ALPINE FLOWERS FOR GARDENS

ROCK, WALL, MARSH PLANTS, AND MOUNTAIN SHRUBS

By W. ROBINSON

AUTHOR OF "THE ENGLISH FLOWER GARDEN"

FOURTH EDITION, REVISED

ILLUSTRATED

1910

Copyright © 2013 Read Books Ltd.
This book is copyright and may not be
reproduced or copied in any way without
the express permission of the publisher in writing

British Library Cataloguing-in-Publication Data
A catalogue record for this book is available from the
British Library

A Short History of Gardening

Gardening is the practice of growing and cultivating plants as part of horticulture more broadly. In most domestic gardens, there are two main sets of plants; 'ornamental plants', grown for their flowers, foliage or overall appearance – and 'useful plants' such as root vegetables, leaf vegetables, fruits and herbs, grown for consumption or other uses. For many people, gardening is an incredibly relaxing and rewarding pastime, ranging from caring for large fruit orchards to residential yards including lawns, foundation plantings or flora in simple containers. Gardening is separated from farming or forestry more broadly in that it tends to be much more labour-intensive; involving *active participation* in the growing of plants.

Home-gardening has an incredibly long history, rooted in the 'forest gardening' practices of prehistoric times. In the gradual process of families improving their immediate environment, useful tree and vine species were identified, protected and improved whilst undesirable species were eliminated. Eventually foreign species were also selected and incorporated into the 'gardens.' It was only after the emergence of the first civilisations that wealthy individuals began to create gardens for aesthetic purposes. Egyptian tomb paintings from around 1500 BC provide some of the earliest physical evidence of ornamental horticulture and landscape design; depicting lotus ponds surrounded by symmetrical rows of acacias and palms. A notable example of an ancient ornamental garden was the 'Hanging Gardens of Babylon' – one of the Seven Wonders of the Ancient World.

Ancient Rome had dozens of great gardens, and Roman estates tended to be laid out with hedges and vines and contained a wide variety of flowers – acanthus, cornflowers, crocus, cyclamen, hyacinth, iris, ivy, lavender, lilies, myrtle, narcissus, poppy, rosemary and violets as well as statues and sculptures. Flower beds were also popular in the courtyards of rich Romans. The Middle Ages represented a period of decline for gardens with aesthetic purposes however. After the fall of Rome gardening was done with the purpose of growing **medicinal herbs** and/or decorating church **altars**. It was mostly monasteries that carried on the tradition of garden design and horticultural techniques during the medieval period in Europe. By the late thirteenth century, rich Europeans began to grow gardens for leisure as well as for medicinal herbs and vegetables. They generally surrounded them with walls – hence, the 'walled garden.'

These gardens advanced by the sixteenth and seventeenth centuries into symmetrical, proportioned and balanced designs with a more classical appearance. Gardens in the renaissance were adorned with sculptures (in a nod to Roman heritage), topiary and fountains. These fountains often contained 'water jokes' – hidden cascades which suddenly soaked visitors. The most famous fountains of this kind were found in the Villa d'Este (1550-1572) at Tivoli near Rome. By the late seventeenth century, European gardeners had started planting new flowers such as tulips, marigolds and sunflowers.

These highly complex designs, largely created by the aristocracy slowly gave way to the individual gardener however – and this is where this book comes in! Cottage Gardens first emerged during the Elizabethan times, originally created by poorer workers to provide themselves with food and herbs, with flowers planted amongst them for decoration. Farm workers were generally provided with cottages set in a small garden—about an acre—where they could grow food, keep pigs, chickens and often bees; the latter necessitating the planting of decorative pollen flora. By Elizabethan times there was more prosperity, and thus more room to grow flowers. Most of the early cottage garden flowers would have had practical uses though —violets were spread on the floor (for their pleasant scent and keeping out vermin); **calendulas** and **primroses** were both attractive and used in cooking. Others, such as **sweet william** and **hollyhocks** were grown entirely for their beauty.

Here lies the roots of today's home-gardener; further influenced by the 'new style' in eighteenth century England which replaced the more formal, symmetrical '**Garden à la française**'. Such gardens, close to works of art, were often inspired by paintings in the classical style of landscapes by Claude Lorraine and Nicolas Poussin. The work of **Lancelot 'Capability' Brown**, described as 'England's greatest gardener' was particularly influential. We hope that the reader is inspired by this book, and the long and varied history of gardening itself, to experiment with some home-gardening of their own. Enjoy.

PLATES

ALPINE FLOWERS AT HOME	*Frontispiece*
	PAGE
MASSIVE ROCKS, FRIAR PARK	12
PART OF ROCK GARDEN	68
STONE PATHWAY	178
GENTLY RAISED ROCK GARDEN, THE HOLT, HARROW WEALD	200

FOREWORDS TO NEW EDITION.

This book is written to dispel a general but erroneous idea, that the plants of alpine regions cannot be grown in gardens. This idea is not confined to the general public; it has been taught by botanists and horticulturists whenever they have had to speak of alpine plants, while the alpine traveller has regretted that we could not enjoy in our gardens these most charming of flowers. The late Duke of Argyll, presiding some years ago at the dinner of the Gardeners' Benevolent Institution, told the company that, though they had overcome almost every difficulty of cultivation, they were beaten by one—that of growing alpine plants.

Any reader of this book may prove for himself that this idea is a baseless one; and that, so far from its being true that these plants cannot be cultivated, there is no alpine flower that ever cheered the traveller's eye which cannot be grown in our island gardens. Instead of being very difficult, they will be found to be among the most easily cultivated of all plants, especially to those who begin modestly and avoid the ugly extravagance of artificial "rocks."

What are alpine plants? The word *alpine* is used to denote the plants that grow naturally on all high mountain-chains, whether they spring from hot tropical plains or from green northern pastures. Above the cultivated land these flowers begin to occur on the fringes of the stately woods; they are seen in multitudes in the vast pastures which clothe many great mountain-chains, enamelling their soft verdure; and also where neither grass nor loose herbage can exist; or where feeble world-heat is quenched and mountains are crumbled into ghastly slopes of shattered rock by the contending forces of heat and cold, even there, amid the glaciers, they spring from Nature's ruined battle-ground, as if the mother of earth-life had sent up her loveliest children to plead with the spirits of destruction.

Alpine plants fringe the vast fields of snow and ice of the high mountains, and at great elevations have often scarcely time to flower and ripen a few seeds before they are again imbedded in the snow; while sometimes many of them may remain beneath the surface for more than a year. Enormous areas of the earth, inhabited by them, are every year covered by a deep bed of snow. Where the tall tree or shrub cannot exist in the intense cold, a deep soft mass of down-like snow settles upon these minute plants, a great cloud-borne quilt, under which they safely rest, unharmed by the alternations of frost and biting winds and moist and spring-like days. It is the absence in our island of this winter rest that is our chief difficulty, in leading to "false starts" in growth, and so injuring certain kinds. But, in spite of this, hundreds of kinds of alpine plants are now grown in the parts of Britain that are most subject to winter's rapid changes.

A reason why alpine plants clothe the ground in these high regions is that no taller vegetation can exist there; were such places inhabited by trees and shrubs, we should find few alpine plants among them; on the other hand, if no stronger vegetation were found at a lower elevation, these plants would make their appearance there. Many plants found on the high Alps are also met with in rocky or bare places at much lower elevations. *Gentiana verna* often flowers late in summer when the snow thaws on a high mountain; yet it is also found on low hills, and occurs in the British Isles. In the struggle for existence upon the plains and tree-clad hills, the more minute plants are often overrun by trees, trailers, bushes, and vigorous herbs, but where, as in northern and elevated regions, these fail from the earth, the choicer alpine plants prevail.

Alpine plants include plants from many divisions of the plant world, embracing endless diversities of form and colour. Among them are tiny Orchids, as interesting as their tropical brethren, though so much smaller; ferns that peep from crevices of high rocky places, clinging to the rocks and not daring to throw forth their fronds with airy grace, as they do on the ground; bulbous plants with all their coarseness gone, and all their beauty retained; evergreen shrubs, perfect in leaf and blossom, yet so small that an inverted glass could cover them; creeping plants, rarely venturing much above mother earth, yet spreading freely over it, and, when they fall over the brows of rocks, draping them with lovely colour; minute plants that scarcely exceed the mosses in size, and quite surpass them in the way in which they mantle the earth with fresh green carpets in the midst of winter; and "succulent" plants in endless variety, though smaller than the mosses of our bogs: in a word, alpine vegetation embraces nearly every type of the plant-life of northern and temperate climes.

ALPINE GARDENING.

As to the merits of "alpine" and like kinds of gardening, as compared with those more in vogue, there can be little doubt in the minds of all who give the subject any thought. Stupidity itself could hardly delight in anything uglier than the daubs of colour that, every summer, flare in the neighbourhood of most country-houses in western Europe. Visit many of our large country gardens, and probably the first thing we shall hear about will be the scores of thousands of plants "bedded out" every year, though no system ever devised has had such a bad effect on our gardens.

Amateurs who cultivate numerous hot-house plants, and who generally have not a dozen of the equally beautiful flowers of northern and temperate regions in their gardens, might grow an abundance of them at a tithe of the expense required to fill a glass-house with costly Mexican or Indian Orchids. Our botanical and great public gardens, in which alpine plants are too often found in obscure corners, might each exhibit a beautiful rock-garden, at half the expense now bestowed on some tropical family displayed in a glass-shed, and there is not a garden, even in the suburbs of our great cities, in which the flowers of alpine lands might not be enjoyed.

This book is written in the hope of showing various *simple* ways in which this may be done. As regards the instructions for cultivation given in it, it will be understood that they can only be applied in a general way, so much being dependent on the difference in conditions, even in our islands, of north, south, east and west; of soil, rainfall, amount of sunshine, and many other considerations not always noticed. The plant that in a garden on a north of England moor might be quite happy and take care of itself, will need care in the sands of Surrey, and plants that thrive with the more copious rainfall on the western coast of Ireland may want much looking after in Kent or Essex. In some cases these difficulties are not easily got over. Even soil is not by any means the simple thing it looks, as that no matter what trouble we take, in certain districts we cannot make soil nearly so good as that which occurs naturally in others. But from this and many other things, we may learn the best lesson of all, as regards rock plants, which is to grow the plants that our *conditions allow us to do best*. We have even seen the hardy Pansies perish in great heats on the south, when in the cool hill-country they were enduring and happy. Therefore in a dry district we should lean more to the southern plants, such as Rock Roses, and in heavy soils, which we cannot easily alter, take up easily-grown plants, like the Candytufts, Rockfoils, Stonecrops, and Houseleeks.

Conditions on the Alps.

If the conditions of plant life in our islands are so varied, how of those of the Alps? In no part of the earth are they so wondrously varied, severe, and even terrible. Valleys that would tempt young goddesses to gather flowers, and valleys flanked with cliffs fit to guard the River of Death; beautiful forest shade for woodland flowers, and vast prairies without a tree, yet paved with Gentians; sunburnt slopes and chilly gorges; mountain copses with shade and shelter for the taller plants, and uplands with large areas of plants withered up, owing to the snow lying more than a year. Plants rooted deep in prime river-carried soil, and others living and thriving in little depressions in the earthless rock. Lakes and pools at every elevation, torrents, streams splashing from snowy peaks; pools, bogs, and spring-fed rills at every altitude; long melting snow-fields, giving the plants imprisoned below them their freedom at different times, and so leading to a succession of alpine flower life.

Most noticeable of all, for us, however, is the great winter rest under the snow which keeps the plants asleep. The absence of snow in our country is the cause of the greatest difficulty we have with alpine plants. Constant change of weather, and the occurrence of mild weather in winter when all the high mountain plants are at rest, should lead us to think more of southern plants and shrubs, which are not subject to this high alpine sleep.

But there is one fact that should make all Britons rock-gardeners, namely, that the climate of our grey islands *corresponds with that of an immense range of mountain ground in central Europe*. The plains of France and of Lombardy are hot, and the alpine passes ice-cold, while the nightingales are singing in millions of acres of mountain pasture set with islets of Wild Rose, Hazel, and Aspen. And these conditions of cool mountain ground between hot valley-land and high frozen passes obtain over vast regions in central and eastern Europe. Even in the south, the same thing occurs. If asked to name two of the most enduring rock-flowers, I could not name any better than the blue Greek Anemone (*A. blanda*) and the purple Rock Cress (*Aubrietia*), which we see in quantity on the hills near Athens. I have never seen the mountains of northern Greece nor the mountainous regions near, but we should expect no less from their flora, as their hillside climate would be more like our own. If we go into Savoy to see its rich alpine flora, we are often struck with the likeness to the conditions of our own land. This is why such large numbers of rock-plants are so easily grown in Britain, we having the same cool summer as in the high mountain ground. And the plants that will enjoy these conditions are far more numerous than those that inhabit the flank of the moraine or the high mountain crest, with often a few weeks of summer only. Hence the summer that burns up the Roses on the plains of Italy or of Southern Germany or France, leaves us cool in the plains of Britain, not to speak of our mountain ground, so admirable for the growth of alpine plants and mountain shrubs. And we may be sure that it is only certain groups of plants inhabiting very high ground, like *Androsace*, that will offer us any difficulty.

It is for these reasons I have brought a greater variety of plants into this edition; hardy mountain shrubs mainly, and those accustomed by nature to a great variety of conditions, including plenty of sun and an "open" winter. It is not only for their own sake that the mountain shrubs are a gain; it is for the gentle shelter and shade they give to plants that grow naturally in woods and copses. Some of these plants, like Lily-of-the-valley, thrive in the open with us; but we lose plants of rare beauty, owing to exposure on the bare rock-garden of plants that in nature live among bushes and in copses and in open and moist woods.

EXTRAVAGANCE

has had a free sway in rock-garden formation, and has always ended in ugliness. Much harm is done by rock-makers, their extravagant plans leading to great cost, of which some startling instances could be given. This is more especially the case in the artificial rock-garden, which is formed of bricks and like material, covered with cement. Even if we got such ugly things at little cost it would stop progress. They are rarely artistic, and they are bad for the growth of plants. If we spend much in preliminary effects, such as these rock-gardens give, there may be little left for the main thing—the plants and their care.

People who have natural rocks in their own property, or near it, are not likely to make such mistakes, and the true way is to begin modestly with a few natural stones. A man who has seen the mountains, and has his heart in the matter, ought to do better with a few loads of natural stone than with five hundred tons of artificial rubbish. In many parts of the Alps the prettiest effects are obtained from plants clustered round a lichen-covered stone, with, it may be, a yard only of its point exposed. Such stones not only look well, but are best for the plants, the roots of which find all they require beneath and near the cool stone. In that way, in many districts, even where the natural stone has to be carried home, such a beginning need cost very little. Where the stone is on the ground, as often happens in the north and west, it might become a question of planting only; but the idea is so much in peoples' heads that they must make some kind of "rock" work, that even in the Alps I have seen men making little artificial arrangements, reminding one of what used to be seen in villa gardens at home, instead of planting the rocky ground ready to their hands.

If we are to make artificial rock, it should be as a last resort, and for effect only, as it never allows us to grow plants half as well as the natural stone or even the level soil.

Much improvement, both in design and cultivation of rock-gardens and rock plants, has taken place within the past twenty years or so, and some effects on these rock-gardens are now seen that were impossible on the old form of "rock-work," with its dust-dry pockets and hopeless ugliness. At the Friar Park, Henley-on-Thames, South Lodge, Leonardslee, Warley Place, Batsford, and many other places, we may see not only the rarest Alpine plants admirably grown, but effects and colour not unworthy of the Alpine fields. Even the public gardens where the most grotesque arrangements were common have changed much for the better. I wish one could say there was the same improvement in the nurseries devoted to these plants. There are fuller collections, but the needlessly costly way of offering single plants at a high price tends to prevent any artistic grouping or massing of the plants such as a beginner might seek. Many alpine plants, like the Houseleeks, Stonecrops, and Rockfoils, are almost too facile in increase, and many others distinct from these are easily raised from seed, while the mountain perennials, like the Globe Flowers and Harebells, are easily increased by division. So that there should be no difficulty for any one with a piece of even poor ground in treating the public more liberally than in the usual way of offering single plants. It would be better both for gardens and the trade if the bolder way were followed of offering plants by the dozen or hundred and at reasonable rates.

The plants in this book are not treated in any one or regular way, for the reason that they differ so much in value. In nature, all plants may be said to be of equal value, but in gardening the difference in their values is enormous, both in degree and in every other way. Therefore, in a purely garden book like this, the only helpful way is to treat plants in some relation to their value in the garden. A great many plants, also, are truly Alpine, but have little or no use or beauty in the garden, and these are not included in this book. Nor can we even in such a vast theme include all the claims to beauty, not to speak of the fact that many of the regions from which these plants come are not yet half explored, and many of the plants that are known are not yet introduced.

Here I leave the Alpine garden to the young enthusiasts of the future; they can never exhaust its variety, but can do much for it, by simple plans and good culture. Done in the worst way and most adverse conditions it is interesting, but, with care and thought in the best, the Alpine garden may be the fairest ever made by the hands of man.

W. ROBINSON.

GRAVETYE MANOR,
January 1903.

ALPINE FLOWERS

PART I

CULTURAL.

In treating of the culture of alpine plants, the first consideration is that much difference exists among them as regards constitution and vigour. We have, on the one hand, many plants that merely require to be sown or planted in the simplest way to flourish—Arabis and Aubrietia for example; but, on the other, there are many kinds, like the Primulas of the high Alps, with many of their companions, which demand some thought and care. Nearly the whole of the misfortunes which these little plants have met with in our gardens are to be attributed to the usual conception of what a rock-garden ought to be, and of what the alpine plant requires. These plants live on high mountains; therefore it is erroneously thought they will do best in our gardens if planted on such ugly heaps of stones and brick rubbish as we frequently see piled up and dignified by the name of "rockwork." Rocks are often "bare," and cliffs are devoid of soil; but we must not conclude from this that the choice jewellery of plant life scattered over the ribs of the mountain, or growing out of the crag and crevice, lives upon little more than the mountain air and the melting snow. Where shall we find such a depth of well-ground stony soil, and withal such perfect drainage, as on the ridges of *débris* flanking some great glacier, stained all over with tufts of crimson Saxifrage? That narrow chink, from which peep tufts of the beautiful Androsace helvetica, has for ages gathered the crumbling grit and scanty soil, into which the roots enter far. If we find plants growing from mere cracks without soil, the roots simply search farther into the heart of the flaky rock, so that they are safer from any want of moisture than in the deepest soil.

We find on the Alps plants not more than an inch high, and so firmly rooted in crevices of half-rotten slaty rock that any attempt to take them out would be futile. But, by knocking away the sides from some isolated bits of projecting rock, we may lay bare the roots and find them radiating in all directions against a flat rock, some of them a yard long. We think it rapacious of the Ash, a forest tree, to send its roots under the walls of our gardens and rob the soil therein; but here is an instance of a plant one inch high, penetrating into the earth to a distance many times greater than its foliage ventures into the alpine air. And there need be no doubt whatever that even smaller plants descend quite as deep, though it is rare to find the texture and position of the rock such as will admit of tracing their roots. It is true, we occasionally find hollows in flat, hard rock, into which moss and leaves have gathered for ages, and where, in a sort of basin, without an outlet of any kind in the hard rock, plants grow freely; but in exceptional droughts they are just as liable to suffer from want of water as they would be in our plains. On level or sloping spots of ground in the Alps, the

Mountain flank in process of degradation.

earth is often of great depth, and if it be not all *earth* in the common sense of the word, it is more suitable to rock plants than what we commonly understand by that term. Stones of all sizes broken up with the soil, sand, and grit, greatly tend to prevent evaporation. The roots lap round them and follow them deeply down while in such positions, they never suffer from want of food and moisture, or weather. Stone is a great preventive of evaporation, and shattered stone forms the soil of the mountain flanks where the rarest alpine plants abound, while the degradation of gritty soil, so continually effected by melted snow water and heavy rains in summer, serves to earth up, so to speak, many alpine plants. I have torn up tufts of them, showing the remains of generations of the old plants buried and half buried in the soil beneath their descendants. This would be effected to some extent by the decaying of the plants themselves, but frequently grit and peat are washed down among them; and, in cases where the washings-down do not come so thickly as to overwhelm the plants, they thrive with unusual vigour.

Now, if we consider how dry even our English air often becomes in summer, and that no natural positions in our gardens afford such cool rooting-places as those described, the need of giving to alpine plants a soil quite different from what has hitherto been in vogue will be seen. The only good principle generally followed is that of raising the plants above the level of the ground. But this raising of the plants above the level should in all cases be accompanied by the more essential way of giving the plants means of rooting deeply into good and firm soil—sandy, gritty, peaty, or mingled with broken stone, as the case may be.

How *not* to do this is shown by persons who stuff a little soil into a chink between the stones in a rockery, and insert some small alpine plant in that. There is usually a vacuum between the stones and the soil beneath them, and the first dry week sees the death of the plant—that not being usually attributed to the right cause. Precisely the same end would have come of it if the experiment had been tried on some alp bejewelled with Gentians. We should not pay so much attention to the stones or rocks as to the earth for the plants. There are certainly alpine plants that do not require a deep soil, or what is usually termed soil at all; but all require a firm medium for the roots.

In numbers of gardens an attempt at "rockwork" of some sort has been made; but in most cases the result is ridiculous; not because it is puny when compared with Nature's work in this way, but because it is so arranged that rock-plants cannot exist upon it. In many places a sort of sloping stone or burr wall passes as "rockwork," a dust of soil being shaken in between the stones. In others, made upon a better plan as regards the base, the "rocks" are all stuck up on their ends, and so close that soil, or room for a plant to root or spread, is out of the question. The best thing that usually happens to a structure of this sort is that its nakedness gets covered by some friendly climbing shrub, or some rampant weed, to the exclusion of true rock-plants.

In moist districts, where frequent rains keep porous stone in a continually humid state, this too showy "rockwork" may manage to support a few plants; but in by far the larger portion of the British Isles it is useless, and always ugly. In the southern and eastern counties, where of late years the rainfall is often very low, the need is all the greater to see that alpine plants are so placed that they will not suffer from drought. It is not alone because the mountain air is pure and clear and moist that the Gentians and like plants prefer it, but because the elevation is unsuitable to the coarser-growing vegetation; and the alpines have it all to themselves. Take a healthy patch of Silene acaulis, by which the summits of some of our highest mountains are mossed with rosy crimson, and plant it two thousand feet lower down in suitable soil, keeping it moist enough and free from weeds, and we may grow it well; but leave it to Nature in the same neighbourhood, and the strong grasses and herbage will soon run through and cover it, excluding the light, and finally killing the diminutive Moss Campion.

It is not only those who make their "rockwork" out of spoilt bricks, cement, and perhaps clinkers, that err in this respect, but the designers of some of the most expensive works of this kind. At Chatsworth, for instance, and also to some extent at the Crystal Palace, we see rocks not offensive so far as distant effect in the landscape is concerned; but, when examined closely, it might well be imagined that rocks and rock-plants were never intended for each other's company, so bare are these of their best ornaments. They are, for the most part, pavements of small stones, huge masses of stone, or imitation rock, formed by laying cement over brickwork, and in none of these cases are they adapted for the cultivation of mountain plants.

It is possible to combine the most picturesque effects of which rocks are capable, with all the requirements for plant-growing; and it is easy to use the large stones and make bold effects, and leave at the same time level intervening spaces of rocky ground in which rock-plants may thrive almost as well as on the many mountain pastures where we see them happy in the mountain turf.

Part of the Rock Garden at Brookfield, Hathersage, Sheffield.

POSITION FOR THE ROCK-GARDEN.

The position selected for the rock-garden should not, as a rule, be near walls, or very near a house; never, if possible, within view of formal surroundings of any kind; and generally be in an open situation; and no effort should be spared to make all the surroundings as graceful, quiet, and natural as they can be made. The part of the gardens around the rock-garden should be picturesque, and, in any case, display a careless grace, resulting from the naturalisation of beautiful, hardy herbaceous plants, and the absence of too formal walks and beds. The roots of forest trees would be almost sure to find their way into the masses of good soil provided for the choicer alpine plants, and thoroughly exhaust them. Besides, as alpine flowers are usually found on treeless and even bushless wastes, it is certainly wrong to place them under trees or in shaded positions, as has generally hitherto been their fate. It need hardly be added that it is an unwise practice to plant pines on rockwork, as has been lately done in Hyde Park and many other places. It will, however, generally be in good taste to have some graceful young pines planted near, as this type of vegetation is usually to be seen on mountains, apart altogether from their great beauty and the aid which they so well afford in making the surroundings of the rock-garden what they ought to be. In small places, and in those where, from unavoidable circumstances, the rock-garden is made near a group of trees, the roots of which might rob it, it would be found a good plan to cut them off by a narrow drain, descending as deep as, or somewhat deeper than, the roots of the trees; this should be filled with rough concrete, and it will form an effectual barrier.

MATERIALS.

As regards the kinds of stone to be used, if one could choose, sandstone or millstone grit would perhaps be the best;

but it is seldom that a choice can be made, and happily almost any kind of natural stone will do, from Kentish rag to limestone; soft, slaty, and other kinds liable to crumble away should be avoided, as also should magnesian limestone. Stone of the district should be adopted for economy's sake, if for no other reason. Wherever the natural rock crops out, as it often does in many hilly parts of our islands, it is sheer waste to create artificial rockwork instead of embellishing that which naturally occurs. Something of the same kind might be said of many of our country seats. In many cases of this kind nothing would have to be done but to clear the ground, and add here and there a few loads of good soil, with broken stones, etc., to prevent evaporation; the natural crevices being planted where possible. Cliffs or banks of chalk, as well as all kinds of rock, should be taken advantage of in this way; many plants, like the dwarf Campanulas and Rock Roses, thrive in such places.

No burrs, clinkers, vitrified matter, portions of old arches and pillars, broken-nosed statues, etc., should ever obtain a place in a garden devoted to alpine flowers. Stumps and pieces of old trees are quite as bad as any of the foregoing materials; they are only fitted to form supports for rough climbers, and it is rarely worth while incurring any expense in arranging them. It is best to begin without attempting much. Let your earliest attempts at "the first great evidences of mountain beauty" be confined to a few square yards of earth, with no protuberance more than a yard or so high, and be satisfied that you succeed with that, before trying anything more ambitious. The stones should usually all have their bases buried in the ground, and the seams should not be visible; whenever a vertical or oblique seam of any kind occurs, it should be crammed with earth, and the plants put in this will quickly hide the seams. Horizontal fissures should be avoided as much as possible; they are only likely to occur in vertical faces of rock, and these should be avoided except where distant effect is sought. No vacuum should exist beneath the surface of the soil or surface-stones. Myriads of alpine plants have been lost from want of observing this precaution, the open crevices and loose texture of the soil permitting the dry air to destroy them in a very short time.

Mound of earth, with exposed points of rock.

In all cases where elevations of any kind are to be formed, *the true way is to obtain them by means of a gentle mound of soil, suitable to the plants, putting a stone in here and there as the work proceeds;* frequently it would be desirable to make these mounds without any "strata." The wrong and the usual way is to get the desired elevation by piling up arid and ugly masses of "rockwork."

HIDDEN NATURAL ROCK.

While many go to great expense in forming masses of artificial rock, made of bricks and cement, and others are

Unearthed Rocks in a Sussex Garden.

satisfied with the old bricks themselves, accompanied by clinkers and other offensive rubbish, few trouble themselves about the rock treasures that often lie beneath the sod. Considering the large sums that are spent in sham rocks, and the greater value in every way of natural rock, masses of it are most valuable to those who care for the picturesque in garden scenery. The illustration on the opposite page gives a feeble notion of one of the rocks that a friend of mine has succeeded in unearthing. His place was somewhat liberally strewn with rock on the surface; but the owner was anxious for more; and by digging out the earth, he has formed a beautiful gorge between two flanks of rock; and by clearing away the earth from the flank of a nose of rock that just projects above a grassy knoll, he has discovered beautiful wrinkles, crevices, and other charms in it. Thus by a little persevering searching and digging, has been produced a scene as interesting as in an alpine country, and one which offers such a variety of aspects that one could desire for a rock-garden. Many kinds of rock plants may be grown on it in the best manner, and arranged on it with the happiest effect.

Stone Pathway in Rock Garden at Warley Place, Essex.

It would seem redundant to advise country folk to develop the beauty of natural rocks—where they happen to have any—but it is not so, as I have seen artificial rock being formed in places where there were acres of beautiful rocks hidden away in the underwood! Even where no desire is felt for the cultivation of alpine plants, the effect of the rock on the landscape should be thought of, as it is often very precious.

I have myself made visible throughout the country-side a quarter of a mile of rocks, which were once hidden in the underwood.

Ascending Pathway in Rock Garden (Warley Place).

As we see too clearly that the rock-gardens too commonly made by those who profess to make them, are not based upon observation of natural form, it is well to show all we can of the way rocks come out of the earth, and of their structure and often beauty of colour and form.

PATHWAYS, ETC.

No walk with regularly-trimmed edges of any kind should pass through, or even come near, the rock-garden. This need not prevent walks through or near it, as, by allowing the edges of the walk to be a little free and stony, and by permitting dwarf Stonecrops, Linaria alpina, and the lawn Rockfoils to crawl into the walk at will, a good effect will arise. In every case where walks pass through rock-gardens, a variety of little plants should be placed at the sides, and allowed to crawl into the walk in their own way. There is no surface whatever of this kind that may not be thus planted: Violets, Ferns, and Myosotis will answer for the moister and shadier parts, and the Stonecrops, Rockfoils, Sandworts, and many others, will thrive in more arid parts and in the full sun. *The whole of the surface of the alpine garden should be covered with plants*, except the projecting points of rocks; and even these should be covered, as far as possible, without concealing them. In moist districts, such plants as Erinus alpinus and Arenaria balearica will grow wherever there is a resting-place for a seed on the face of the rocks; and even vertical faces of rock may be half covered with a variety of plants; so that there is no reason why any level surfaces of ground should be bare.

Rocky Steps.

A propos of simple ways of getting good effects, I may mention what took place in a garden in Sussex, where stone steps had been placed in the rock garden just as a pathway. The plants inserted between the rough stones—Gentians and Stonecrops in a varied collection—gave the prettiest effect, and showed the finest health of any plants in the place; and with good reason, because they were protected from the heat much more effectually than the plants in the rock garden near, as

Rocky Path at Lydhurst, Sussex.

they could spread their roots under the great stones. The result was quite a picture, and got in the most simple way.

CONSTRUCTION.

In no case should regular or mason-built steps be permitted in or near the rock-garden. Steps may be made irregular, and even beautiful, with violets and other small plants jutting from every crevice. No cement should be used in connection with the steps. The woodcut on page 10 is from a photograph of the lower part of rude steps ascending from a deep and moist recess in a rock-garden. It shows imperfectly—no engraving could show it otherwise—the crowds of lovely plants that gather over it, except where worn bare by feet. In cases where the simplest type of rock-garden only is attempted, and where there are no rude walks in the rock-garden, the very fringes of the gravel walks may be graced by the dwarfer Stonecrops. The alpine Linaria is never more beautiful than when self-sown in a gravel walk. "Rockwork," which is so made that its miniature cliffs overhang, is useless for alpine vegetation; and all but such wall-loving plants as Corydalis lutea, perish on it. The tendency to make it with overhanging brows is everywhere seen in cement rock-gardens. Into the alpine garden this kind of construction should never be admitted, except to get the effect of bold cliffs. When this system is admitted, the designer should be requested to obtain his picturesque effect otherwise than by making all his "cliffs" overhang. It is erroneous to suppose that heaps of stones or small rocks are necessary for the health of alpine plants. The great majority will thrive without their aid if the soil be suitable; and though all are benefited by them, if properly used as elsewhere described, it is important that it should be generally known how needless is the common system of inserting mountain plants among loose stones. Half burying rocks or stones in the earth round a rare species, which it is intended to save from excessive evaporation, and which has a deep body of soil to root into, is, however, a different and a good practice.

MASSIVE ROCKS (MILL STONE GRIT), FRIAR PARK. *June* 1910.

SOIL.

The great majority of alpine plants thrive best in deep, cool, and gritty soil. In it they can root deeply, and when once they are so rooted, they will not suffer from drought, from which they would quickly perish if planted in the usual way. Two feet deep is not too much for most species in dry districts, and it is in nearly all cases a good plan to have plenty of broken sandstone or grit mixed with the soil. Any good free loam, with plenty of sand and grit, will be found to suit many alpine plants, from Pinks to Gromwells. But peat is required by some, as, for example, various small and brilliant rock-plants like Menziesia, Trillium, Cypripedium, Spigelia marilandica, and other mountain and bog plants. Hence, though the general mass of a rock garden may be of an open loam nature, it will be desirable to have a few masses of sandy and leaf soil and peat here and there. This is better than forming all the ground of good loam, and then digging holes in it for the reception of small masses of peat. The soil of one or more portions might also be chalky or calcareous, for the sake of plants that are known to thrive best on these formations, as Polygala calcarea, the Bee orchis, and Rhododendron chamæcistus. Any other varieties of soil required by individual kinds can be given as they are planted.

Much consideration has been given by botanists to the plants that grow on the different formations, but we have evidence in British gardens that the good soils common in them will sustain in health a great number of kinds well, that in Nature are found on soils of a special character.

Mr Correvon, who has given much thought to the matter, writes as follows in *The Garden:*—

The flora of the Alps depends in a much greater degree than that of the plains on the chemical nature of the soil. We know that from the point of view of chemistry, the mountains are divided into two main classes, namely, the calcareous and the granitic, otherwise the sedimentary and the igneous.

All the mountain ranges of the Alps are either of limestone or of granite. The vegetation that adorns them is directly subject to their influence, and hence becomes a flora either calcareous or silicious. Thus, also, there is among the alpine plants that we have in cultivation, some that desire or actually require lime, just as there are others that avoid it, and must have silica. It is important to know to which category the various plants belong, in order to combine them rightly. There are, notwithstanding, a great number —indeed the larger number—of mountain plants whose distribution is general, and which do equally well in either soil. It is just these, of all the plants of the Alps, that submit most readily to cultivation, and that have long been established in gardens.

But there are great numbers of other species which, though easy to grow at Geneva, where the soil, the water, and the stone contain lime, are by no means so accommodating in the west of France or in the parts of England that are granitic; while there is a whole range of other species that are readily grown in these regions, and that we cannot persuade to feel at home in our lime-impregnated garden.

One of my friends, Dr A. Rosenstiehl, a chemist, who is also an excellent botanist, has gone deep into the subject, and, thanks to a system of watering with distilled water, has arrived at some excellent results. He set to work with all the necessary care and precaution, keeping his granite rock free from contact with lime, and the results he has obtained prove that those botanists are right who class some plants as lovers, and others as haters, of lime, and others again as inimical to granite.

The juices of plants are acid; these acids, when brought into contact with the carbonate of lime absorbed by the plant, become saturated and neutralised. There are formed therefore in the plant certain salts of lime, which, if they are soluble in water, can circulate in its organism; but if they are insoluble, as is often the case, the channels of circulation become choked, and nutrition is impeded. Their presence, therefore, is a mechanical impediment to the well-being of the plant. Dr Rosenstiehl has verified the presence of such acids in the lime-hating plants he has examined, and it is certain that these plants, if grown in soil containing lime, will sooner or later become poisoned. He has shown me in his garden examples of Sphagnum and Vaccinium, plants essentially lime-hating and granite-loving, whose leaves were throwing out small calcareous crystals and were dying. All plants, however, require lime in a certain proportion for the building up of their tissues, and it is found in the ashes of even the most lime-hating of plants. Each species must have a certain amount, but cannot endure too strong a dose, and on these a little too much acts as poison. The careful cultivator must therefore learn exactly how much must be given to each species.

Dr Rosenstiehl grows Asplenium germanicum in soil containing

0.293 per cent. of carbonate of lime, while the earth in which Edelweiss is growing contains a great deal more. This plant is, as is well known, essentially lime-loving and its flower-bracts are just so much more white and woolly in proportion as the soil it grows in is rich in lime.

Here is a list of the principal alpine plants that need one or other of the two soils containing respectively either lime or granite:

Calcareous.	*Granitic.*
Achillea atrata	Achillea moschata
Aconitum Anthora	Aconitum septentrionale
Adenostylis alpina	Adenostylis albifrons
Androsace chamæjasme	Androsace carnea
„ arachnoidea	„ lactea
„ helvetica	„ glacialis
„ pubescens	„ imbricata
„ villosa	„ vitaliana
Anemone alpina	Anemone sulphurea
„ narcissiflora	„ baldensis
„ Pulsatilla	„ montana
„ Hepatica	„ vernalis
Anthyllis montana	Arnica montana
Artemisia mutellina	Artemisia glacialis
Braya alpina	Astrantia minor
Campanula thyrsoidea	Azalea procumbens
„ cenisia	Braya pinnatifida
Cephalaria alpina	Campanula spicata
Cyclamen europæum	„ excisa
Daphne alpina	Daphne petræa
„ Cneorum	„ striata
Dianthus alpinus	Dianthus glacialis
Draba tomentosa	Draba frigida
Erica carnea	Ephedra helvetica
Eryngium alpinum	Eritrichium nanum
Erinus alpinus	Gentiana brachyphylla
Gentiana alpina	„ Kochiana
„ angustifolia	„ frigida
„ Clusii	„ pneumonanthe
„ ciliata	„ pyrenaica
„ asclepiadea	Geranium argenteum
Geranium aconitifolium	Gnaphalium supinum
Globularias	Linnæa borealis
Gnaphalium Leontopodium	Lychnis alpina
Gypsophila repens	Meum athamanticum
Lychnis Flos-jovis	Oxytropis campestris
Moehringia muscosa	Papaver rhæticum

Calcareous.

Oxytropis montana
Papaver alpinum
Primula Auricula
" Clusiana
" integrifolia
" minima
" spectabilis
Ranunculus alpestris
" Seguieri
Rhododendron hirsutum
Ribes petræum
Saussurea discolor
Saxifraga longifolia
" cæsia
" diapensioides
" burseriana
" tombeanensis
" squarrosa
" media
" aretioides
Senecio abrotanifolius
" aurantiacus
Sempervivum dolomiticum
" hirtum
" Neilreichii
" Pittoni
" tectorum
Silene acaulis
" alpestris
" Elizabethæ
" vallesia
Valeriana saxatilis
Viola cenisia

Granitic.

Phyteuma hemisphæricum
Phyteuma pauciflorum
Primula hirsuta
" glutinosa
" wulfeniana
" Facchinii
" longiflora
Ranunculus crenatus
" glacialis
Rhododendron ferrugineum
Ribes alpinum
Saussurea alpina
Saxifraga Cotyledon
" Hirculus
" Seguieri
" moschata
" aspera
" bryoides
" ajugæfolia
" exarata
" retusa
Senecio uniflorus
" carniolicus
Sempervivum arachnoideum
" acuminatum
" debile
" Gaudina
" Wulfeni
Silene exscapa
" rupestris
" pumilio
" quadrifida
Vaccinium uliginosum
" oxycoccus
Valeriana celtica
" Saliunca
Veronica fruticulosa
Viola comollia

Ferns.

Cystopteris alpina
" montana
Aspidium Lonchitis
Asplenium Selovi
" fontanum
" viride

Woodsia hyperborea
" ilvensis
Blechnum spicant
Allosorus crispus
Asplenium germanicum
" septentrionale

This list is necessarily incomplete, and comprises only the most characteristic examples of the plants special to the limestone and to the granite, and those which we have actually tried and proved, either at Geneva, at the alpine garden of the Linnæa at Bourg St Pierre, which is essentially granitic, or at the one at the Rochers de Naye, which is of limestone. The names of the plants are so placed in the two columns that related species are opposite one another, so that readers may see at a glance the part that is played by the presence or absence of lime.

While in our garden on the Rochers de Naye above Montreux, which is essentially calcareous, we have never been able to establish species essentially granitic; in that of Bourg St Pierre, which is granitic, we are unable to cultivate Primula Auricula, Campanula thyrsoides, Gentiana lutea, alpina, angustifolia, and Clusi, and other calcareous plants.

It is always well, however, in considering alpine plants in relation to soil in their native homes, to remember that the nature of the rock is but one of the conditions that may lead to the presence or absence of plants in any given situation; rainfall, altitude, temperature, length of growing season, presence or absence of snow, and the absence of more vigorous plants, having all to be counted with, and other conditions not so clear to us.

Need of poor soil for certain plants.—The tendency of gardeners is to overrich earth in almost everything, and among alpine flowers we often see the effects of this in too rank a growth, making some plants less able to endure our winter and early spring weather. Deep soil is not against us, but it would be better in many cases without any humus, but formed of grit, broken sandstone, or other stones, as the case may be. On such earth plants that fail in the ordinary borders or banks might often be grown in a firm and healthy state.

I mean simply heaping up banks of rough sand or decayed stone, and so as to secure various aspects. In certain cases there should be no rich soil whatever, so as to get the dwarf, wiry growth that we often find on the more arid and stony parts of the Alps.

Grit.—A gritty soil, or pure grit, are often very useful in the rock-garden, and where there is a large collection of plants difficult to grow and keep, heaps and banks of grit will help much. The detritus of millstone grit and granite are among the best, and in some districts sharp river sand, but sea sand does not, as a rule, take the place of these grits, granite grit being for plants of granitic formations. These banks would be all the better having different aspects, some cool and moist.

It is, however, a mistake to suppose that all rock-plants will not endure drought. Many, such as the Rock-roses (*Cistus*), Sun-roses (*Helianthemum*), Stonecrops (*Sedum*), Sandworts (*Arenaria*), the rock Bindweeds, Heaths, and many other rock-plants, supporting drought and sunshine bravely.

VARIOUS ROCK-GARDENS.

We will now enter into particulars as to the various ways in which alpine plants may be grown, beginning with the best type of rock-garden—that in which, in addition to the low-lying, stony, and rocky banks and slopes, where numbers of hardy and vigorous species may be grown, there are miniature cliffs and ravines, with perhaps bog and water. The most usual of the faults in setting rocks is that of so placing the stones that they seem to have as little connection with the soil of the spot as if thrown out of a cart. Instead of allowing what may be termed the foundations of the rock-garden to barely show their upper ridges above the earth, and thereby suggesting much more endurable ideas of "rock" than those arising from the contemplation of the unnatural-looking masses usually seen, the stones are often placed on the ground much as a bricklayer places bricks

Half-buried Stone in Rock-Garden.

The surface of every part of the rock-garden should be so arranged that all rain will be absorbed by it; here, again, the objection to overhanging faces holds good. If the elevations

Part of the Rock-Garden at Elmet Hall, Leeds.

are obtained, as they should be, by gradually receding, irregular steps, rather than by abrupt "crags," walls, etc., all the plants on the surface will be refreshed by rains. The upper surfaces of crags and mounds should in all cases be of earth, broken stones, grit, etc., as indeed should every spot where projecting stones or rock are not required for the sake of effect. All the soil-surfaces of the rock-garden should be protected from excessive evaporation by finely broken stones, pebbles, or grit scattered on the surface, or by means of small pieces of broken sandstone or millstone half buried in the ground.

If we merely want a certain surface of rock disposed in a picturesque way, such details as these may not be worthy of attention, but if we wish our rock-gardens to be faithful

Well-formed Sloping Ledges.

Artificial Rock on which Plants do not Thrive.

miniatures of those wild ones which are among the most exquisite of Nature's gardens, then they are of much importance.

In dealing with the construction of the bolder masses of rockwork, we cannot have a better guide than the late Mr James Backhouse, of York, who wrote:—

"Comparatively few alpines prefer or succeed well in horizontal fissures. Those, however, which, like *Lychnis*

Viscaria and *Silene acaulis,* form long tap roots, thrive well in such fissures, provided the earth in the fissure is continuous, *and leads backward to a sufficient body of soil.* Where the horizontal fissures are very narrow, owing to the main rocks being in contact in places, and leaving only irregular and interrupted fissures, such plants as *Lychnis Lagascæ, Lychnis pyrenaica,* and others, bearing and preferring hot sunny exposures, do well. But many plants that would bear the heat and drought, *if* they could get their roots far enough back, would quickly die if placed in such fissures, from the want of soil and moisture near the front; therefore it is usually better, in building rockwork with these fissures, to keep the main rocks slightly apart by means of pieces of very hard stone (basalt, close-grained 'flag,' etc.), so as to leave room for a good intermediate layer of rich loam, stones, or grit, mingled with a little peat. The front view of such a structure would be as above—the dark spaces being firmly filled with the appropriate mixture of soil *before* the upper course of large rocks is placed.

Lychnis and Silene in Fissures.

Horizontal Fissure.

"As a rule, oblique and vertical fissures are both preferable to horizontal ones; but care should be taken with oblique fissures that the upper rock does not overhang. A plant placed at J will often die, when the same placed at H will live, because

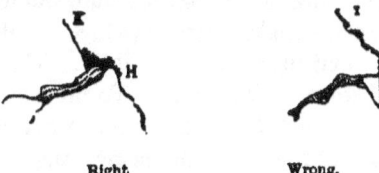

Right.   Wrong.

the rain falling on the sloping face of rock at I will drop of at J, and miss the fissure J altogether, while that falling on the sloping face of rock at K will all *run into* the fissure H. There are, however, some plants, like *Nothochlæna Marantæ* and *Androsace lanuginosa*, which so much prefer positions dry in winter that a fissure like J would suit them better than one like H. Such, however, are rare exceptions.

"The best and worst general forms of steep rockwork we have tried are those indicated in the following figures. By making each rock slightly recede from the one below it, the rain runs consecutively into every fissure. Where the main fissures reverse this order, almost everything dies or languishes. Care should be taken to have the top made of mixed earth and stones —*not* of rock, unless use is intentionally sacrificed to scenic effect.

"Vertical fissures (which suit many rare alpines best of all) should always, so far as possible, be made narrower at the

Right. Wrong.

bottom than at the top. If otherwise, the intervening earth, etc., leaves the sides of the rock as it 'settles,' instead of becoming tighter. In figure A, as the total mass of soil sinks, it becomes compressed against the sides of the rock; while in B, the soil *leaves* the sides of the fissures more and more as the mass sinks, and almost invariably forms distinct 'cracks' (separations between the soil and rock) sooner or later. The same principle applies to small stones and fissures. To prevent undue evaporation in the case of such fissures, stones, larger or smaller, may be laid on the *top* of the soil, care being taken not to cover too much of it, to the exclusion of rain.

"Where a large fissure exists, the smaller pieces of stone *in* it are on this account best placed with the narrowest edge or point upwards (fig. c)—not downwards. It will easily be seen that the tendency of the mixed soil, both as a whole and in each

(A) Right.

(B) Wrong.

(C) A properly formed large vertical fissure.

of its subdivided parts, is to become more and more compressed by its own weight and by the action of rain."

In the construction and planting of every kind of rock-garden, it should be remembered that every surface may and should be embellished with beautiful plants. Not alone on rocks or slopes, or favourable ledges or chinks, or miniature valleys, should we see this kind of plant-life. Numbers of rare mountain species will thrive on the less trodden parts of footways; others, like the two-flowered Violet, seem to thrive best of all in the fissures between rude steps of the rockwork; many dwarf succulents delight in gravel and the hardest soil, and various other plants will run free in among low shrubs near the rock-garden.

As a rule, much more vegetation than rocks should be seen. Where vast regions

Showing ascending rock with base buried.

are inhabited by alpine plants, acres of crags, with a stain of flowers here and there, are attractive parts of the picture;

but in gardens, where our creations in this way can only be Lilliputian, a different method must be pursued, except in places where great cliffs are naturally exposed; and even in this case much vegetation is best. Frequently masses of stone with an occasional tuft of vegetation, are met with under the name of "rockwork," every chink and joint between the stones being quite exposed. This should not be so; every minute chink should have its little line of verdure. Where the ground is low, there is not the slightest need for placing stones all over the surface; an occasional one cropping up here and there from the mass of vegetation will give the best effect. Alpine flowers are often seen in multitudes and in their loveliest aspect in some little elevated level spot, frequently without rocks being visible through it, and when they do occur, merely peeping up here and there. They are lovely, too, in the awful wastes of broken rock, where they cower down between the great stones in lonely tufts, but it is only when Gentians and silvery Cudweeds, and minute white Buttercups, and strange large Violets, and Harebells that waste all their strength in flowers, and fairy Daffodils that droop their heads as gracefully as Snowdrops, are seen, forming a dense turf of living enamel work, that they are seen in all their beauty. Fortunately, the flowery turf and gentle mound are much more possible to us than the moraine ruin or arid cliffs.

Ledge of Alpine Flowers (a Garden Sketch).

In cultivating the rarest and smallest alpine plants, the stony, or partially stony, surface is to be preferred. In their case, we cannot allow the struggle for life to have its own relentless way, or we should often have to grieve at finding the Eritrichium from the high Alps of Europe overrun and exterminated by an alpine American Phlox. Full exposure is also necessary to complete success with very minute plants, and the stones prevent excessive evaporation from the roots. A great number of alpine plants may be grown on exposed level ground as readily as the common Chamomile; but there are, on the other hand, not a few that require care to establish them, and there are usually new kinds to be added to the collection, which, even if vigorous ones, should be kept apart for a time. Therefore, in every place where the culture of alpine plants is entered into with zest, there ought to be a select spot on which to grow the

Alpine Plants growing on the level ground.

delicate, rare, and diminutive kinds. It should be fully exposed, and while sufficiently elevated to secure perfect drainage and all the effect desirable, should not be riven into miniature cliffs.

SLUGS.

Alpine plants will not perish from cold or heat or wet, if properly planted, but many of them are so small that they hardly afford a full meal to a browsing slug, and often disappear during a moist night. Now, as our gardens abound with slugs that play havoc with many things colossal compared with our alpine friends, it is clear that one of the main points is to guard against slugs and snails, and, as far as possible, against worms. Mr Backhouse fenced off the choicest parts of his rock-garden from them by a very irregular little canal, so arranged that, while not an eyesore, it is water-tight, and no slug can cross it. It thus becomes an easier task to guard the plants from slugs than when they are allowed to crawl in from all points of the compass. But even with this precaution, it is necessary to search continually for snails and slugs; and in wet weather the choicest plants should be examined in the evening, or very early in the morning; with a lantern, if at night. Sir Charles Isham, an enthusiastic cultivator of rock-plants, says that he not only protects toads, but does not forget to lay stones, so as to form little retreats for them underneath. They prefer a stone just sufficiently raised to crawl under, and do a deal of good by destroying slugs. He also protects frogs and all carnivorous insects. Ceaseless hand-picking, however, is the best remedy for slugs, and where this is not done, there is little hope of succeeding with some plants, at least where slugs are as abundant as we usually find them in gardens.

GEOGRAPHICAL ARRANGEMENTS OF ROCK-PLANTS.

I have seen in the Berlin Botanic Gardens an interesting essay to grow alpine flowers as distributed over the various ranges of mountains in central Europe; keeping the plants on such rocks, stones, and soil as they are found upon. While such a plan may be pursued with some reason in a botanic garden, it is doubtful, generally, for private places, and not an artistic plan to pursue in a botanic garden, as the more we find such ideas pursued, the less beauty we see, and beauty should be the first *raison d'être* of a garden. The so-called "natural" arrangements of plants in botanic gardens were most wearisome, and still uglier were the "Linnæan" arrangements of living plants in botanic gardens. If the mind is fixed much on any book system of setting out plants in gardens, the precious gift of beauty is often lost. Therefore attempts to imitate the particular mountain ranges and their flora is not likely to lead to so good results as where we are free to get the best result our conditions will allow of.

One exception, however, I would make in our country, and that would be a British Alpine and Moor Garden. We have our own mountains, and many of them—Welsh, Irish, Scotch, and North English—with many beautiful plants on them. It would however, be an instance of hyper-refinement to grow separately the plants of each of our own islands; the effort should be rather to show their unity and connection. So many people buried in cities do not know that we have beautiful alpine flowers, natives of our own land, that it might be well to let them see in a garden of British Alpine and Moor plants.

CULTURAL

CASCADES, BRIDGES, ETC.

Where water occurs near the rock-garden, bridges here are often seen; but some such arrangement as that suggested would be better. It is, however, introduced here chiefly for the

Stepping-Stone Bridge, with Water-Lilies and Water Plants.

purpose of showing how well it enables one to enjoy various beautiful aquatic plants, from the fringed and crimson-tipped Bog-bean and graceful *Carex pendula* at the sides, to the golden Villarsia and Water Lilies sailing among the stones. Care is required to arrange it so that it may satisfy the eye, offer free passage to the water, and an easy means of crossing it at all times.

Plan of preceding figure.

Rock-gardens made on the margin of water are very often objectionable—rigid, abrupt, unworn, and absurdly unnatural. In no position is an awkwardness more likely to be detected; in none should more care be taken not to offend good taste. Good effects may be obtained on rocky mounds near water, by planting with moisture-loving rock plants; but even the grace and beauty of the finest of these will not relieve the hideousness of the masses of brick-rubbish and stone that are frequently placed by the margins of water.

Rock, near water, suited for bold vegetation.

The next figure, showing the fringe of a little island in one of the lakes of Northern Italy, may serve to show how irregularly and prettily the waves carve the rocky shore.

Margin of Island in Lake Maggiore.

Frequently in such places diminutive islands from a few feet to a few yards across are seen, and, when tufted with Globe-flowers, Ivy, and Brambles, are very pretty.

Rocky Water-margin (Oak Lodge).

THE ROCK-GARDEN FERNERY.

It is the fashion to make the hardy fernery in some obscure and sunless spot, in which it would be difficult to grow alpine plants, but there is no reason why it should not be made in more open positions, and as part of the rock-garden. No plants adhere more firmly to vertical rocks, or better sustain themselves in health without any soil, than some ferns. In a wild state we find the Maidenhair Fern and many other species rooted into little fissures in hard rocks. Some of our own small British Ferns are found on the face of dry brick walls, when they are not to be found growing on the ground, in the same neighbourhood.

The general idea is that Ferns want shade, humidity, and sandy vegetable earth; but, though these suit a great number of Ferns, others thrive under conditions the very opposite. The late M. Naudin, of the Institute, told me that the pretty little sweet-scented Fern, *Cheilanthes odora*, is found, even in that warm and sunny region, on the south side of bare rocks and walls, where it is exposed to the full rays of the sun, and is sought for in vain on northern exposures. In the middle of winter it is in full vigour, by the end of spring the fronds begin to dry, and through the torrid summer, when the stones of the walls are burning hot, its roots, fixed between the hot stones, are the only parts with life. In humid valleys and recesses it is not found. Other Ferns show like tendencies. This, by way of proof that some of the choice Ferns may not only be grown well in sunny positions, but better on them than elsewhere.

I was informed by Mr Atkins, of Painswick, who was the first to bring the little *Nothochlæna Marantæ* alive into this country, that he has had it in health on a sunny rock for many years, and without protection. It is reasonable to assume that many Ferns, which in a wild state are found in half-shady spots, would, in our colder clime, flourish best if permitted to enjoy the sun, while Ferns that inhabit rocks in countries not much warmer than our own, should always have the warmest positions we can give them. And in the case of the species that require shade, it is quite possible to grow them in recesses in the rock-garden, and in deep passages leading through it, even if a portion be not specially designed as a fernery. Some small species and varieties may be used in any aspect as a graceful setting to flowering plants. Among the select lists, that of the Ferns that thrive best in open exposed places may meet the wants of some, but where the fernery is specially designed as a part of the rock-garden, there is no necessity for any selection, as all hardy kinds may then be grown.

Entrance to Cave for Killarney Fern.

Even the rare Killarney Fern, usually kept in houses, may be grown successfully in a cave in the rock-garden. The illustration shows the entrance to Mr Backhouse's cave for growing this plant. It is in a deep recess, perfectly sheltered and surrounded by high rocks and banks clothed with vegetation. Here in the darkness grows the Killarney Fern, tufts of Hart's Tongue guarding the entrance.

ROCKS FORMED OF CONCRETE.

Picturesque effects may be effected in this way, and may be graced with shrubs and vigorous trailing plants, but it is unsuitable for alpine plants. When properly constructed, care is taken to make the interior of the cemented masses of deep beds of earth, leaving holes here and there in the face of the structure, from which plants can peep forth, while the top is left open, and may be planted with shrubs or trees, but the stony mound, free in every pore, or constructed of separate pieces of stone, is infinitely the best for the flora of the rocks. The plants that thrive on walls, and send their roots far into their crevices, cannot get the slightest footing on these large masses coated with cement; and little plants stuck in the "pockets," which the constructors leave here and there on the face of the edifice, rarely thrive or look happy. They should never be placed in such positions, and the rock-gardens of natural stone should be preferred at any sacrifice. Where, however, natural stone cannot be obtained, the cemented work may be used, and in positions where only the picturesque effect of rocks is sought. In places where it already exists, some improvement may be effected by banks of true alpine garden in open spots near, covering the artificial rock gracefully with low shrubs and hardy climbers, and coniferæ like the Swiss Pine, and Mountain Pine, and the Junipers.

Rocky Bank at Oak Lodge.

THE SMALL ROCK-GARDEN.

One of the simplest ways of enjoying alpine plants is in small rocky beds, arranged on the turf of some parts of the garden, cut off by trees or shrubs from the ordinary flower-beds. One of these will give more satisfaction than many a pretentious "rockwork," and by the exercise of a very little judgment is readily constructed, so as not to offend the nicest taste. I once induced the owner of a garden in the northern suburbs of London to procure a small collection of alpines, and try them in this way, and the result was such, that a few words as to how it was attained may be useful.

A little bed was dug out in the clay soil to the depth of two feet, and a drain run from it to an outlet near at hand; the bed was filled with sandy peat and a little loam and leaf-mould, and, when nearly full, worn stones of different sizes were placed around the margin, so as to raise the bed one foot or so above the turf. More soil was then put in, and a few rough slabs, arranged so as to crop out from the soil in the centre, completed the preparation for Sedums and Sempervivums, such Saxifrages as *S. cæsia* and *S. Rocheliana*, such Dianthuses as *D. alpinus* and *D. petræus*, Mountain Forget-me-nots, Gentians, little early bulbs, *Hepatica*. They were planted, the finer and rarer things getting the best positions, and, when finished, the bed looked a nest of small rocks and alpine flowers.

In about eight weeks the plants had become established, and the bed looked quite gay from a dozen plants of *Calandrinia umbellata*, that had been planted on the little prominences, flowering profusely. Another was made in the same manner, with more loam, however, and planted with subjects as different

Small Bed of Alpine Flowers.

from those in the other bed as could be got; confining them, however, to the choicest alpines, except on the outer side of the largest stones of the margin, where such plants as *Campanula carpatica bicolor* were planted with the best results.

The only attention these beds have required since planting has been to keep a free-growing species from overrunning plants, like *Gentiana verna*, to water the beds well in hot weather, and to remove the smallest weeds. With the exception of the fine *Gentiana bavarica*, every alpine plant grew well, and the beds showed fresh interest every week from the dawn of spring till late in autumn.

In such little-exposed beds some may fear the sun burning up their plants; yet the sun that beats down on the Alps and Pyrenees is fiercer than that which shines on the British garden. But, while the Alpine sun cheers the flowers, it also melts the snows above, and water and frost grind down the rocks into earth; and thus, enjoying both, the roots form healthy plants. Fully exposed plants do not perish from too much sun, but from want of moisture. Therefore, for the greater number of rock-plants, full exposure is one of the first conditions of culture — abundance of free soil under the roots and such a disposition of the soil and rocks that the rain may permeate through all, being also indispensable.

Alpine Plants growing in a level border.

An open, slightly elevated, and, if possible, quite isolated spot should be chosen, and a small rock-garden so arranged as to appear as if naturally cropping out of the earth. With a few cart-loads of stones and earth, good effects may be produced in this simple way.

Having determined on the position of the bed, the next thing to do is to excavate the ground to a depth of two feet, or thereabouts, and to run a drain from it if very wet. If not, it is better let alone, as with many kinds success depends upon the beds being continually moist; and in dry soils, instead of draining, it would be better to put in a subsoil of spongy peat, so as to retain moisture. As to soil, rock-plants are found in all sorts, but a turfy loam, with plenty of river sand added, will be found to suit a greater number of kinds than any other. If not naturally free and open, it should be so made by the addition of leaf-mould, cocoa-nut fibre, or, failing these, peat.

With the soil should be mixed the smallest and least useful stones and *débris* among those collected for the work, so that the plants to adorn the spot may send down their roots through the mixture of earth and stone. When this is well and firmly done, the larger stones may be placed—half in the earth as a rule, and on their broadest side, so that the mass, when completed, may be perfectly firm. Have nothing to do with tree-roots or stumps in work of this kind; they crumble away, and are at best a nuisance and a disfigurement in a garden. The intervening spaces may then be filled up, half with the compost and half with the stony matter, and the smaller blocks placed in position—the whole being made as diversified as may seem desirable, but without much show of "rock." When finished, it should look like a bit of rocky ground, and in no way resemble the "rockwork" of books and most gardens. Two or three feet will, as a rule, be high enough for the highest stones. In some of our public and private gardens want of means is given as an excuse for the presence of the hideous masses of rockwork that disfigure them. The plan here recommended is as much less expensive than these, as it is less offensive!

ROCK AND ALPINE FLOWERS IN BORDERS AND BEDS.

The most uninviting surfaces often afford a home to various forms of plant life: pavements, the stone roofs of old buildings, the stems and branches of trees, the faces of inaccessible rocks, and ruins, are all frequently adorned with ferns and wild flowers, and we are far from the end of simple ways of growing our Alpine favourites. The mixed-border system rightly done enables us to cultivate, with little trouble, many of the more vigorous alpine plants as edgings and carpets beneath the taller and more stately plants: dwarf Hairbells, Pinks, Phlox,

Cinquefoils, dwarf silvery Yarrows, purple rock Cresses, Rockfoils, Stonecrops, and Gentianella, all helping well in this way. In many positions the best of all edgings are those of natural stone, such as that shown in the wood-cut on this page. The cool soil below and behind the stones is the very place for rockplants that suffer in a hot season in dry soils, and many kinds may be grown in this way, as well and even better than in the most costly rock-garden.

Rough stone-edging to border, with Rock-Plants set behind it. In this simple way many of the most beautiful kinds may be admirably grown. (Engraved from a Photo by George Champion, in my garden.)

The common way of repeating the same plants at intervals is fatal to good effect here as elsewhere. The reverse of that is the true system for the best kind of mixed border. In a well-arranged one, no six feet of its length should resemble any other six feet of the same border. Certainly, it may be desirable to have several of a favourite plant; but any approach to planting the same thing in numerous places along the same line should be avoided. I should not, for instance, place one of the neat Saxifrages along in front of the border at regular intervals, fine and well suited as it might be for that purpose; but, on the contrary, attempt to produce in all parts totally distinct types of vegetation.

It is a great mistake to *dig* among choice rock plants, and therefore no pains should be spared in the preparation of the ground at first. If thoroughly well made then, there will be no need of any digging of the soil for a long time.

Many alpine plants, when grown in borders, are benefited

by being surrounded by a few half-buried rugged stones or pieces of rock. These are useful in preventing excessive evaporation, in guarding the plant when small and young from being trampled upon or overrun by coarse weeds or plants, and in keeping the ground firmer and cooler.

Alpine Plant on border surrounded by half-buried stones.

A few barrowfuls of stones—the large flints of which edgings are often made will do well, if better cannot be obtained—will suffice for many plants; and this simple plan will be found to suit many who cannot afford the luxury of a rock-garden. Lists of alpine plants suitable for the mixed border will be found in the selections at the end of the book.

ROCK-GARDENS ON LEVEL GROUND.

Mr F. Lubbock has been most successful in the cultivation of alpine flowers, in modest and simple ways, that so many may follow in any open spot of ground, and, acceding to my request, he writes of it as follows:—

"My experience is, that most alpine plants can be more easily and conveniently grown in the open ground, with little hillocks and ridges thrown up, so as to provide different aspects, and dryer or moister positions, than in the more imposing artificial 'rockery' constructions—the latter, if well made, do, no doubt, show off some plants to advantage, and are better suited to a few of the most difficult sorts; but they are expensive to build, and if, as usually happens, some spreading intruder establishes itself, it is far more troublesome to dislodge it. Then it is much more difficult to put in a plant properly in a rock crevice, and, with most alpines, it is of the greatest importance to plant them well and firmly at the outset. Moreover, it frequently happens that a mistake is made in the position given to a plant, and it is far easier to move it from the open ground than to pull it out of a rock crevice.

"I find it most convenient to grow the smaller and choicer plants in a separate part, where they can be more carefully tended. In another part I grow the stronger sorts which can hold their own, and this part I allow to be overrun by red and white wild Thyme, under which a number of small bulbs—several species of Anemones, Campanulas, and many other sorts—are quite happy.

"A few large weathered stones, judiciously placed, look well, and are often of advantage in giving a plant the aspect that suits it. It is usually recommended that such stones should be half buried, and no doubt many plants like to spread their roots down the side of a stone. On the other hand, this is just where some aggressive weed will run underground, and it often

Part of Rock-Garden on level ground at Emmots, Ide Hill, Sevenoaks, Kent.

happens that the only way to eradicate it effectually is to pull up the stone, causing a considerable upheaval. To obviate this, I find it generally more convenient to sink the stones about an inch only, and they can then be lifted and put back with very little disturbance.

"There are many disappointments in growing alpines, as in everything else, but they afford a constant and daily interest, and given a breezy open situation and a deep light soil, there should be many more successes than failures."

WALL GARDENS OF ROCK AND ALPINE FLOWERS.

Many plants that in gardens have carefully prepared soil grow naturally on the barest and most arid surfaces. Most of those who are blessed with gardens have usually a little wall surface at their disposal; and all such may know that some plants will grow thereon better than in the best soil. A mossy wall affords a home for some dwarf rock plants which no specially-prepared situation could rival; and even on well-preserved walls we can establish some little plants, which year after year will repay for the slight trouble of planting or sowing them. Now, numbers of alpine plants perish if planted in the ordinary soil of our gardens, and even do so where much pains is taken to attend to their wants. This often results from over-moisture at the root in winter, the plant being injured by our green winters inducing it to grow in the bitter winter and spring, when it ought to be at rest. By placing many of these rock plants where their roots enjoy a dry spot, they remain in perfect health. Many plants from mountains a little further south than our own, and from alpine regions, find on walls and ruins that stony firmness of "soil" and dryness in winter which make them at home in our climate. There are many alpine plants now cultivated with difficulty in frames, that any beginner may grow on walls.

Nor must it be supposed that a moist district is necessary, for the illustrations on pages 39 and 42 are engraved from photographs of walls built and covered with plants in a southern county in one year.

Sloping wall of local sandstone, supporting banks on each side of path, rock plants placed between each line of stones as wall was built. (Engraved from a photograph taken in my garden, by G. A. Champion.)

WALL PLANTS FROM SEED.

A good way to establish rock plants on walls is by seed. The Cheddar Pink, for example, grows on walls at Oxford much better than on the level ground, on which it often dies. A few seeds of this plant, sown in a mossy or earthy chink, or even covered with a little fine soil, would soon take root and grow, living, moreover, for years in a healthy state. So it is with most of the plants enumerated; the seedling roots vigorously into the chinks, and gets a hold which it rarely relaxes. But of some plants seeds are not to be had, and therefore it will be often necessary to use plants. In all cases, young plants should be selected, and as they will have been used to growing in fertile ground, or good soil in pots, and have all their little feeding roots compactly gathered up near the surface, they must be placed in a chink with a little moist soil, which will enable them to exist until they have struck root into the interstices of the wall. In this way several interesting species of Ferns are established, and also the silvery Rockfoils, and the appearance of the starry rosettes of these little rock plants (the kinds with incrusted leaves, like S. longifolia and S. lingulata) growing flat against the wall is strikingly beautiful.

While few have ruins and walls on which to grow alpine plants, all may succeed with many kinds by building a rough stone wall, and packing the intervals as firmly as possible with soil. A host of brilliant plants may be thus grown with little attention, the materials of the wall affording precisely the conditions required by the plants. To many species the wall would prove a more congenial home than any but the best constructed rock-garden. In very moist places, natives of wet rocks, and trailing plants like the Linnæa, might be interspersed here and there among the other alpines; in dry ones it would be desirable to plant chiefly the Saxifrages, Sedums, small Campanulas, Linarias, and plants that, even in hotter countries than ours, find a home on the sunniest and barest crags. The chief care in the management of this wall of alpine flowers would be in preventing weeds or coarse plants from taking root and overrunning the usually dwarf rock plants. When these intruders are once observed, they can be easily prevented from making any further progress by continually cutting off their shoots as they appear; it would never be necessary to disturb the wall even in the case of a thriving Convolvulus. The wall of alpine plants may be placed in any convenient position in or near the garden: there is no reason why a portion of the walls usually devoted to climbers should not be prepared as described. The boundary walls of multitudes of small gardens would look better if graced by alpine flowers, than bare as they usually are.

DRY STONE WALLS FOR ROCK PLANTS.

In garden formation, especially in diversified ground, what is called a "dry" wall is often useful, and may answer the purpose of supporting a bank or dividing off a garden quite as well as an expensive brick or masonry wall. Where the stones can be got easily, men used to the work will often make gently "battered" walls which, while fulfilling their first use in supporting banks, will make homes for rock plants which would not live one winter on a level surface in the same place. Blocks of sandstone laid on their natural "bed," the front of the stones almost as rough as they come out, and chopped nearly level between, so that they lie firm and well, no mortar being used, do well. As each stone is laid, slender-rooted rock plants are placed along in lines between with a sprinkling of fine earth, enough to slightly cover the roots and help them in getting through the stones to the back, where, as the wall is raised, the space behind it is packed with earth. This the plants soon find out and root firmly into. Even on old walls made with mortar, rock and small native Ferns often establish themselves, but the "dry" walls are more congenial to rock plants, and we may have any number of beautiful alpine plants in perfect health in them.

One charm of this kind of wall garden is that little attention is required afterwards. Even in the best-made rock-gardens things get overrun by others, and weeds come in;

but in a well-planted dry wall against an earth bank, we may leave plants for years untouched, beyond pulling out any weed that may happen to get in. So little soil, however, is put with the plants, that there is little chance of weeds, while moles—a nuisance in England—worms, and slugs are not such a trouble as on the level. If the stones were separated with much earth, weeds would get in, and it is best to have the merest dusting of soil with the roots, so as not to separate the stones, but let each one rest firmly on the one beneath it. The roots soon run back to the good earth behind, and it is surprising how soon good effects arise by this simple plan. It may be noticed that there is no pretension of "design" about these walls, made simply to do their work in supporting the bank.

Dry wall of sandstone blocks, supporting earth banks; plants placed as the wall is built; wall trellised with Bamboos for Roses and other climbers.

PLANTS FOR "DRY" WALLS.

Arabis, Aubrietia, and Iberis are among the easiest plants to grow; but as such things can be grown without walls, it is hardly worth while to put them thereon. Between these stones is the very place for Mountain Pinks, which thrive better there than on level ground; the dwarf alpine Hairbells, while the alpine Wallflowers and creeping rock plants, like the Toad Flax (Linaria), and the Spanish Erinus, are quite at home there. The Gentianella does very well on the cool sides of such walls, and we get a different result according to the aspect. All our little pretty wall Ferns, now becoming so rare where hawkers abound, thrive on such walls, and the alpine Phloxes may be used, though they are not so much in need of the comfort of a wall as the European alpine plants, the Rocky Mountains dwarf Phloxes being very hardy and enduring on level ground. The Rockfoils are charming on a wall, particularly the silvery and mossy kinds, and the little stone-covering Sandwort (Arenaria balearica) will run everywhere over such walls. Stonecrops and Houseleeks do well, but are easily grown in any open spot of ground. In many cases the rare and somewhat delicate alpines, if care be taken, would do far better on such a wall than as they are usually cultivated. Plants like the Thymes are quite free in such conditions, also the alpine Violas, and any such rock trailers as the blue Bindweed of North Africa. I have hundreds of plants of Gentiana acaulis thriving on such walls, to the surprise of all who see them in bloom.

We have spoken of "dry" walls, which are necessary, apart from their flower life, that is to say essential, for the support of banks by the side of "cuttings," or where terraces are cut out of steep ground; the sides of steps, ascending banks, and a variety of positions which will occur in diversified and in hilly districts.

Rock plants established on an old wall.

ALPINE FLOWERS [Part I.

These are by far the best positions, as in nearly every case we place our stone against the bank, ensuring moisture and food behind. Often walls are made straight against terraces which would be quite as well made in this way, with a gentle "batter" or slope backwards, and built with earth between the stones; they would be as good for shelter and for supporting terrace banks, and even for climbers, when the shade of Tea Rose foliage and other plants would not prevent Ferns and many plants from growing well. In fact—in the case of walls facing due south in dry seasons—the shade of the creepers above would help the plants a little against the power of the sun.

On level ground there is no need for any dry walls support-

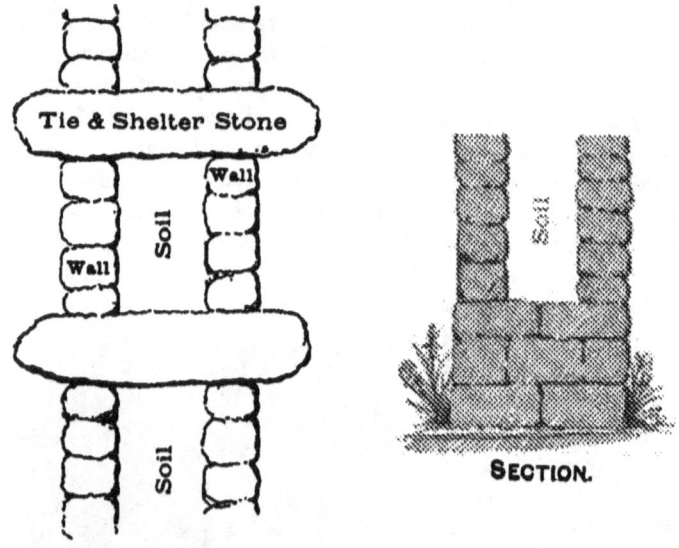

PLAN
Hollow wall for rock plants, forming dividing line round yard. (See page 46.)

ing banks, and where rock flowers on walls are desired, we may have to make a wall away from all support of earth banks, but which also will suit the cultivation of rock plants. Here a *hollow* wall and a variety of plants may give us a good result, the principal being to get our mass of soil in the centre of the wall, and make it very firm, but so that rain will refresh it. It is clear such a wall might take the place of the dividing lines we often have in gardens, separating different gardens or plots, and the following is a case in which such a wall was made, with good results.

"We are told that Solomon knew all 'green things,' from the Cedar of Lebanon to the Hyssop which sprang from the wall, and there is no doubt that wall gardening began soon after walls themselves were made. The beautiful wall garden which Nature had made on the ruins of the Colosseum is now destroyed, but the Wallflowers and Catchfly yet linger on the sunny castle rock at Nottingham, and the ruins at Conway are a study every summer, so beautiful is the Centranthus, which sows itself among the stones. At Dinan the top of an old entrance doorway is draped with Ferns and weeds, with delicately poised Bellflowers and Yarrow-heads, white as the sea foam. Wherever old walls or ruins exist in gardens or pleasure-grounds, it is easy to beautify them by sowing seeds of the many beautiful flowers which luxuriate in such positions. Wallflowers, Snapdragons, Erinus, and some species of Dianthus grow perfectly well, naturalise themselves, in fact, on sunny walls, while on shady damp ones many Ferns grow equally well, often better on a wall than elsewhere. A good plan to get Ferns to grow on a damp, shady, old wall is to wash off the spores from Asplenium, Scolopendriums, Ceterach, and Wall Rue, into a pail of tepid water, which may then be dispersed over the wall by means of a syringe. It is something for us to know that a broken stone or the crumbling edge of a brick may nourish in sunshine flower beauty of the highest, or that in shade it may yield us feathery drapery of tenderest Fern fronds. A rough stone-topped wall may become a garden of Sedum, Saxifrage, Erinus, both purple and white, and of many other rock plants. There are some mountain plants that never grow better or look more beautiful than when grown on rough-topped walls or in the interstices of stony edgings. The Erinus is one of the best wall-plants, and sows itself every year. Ramondia and Edelweiss both love to grow wedged tightly in among the stones.

"A WALL MADE FOR ROCK PLANTS.

"Having that worst of all things in a garden—viz. a rubbish and manure-yard somewhat exposed to a public road—it became a necessity to erect a shelter wall, so as to secure more privacy and to conceal from the public gaze a sort of laboratory necessary in every garden. Having that old proverb about 'two birds with one stone' in my mind's eye, I resolved to make the wall not only a shelter, but also an object pleasant to the eye as well. This has now been done fairly well, as I imagine, by the building of a hollow wall topped with tie or binding stones, and pocketed for the reception of soil and plants, as shown in the diagrams on page 44, made to a scale of half an inch to the foot. In such a plant-wall the principle is everything, and the proportions may be varied to suit any special conditions, circumstances, or surroundings. The wall is a little over 4 feet high and over 2 feet through, and 30 yards or 40 yards in length. Having filled up the hollow centre of the wall with suitable soil, I planted the top with Iris of the I. germanica and the I. pumila sections, with Cloves, Carnations, Pinks, Linarias, Aubrietias, Stonecrops, Edelweiss, and Sempervivums; but I am especially anxious to see established on its face a group of the Californian Zauschneria, which does not always flower well with me on the ground level, except during very hot, dry summers.

"A wall of the above size may be made by any man handy at stonework, and at no great cost. The stones I was very fortunate in procuring almost free of charge, and every one of them is precious, as having originally formed a portion of the Trinity College Library, removed during alterations. They have come from the world of books into a world of flowers, and in a short time they will, I hope, be crowned with blossoms and green leaves.—F. W. B."

MOUNTAIN SHRUBS FOR THE ROCK-GARDEN.

It might well be borne in mind that there is in Nature no hard-and-fast line, like the little divisions we make for our convenience in books, and though the most alpine of plants are very

Rhododendrons among natural rock at Howth, Co. Dublin. (Engraved from a Photograph sent by Mr Geo. E. Low.)

tiny evergreen herbs on all hilly and mountain ground, there is yet much beauty of shrub life on the mountains, from that of the

Heaths of our own land to the Rhododendrons of the European and other alpine regions. Therefore, it is right in all ways to associate shrubs with the rock-garden, and on its outer parts or in groups near it, seek beauty from such shrubs as are here named, and others to come, as the flora of northern regions becomes better known to us. Danger from the association of shrubs of a spreading nature with little rock plants may be avoided by grouping all shrubs on banks, or groups, apart from the place of the Gentians, Androsaces, and other fairies of the high rocks and alpine meadows. Even without any attempt at a rock-garden made in the ordinary garden-way, there are many places in various parts of our islands where lovely rock-gardens may be made by merely planting the natural rocks as they come out in their own beautiful way whether on the often bare hills of Wales, the many lovely rocky sites on the fringe of mountains around Ireland, Scotland, northern and southern England, and even on the sandstone rocks—quite near London —in Sussex and Kent. In such places, without set design or much care, we may enjoy the most enduring and the easiest to form of rock-gardens. Another reason for making bush rock-gardens about natural rocks cropping out of the ground is that the soil about is often the sort we seek for evergreen shrubs of the choicer kinds, being decayed rock, often of a peaty or sandy kind, and the best for Rhododendrons, Azaleas, dwarf Kalmias, Heaths, and many shrubs that in Nature inhabit the mountains, so that where the natural rock breaks out, the very conditions so very difficult to secure in the stoneless lowland country exist. As an example of good work on such ground, we quote this about planting rocky ground at Howth, near Dublin, by Mr Burbidge, in the *Field* :—

"Coming upon them rather suddenly, the flashes of colour amongst the grey crags are startling in their intensity. A shower had just passed over the hillside, and a gleam of sunshine illumined the flowers, which shone out in all shades of crimson and purple, and of orange and vermilion, softening down in shady corners into the richest of old gold. Great rocks, like the moraine of some old glacier, are piled and scattered on a sloping surface, above which great masses of old Cambrian formation tower seemingly into the sky.

A rocky path leads one up and down, now closed in overhead by Hawthorns embowered in Honeysuckle, Vine, and now open and clear, and as you pick your way over matted tree-roots or past slippery rocks, the acres of Azaleas and Rhododendrons flash out in the evening sunshine, each cluster glowing like jewelled lamps full of coloured light. They are mostly garden kinds or hybrids, but there are noble plants of the Himalayan R. Thomsonianum, R. Falconeri, and R. Edgworthii amongst them. The colours vary from white and soft lilac-purple through all shades of red and crimson, the complimentary shades of yellow, orange, and ivory-white being supplied by occasional groups of coloured Azaleas, with their sunrise and sunset shades and hues. There are, no doubt, far finer collections of Rhododendrons in Ireland, as also in Cornwall and Devon and elsewhere, but the great charm at Howth is that the picturesque position and the grouping of the Rhododendrons form such a succession of pictures, no two alike. An old traveller, whom we met here, told us: 'I have seen far finer Rhododendrons and far more noble rocks, but I must say I have never seen such glorious masses of colour and such picturesque rocks associated as they are here.' The rocky slopes and rocky scarps, on which the shrubs are now so beautiful, formed originally a sheltered little wood of Birch, Larch, Scots Fir, Oak, Mountain Ash, and Hawthorn, overrun with Woodbine, and in the more open spots by Gorse and Brambles. The floor of the little forest then, as now, was carpeted with Bluebells and Primroses, Stitchwort, Anemones, Wood Sorrel, and Ferns of a stature not often seen, even in Ireland. There was but scant root room in many places, and little or no soil, but men brought down and up peat, earth, and leaf-mould to chink and cleft, or rocky hollows and crevices, and to-day the result is seen and felt by all who, like the Japanese, come here on a June-day pilgrimage to see the flowers."

Though such natural situations are impossible to many, they are not at all essential for the cultivation or the good effect of mountain-shrubs, as we have proof in the garden at Warley Place, and other lowland gardens, where the rock shrubs are such a feature, garlanding the outer parts of the rock-gardens—Wild Rose, Azalea, Furzes, Sun-Roses, Brooms, Daphnes, and many other shrubs clustering about the banks and often grouped on the turf.

Whatever difficulty the cultivation of true alpine plants may present in certain conditions, there is little or none in connection with the mountain shrubs, and many of them are among the hardiest shrubs of the mountains of N. America and Asia.

TREES AND ALPINE GARDENS.

We often see trees, more or less suitable, planted about the alpine garden, and sometimes above the level of the plants. If possible, this should be avoided. Although alpine and rock-plants and shrubs *may* sometimes occur in woods, yet, as a

general rule, the trees cease from the hills before we come to the true dwarf alpine plants. If any shelter or dividing mass of trees is desired near the alpine garden, the trees chosen should always be mountain kinds, such as the Swiss Pine (P. Cembra), Juniper, Savin (also a Juniper), dwarf rock Pine (P. Montana), interspersed, if desired, with a few summer-leafing northern trees, like the Beech, Birch, and Mountain Ash. The Spruces and Pines of the Rocky Mountains of N.W. America might also be used, "holding them together" in groups where possible.

JAPANESE DWARFED TREES FOR THE ROCK-GARDEN.

There has been much talk of late years of these, of which numbers have been brought to this country and, still more, to America, some of the plants very unworthy of a place in a good garden, as they too often resemble the refuse of the nurseries. Among the best, however, there are some really interesting things, especially plants of the Cypress tribe, which occasionally retain their picturesque forms, although on such a small scale, and some graceful deciduous plants and shrubs like Wistaria, which are pretty grown in that way. Now, this curious and ancient way of growing plants, which seems so strange and new to many of us, is undoubtedly based on facts of Nature, and has its origin in the habits of plants on the high mountains often starved and dwarfed. We may see such dwarf and often distorted trees and shrubs on high rocks or mountains, or otherwise starved out of their natural vigour and habit by unnatural exposure, cold, or drought. We see it in the Alps occasionally, and even in the stately cedars of Lebanon and Atlas we see them in many different shapes, dwarf and stunted, and yet always beautiful in form. This being so, the true place for these quaint shrubs is the rock-garden, where they might be grouped together near a little streamlet on a modest bank of rocks. They are arranged in this way prettily at Warnham Court, and where rocks and shrubs are associated with the true alpine plants (as I think they should always be where there is room enough), there these quaint little trees come in very well.

Mr Alfred Parsons writes:—"The Japanese dwarf trees in their gardens, which are essentially rock-gardens, are planted among stones, which probably helps to stunt their growth, but besides this, they are most carefully trimmed to keep them to the desired size and shape—sometimes this form is quite stiff and symmetrical, especially in the case of Azalea bushes; more often it is a miniature of the characteristic shape of the tree in Nature under similar conditions, or a suggestion of some celebrated tree of the kind grown."

THE ALPINE MARSH-GARDEN.

In the great mountain regions, marshy ground and boggy places are frequent, and some of the fairest of the mountain's flowers adorn them, and may only be well grown in like conditions, happily easy to imitate. Therefore, while water as a separate element is not a necessity of even a noble rock-garden, some little place for marsh plants is needed, if we are to see the beauty in our gardens of many singularly pretty and some brilliant plants.

THE MARSH-GARDEN.

The marsh-garden is a home for the numerous children of the wild that will not thrive on our harsh and dry garden borders, but must be cushioned on moss or grown in moist peat. Many beautiful plants, like the Wind Gentian and Creeping Harebell, grow on our own marshes, much as these are now encroached upon. But even those acquainted with the beauty of our bog-plants have but a feeble notion of the multitude of charming plants, natives of northern and temperate countries, whose home is the open marsh or boggy tract. In our own country, we have been so long encroaching upon the wastes that we come to regard them as exceptional tracts all over the world. But when one travels in northern climes, one soon learns what a vast extent of the world's surface was at one time covered with bog. In North America, day after day, even by the side of the railroads, we may see the vivid blooms of the Cardinal-flower springing from the wet peaty hollows. Far under the shady woods stretch the black bog-pools, the ground between being so shaky that we move a few steps with difficulty. One wonders how the trees exist with their roots in such a bath, and where the forest vegetation disappears the American Pitcher-plant (*Sarracenia*), Golden Club (*Orontium*), Water Arum (*Calla Palustris*), and a host of other handsome and interesting plants cover the ground for hundreds of acres, with perhaps an occasional slender bush of Laurel Magnolia (*Magnolia glauca*) among them. In some parts of Canada, where the painfully long and straight roads are often made through woody swamps, and where the few scattered and poor habitations offer little to cheer the traveller, he will, if a lover of plants, find much beauty in the ditches and pools of black water beside the road, fringed with Royal and other stately Ferns, and with masses of water-side plants.

Southwards and seawards, the marsh-flowers become tropical in size and brilliancy, as in the splendid kinds of *Hibiscus*, while far north, and west, and south along the mountains, the beautiful Mocassin-flower (*Cypripedium spectabile*) grows the queen of the peat-bog and of hardy orchids. Then in California, all along the Sierras, a number of delicate little annual plants grow in small mountain bogs long after the plains are parched, and vegetation has disappeared from the dry ground. But who shall record the beauty and interest of the flowers of the wide-spreading marsh-lands of this globe of ours, from those of the vast wet woods of America, dark and brown, where the fair flowers only meet the eyes of water-snakes and frogs, to those of the breezy uplands of the high Alps, far above the woods, where the little mountain-marshes teem with Nature's most brilliant flowers, waving in the breeze? Many mountain-swamp regions are as yet as little known to us as those of the Himalaya, with their giant Primroses and strange and lovely flowers. One thing, however, we may gather from our small experiences—that many plants commonly termed "alpine," and found on high mountains, are true marsh-plants. This must be clear to any one who has seen our Bird's-eye Primrose in the wet mountain-side bogs of Westmoreland, or the Bavarian Gentian in the spongy soil by alpine rivulets. We enjoy at our doors the plants of hottest tropical isles, but many wrongly think the rare bog-plants, like the minute alpine plants, cannot be grown well in gardens. Like the rock-garden, the marsh-garden is seldom seen well made, and with its most suitable plants.

In some places, naturally boggy spots may be found, which may be converted into a marsh-garden, but in most places an artificial is the only possible one. It may be associated with a rock-garden with good effect, or it may be in a moist hollow, or may touch upon the margins of a pond or lake. By the margins of streamlets, too, little bogs may often be made. But the mania for draining springy and marshy spots has in most places left little chance of a natural site, such as might readily be turned into a marsh-garden. A tiny streamlet may be diverted from the main one to flow over the adjacent grass—irrigation on a small scale. Another good kind could be made at the outlet of a small spring. It was in such little bogs around springs that I found the Californian Pitcher-plant in dry parts of California. In some of these positions the ground will often be so moist that little trouble beyond digging out a hollow to give a different soil to some favourite plant will be needed. Where the marsh-garden has to be made in ordinary ground, and with none of the above aids, a hollow must be dug to a depth of at least two feet, and filled in with any kind of peat or leaf soil that may be obtainable. If no peat is at hand, turfy loam with plenty of leaf-mould, etc., must do for the general body of the soil; but, as there are some plants for which peat is indispensable, a small portion of the beds should be of that soil. The bed should be slightly below the surface of the ground, so that no rain or moisture may be lost to it. There should be no puddling of the bottom, and there must be a constant supply of water. This can be supplied by means of a pipe in most places—a pipe allowed to flow forth over some firmly-tufted plant that would prevent the water from tearing up the soil.

As to planting the marsh-garden, all that is needed is to put as many of the under-mentioned plants in it as can be obtained, and to avoid planting in it any rapid-running sedge or other plant, as in that case, all satisfaction with the garden is at an end. Numbers of Carexes and like plants grow so rapidly that they soon exterminate choice marsh flowers. If any roots of sedges, etc., are brought in with the peat, every blade they send up should be cut off with the knife just below the surface; that is, if the weed cannot be pulled up on account of being too near some precious plant one does not like to disturb. All who wish to grow the tall sedges and other coarse bog-plants should do so by the pond-side, or in moist or watery places set apart for the purpose. Given the necessary conditions as to soil and water, the success of the marsh-garden will depend on the continuous care bestowed in preventing rapidly growing or coarse plants from exterminating others, or from taking such a hold in the soil that it becomes impossible to grow any small plant in it. Couch and all weeds should be exterminated when very young and small.

The following are the bog and marsh plants at present most worthy of culture; but there are many not yet in cultivation, equally lovely.

A SELECTION OF MARSH PLANTS.

Anagailis tenella ; Calla palustris ; Caltha in var. ; Campanula hederacea ; Chrysobactron Hookeri ; Coptis trifolia ; Cornus canadensis ; Crinum capense ; Cypripedium spectabile ; Drosera ; Epipactis ; Galax ; Gentiana ; Helonias ; Iris Monnieri, ochroleuca, sibirica ; Leucojum æstivum, Hernandezii ; Linnæa ; Parnassia ; Lycopodium in var. ; Menyanthes trifoliata ; Myosotis dissitiflora, palustris ; Nierembergia rivularis ; Orchis latifolia and vars., laxiflora, maculata ; Orontium aquaticum ; Pinguicula in var. ; Primula rosea, sikkimensis, farinosa ; Rhexia virginica ; Sagittaria in var. ; Sarracenia purpurea ; Saxifraga Hirculus ; Spigelia marilandica ; Swertia perennis ; Tradescantia virginica ; Trillium ; Lastrea Thelypteris.

The above are suitable for the select marsh bed kept for the most beautiful and rare plants ; and among these, as has been stated, should be planted nothing which cannot be readily kept within bounds. To them lovers of British plants might like to add such native plants as Malaxis paludosa ; but it is better, as a rule, to select the finest, no matter whence they come. Among the most interesting plants for the bog-garden are the Pitcher-plants of North America. Some may doubt if the American Pitcher-plant (*Sarracenia purpurea*) would prove hardy in the open air in this country. It certainly is so, as one might expect from its high northern range in America. It will thrive in the wettest part of the bog-garden and in its native country I usually observed the Pitchers half buried in the water and sphagnum, the roots being in water.

As however no natural opportunities occur in many places, the plan followed by a very successful cultivator may be useful here.

MR LATIMER CLARK ON FORMING A BOG-GARDEN.

"*Artificial Bogs—How to make them and what to plant in them.*— All that is requisite to form a bog-garden is to form a hollow space which will contain water. The simplest way is to buy a large earthenware pan or a wooden tub, bury it 6 inches beneath the surface of the ground, fill it full of broken bricks and stones and water, and cover with good peat soil; the margin may be surrounded with clinkers or tiles at discretion, so as to resemble a small bed. In this bed, with occasional watering, all strong-growing bog plants will flourish to perfection; such plants as Osmundas and other Ferns, the Carexes, Cyperuses, etc., will grow to a large size and make a fine display, while the cause of their vigour will not be apparent.

"A more perfect bog-garden is made by forming a basin of brickwork and Portland cement, about 1 foot in depth; the bottom may be either concreted or paved with tiles or slates laid in cement, and the whole must be made water-tight; an orifice should be made somewhere in the side, at the height of 6 inches, to carry off the surplus water, and another in the bottom at the lowest point, provided with a cork, or, better still, a brass plug valve to close it. Five or six inches of large stones, brick, etc., are first laid in, and the whole is filled to the top with good peat soil, the surface being raised into uneven banks and hillocks, with large pieces of clinker or stone imbedded in it, so as to afford drier and wetter spots; the size and form of this garden or bed may be varied at discretion. An oval or circular bed, 5 or 6 feet in diameter, would look well on a lawn or in any wayside spot, or an irregularly formed corner may be rendered interesting in this way; but it should be in an open and exposed situation; the back may be raised with a rockwork of stones or clinkers, imbedded in peat, and the moisture ascending by capillary action will make the position a charming one for Ferns and numberless other peat-loving plants. During the summer the bed should always contain 6 inches of water, but in winter it may be allowed to escape by the bottom plug. It is in every way desirable that a small trickle of water should constantly flow through the bog; ten or twelve gallons per diem will be quite sufficient, but where this cannot be arranged, it may be kept filled by hand. The sides of such a bog may be bordered by a very low wall of flints or clinkers, built with mortar diluted with half its bulk of road-sand and leaf-mould, and with a little earth on the top; the moisture will soon cause this to be covered with Moss, and Ferns and wall plants of all kinds will thrive on it.

"Where space will permit, a much larger area may be converted into bog and rock-work intermingled, the surface being raised or depressed at various parts, so as to afford stations for more or less moisture-loving plants. Large stones should be freely used on the surface, so as to form mossy stepping-stones; and many plants will thrive better in the chinks formed by two adjacent stones than on the surface of the peat. In covering such a large area, it is not necessary to render the whole area water-tight. A channel of water about 6 inches deep, with drain pipes and bricks at the bottom, may be led to and fro, or branched over the surface, the bends or branches being about 3 feet apart. The whole, when covered with peat, will form an admirable bog, the spaces between the channels forming drier portions, in which various plants will thrive vigorously.

"Perhaps the best situation of all for a bog-garden is on the side of a hill or on sloping ground. In this case the water flows in at the top, and the surface, whatever its form or inclination, must be rendered water-tight with Portland cement or concrete. Contour or level lines should be then traced on the whole surface, at distances of about 3 feet, and a ridge of two bricks in height should be cemented on the surface along each of the horizontal lines. These ridges, which must be perfectly level, serve to hold the water, the surplus escaping over the top to the next lower level. Two-inch drain tiles, covered with coarse stones, should be laid along each ridge, to keep the channel open, and a foot of peat thrown over the whole. Before adding the peat, ridges or knolls of rock-work may be built on the surface, the stones being built together with peat in the interstices. These ridges need not follow the horizontal lines. The positions thus formed are adapted both to grow and to display Ferns and alpine plants to advantage.

"There is another way in which a minute stream of water may be turned to advantage, and that is by causing it to irrigate the top of a low wall; such a wall should be built 12 inches high, the top course being carefully laid in Portland cement. A course is then formed by bricks projecting over about 2 inches at each side, with a channel left between them along the centre of the wall, which must be carefully cemented. Small drain pipes are laid along this channel and fitted in with stones. Large blocks of burr or clinker are then built across the top of the wall, with intervals of 12 or 15 inches between them, and these are connected by narrow walls of clinker on each side, so as to form pockets, which are filled with a mixture of peat and sandy loam. The projecting masses of burr stand boldly above the general surface, and, occurring at regular intervals, give a castellated character to the wall, which may be about 2 feet high when finished. Hundreds of elegant wall plants find a choice situation in the pockets, which are kept constantly moist by the percolation of the water beneath them, while Sempervivums and Sedums clothe the projecting burrs. In fact, with Wallflowers, Snapdragon, Cistuses, and Sedums, such a wall forms a garden of blossom throughout the whole spring and summer.

"In addition to true bog plants, almost all the choice alpines will luxuriate and thrive in the drier parts of the bog-garden better than in an ordinary border or in pots. Perhaps the most charming plants to commence with are our own native bog plants—Pinguicula, Drosera, Parnassia, Menyanthes, Viola palustris, Anagallis tenella, Narthecium, Osmunda, Marsh Ferns, Sibthorpia, Linnæa, Primula, Campanula, Saxifraga Hirculus, aizoides, and stellaris; Mimulus luteus, Cardamine, Leucojum, Fritillaria, Marsh Orchises, and a host of plants from our marshes, and from the summits of our higher mountains, will flourish as freely as in their native habitats, and may all be grown in a few square feet of bog; while dwarf Rhododendrons, Kalmias, Gunnera scabra, the larger Grasses, Ferns, Carexes, etc., will serve for the bolder features.

"I have not space to enumerate the many foreign bog plants of exquisite beauty which abound, and which may be obtained from our nurseries, although many of the best are not yet introduced into this country; in fact, one of the great charms of the bog-garden is that everything thrives and multiplies in it, and nothing ever droops or dies, the only difficulty being to prevent the stronger plants from overgrowing and eventually destroying the weaker ones."

Ferns on an old wall.

Mr. F. W. Meyer, an excellent and experienced worker in rock-gardening, writes well of the formation of bog-beds in the Garden:

"Though the term may be suggestive of a formal bed, there should really be no hard-and-fast outline in the rock-garden, and the bog-bed should be harmonised with its surroundings in such a way as to make it impossible to discern its extent. We might have several such beds in different positions regarding light, as some marsh plants thrive in the sun, while others delight in shady nooks, and the wants of the plants must therefore be our first consideration.

"*Bog Beds without Cement* are to be recommended when the water supply is unlimited: if in connection with a pond fed by a streamlet, so much the better. The overflow water of the pond can then be used for feeding the bog-bed, or if the water should only run occasionally, a short pipe fitted with a regulating tap may be let into the side of the pond and connected with the bog-bed, this arrangement having the advantage of enabling us to keep the water supply under control. The construction of such a bed is simplicity itself; dig a pit of the desired size about 18 inches deep, spreading at the bottom a layer of porous stones, brickbats, and a little charcoal, and covering the same with pieces of peat. Peaty soil, mixed with a little leaf-mould, Sphagnum Moss, sand and broken stone, is then added till the pit is filled up. A few larger stones are then placed with some care, partly with a view to effect, and partly to give shade or shelter to the plants to be grown by their side. If the ground is heavy, the bottom of the pit must be drained to get rid of stagnant water; but if of a porous nature, the water will soak away naturally through the bottom of the bed thus prepared.

"*The Cemented Bog Bed.*—Though at first involving a little more expense, this will be found of great advantage in rock-gardens on a small scale, where the supply of water comes through a small pipe. It is an irregular underground pond, made of cement concrete, and filled with soil as well as with water, to a depth of 12 inches to 15 inches. Besides being fitted with a supply pipe and tap, so arranged as to be within easy reach (though hidden from view), it should have an overflow and an outlet pipe fitted with another tap for completely emptying the whole at will. If the bed is large, it would be well to arrange for stepping-stones here and there to ensure easy access to the plants. When space is limited, I often use for this purpose thin flat stones raised a little and supported at each end by a miniature pillar of bricks and cement, thus forming a little bridge, as it were, and admitting of the space between the little pillars and beneath the stones being filled with the proper soil. That every trace of cement-work would be hidden by soil, stones, or plants, goes without saying. One advantage of this sort of bed is that the water supply and drainage can be regulated in the simplest manner by the mere turning of a tap.

"*The Partly Cemented Bog Bed.*—The advantage I claim for this lies in the facility it affords for graduating moisture, which makes it possible to grow plants requiring different degrees of humidity in the same bed. First of all, a bog bed is constructed after the manner described above under the heading of 'Bog Beds without Cement,' but instead of having the sides more or less upright, they are kept gently sloping. A winding trench is then excavated through this bed and secured with cement concrete—a water-tight trench not more than a foot wide and 6 inches or 8 inches deep. The cemented sides should be level, so that, when filled, the water would flow evenly over the sides and into the outer parts of the bed, so giving different degrees of moisture between the cemented centre and portions and the sloping sides, from which the water would drain away naturally. Before the water is admitted, the trench is filled with loose stones and brickbats, and is then bridged over with large pieces of peat, and covered with a few inches of suitable soil. It is then levelled, so as to show no visible difference from the rest of the bed. As soon as the trench is filled with water, however, the latter will rise by capillary attraction not only through the pieces of peat, but also the soil above it, showing even on the surface of the soil the course of the water-trench beneath. But if the soil is filled up to such an extent that the rising water cannot be seen on the surface, it would be well to mark the course of this underground trench with a few sticks projecting through the soil, to guide us when planting, and enabling us to put all plants requiring an extra degree of moisture directly over the water-trench where the roots could help themselves to the water.

"On a steep slope, where the forming of such beds would be difficult, an ordinary lead pipe, a few inches underground and perforated at intervals, will be found useful, and may be regulated so as to supply water trickling through the soil throughout the summer."

WATER-PLANTS IN THE ROCK-GARDEN.

The water-garden has no essential connection with alpine or rock-gardens for this reason (among others), that millions of acres of many countries are covered with beautiful rock plants with no water near. But as some water often occurs in connection with the rock-garden, it may as well be treated rightly. Many beautiful natural alpine gardens are far above all water, except what falls from the clouds as snow or rain. Many alpine plants live on sunny rocks and in high waterless plateaux, and my own wish in the formation of alpine gardens would be to get as near as I could to the same conditions. I would seek exposure to all winds and weathers, and on as elevated and open airy spots as I could, keeping my stream, banks, and water-margins in the vale for other and stouter plants. Of late years a precious aid has come to us in the shape of many beautiful uncommon things for the water-garden, and above all, the hardy water-lilies, raised by M. Latour Marliac, which give us in a cold country such beauty as at one time was thought to be only possible in sub-tropical countries. We now have water-lilies so bright in colour, as hardy as a Dock, and it is impossible to resist such beauties, especially when we may grow them in a small pool, and in close relation to our rock-garden, if such we desire. A skilfully-formed lakelet will be prettier than a stiff tank, but in either it is quite easy to grow water-lilies, the essential thing being to plant in a good depth of mud or soil. There is nothing better than the mud which is washed down by little streams, but any good earth will do, and the result of planting in the soil will be much better than if we had put them in pots or tubs of any kind. The beauty and length of bloom of these water-lilies makes them a very precious aid in the garden, while for the margins of our lakelet we have many graceful plants in the way of Reeds, Rushes, Arrowheads, and many water-plants, such as Day-lilies, tall Irises, Swamp Lilies, Loosestrife, Golden Rod, Cardinal flowers, and the nobler hardy Ferns, like the Royal and Feather Ferns. It is necessary to keep off the common water-rat, which cuts off the flowers and eats them on the bank-side, and also the common water-hen, which picks at and destroys the flowers; and, generally, it may be said that it is not possible to have water-fowl and living creatures if we would grow water-lilies well.

The new kinds, which are now coming out, demand more careful treatment than the well-known ones, and should be kept apart in small tanks. The older and bolder kinds may be put out in the open water with the greatest confidence. I have grown some of them in open ponds fully exposed to storms, and with good results; but always planting in the natural mud, and in a good depth of it if possible, and that is not difficult where mud is washed in freely by streamlets.

For those who desire to go into the question of water-gardening more at length, there is a fuller account in the "English Flower Garden," than we can find room for here. And there are often happy incidents where a natural stream would come near us to give its precious help, and there are various cases in which water—either moving or still water—may be happily associated with marsh and alpine gardens.

The Water Garden at Fota, Co. Cork.

HARDY WATER PLANTS.

Water and water-side plants are often intimately associated with rock-gardens, and much beauty may be added to the margins, and here and there to the surface, of water, by waterplants. Usually we see the same monotonous vegetation all round the margin if the soil be rich; in some cases, where the bottom is of gravel, there is little or no vegetation, but an ugly line of washed earth between wind and water. In others,

The White Water-Lily.

water-plants accumulate till they are a nuisance and an eyesore —I do not mean submerged plants like *Anacharis*, but such as the water-lilies, when they get matted.

One of the prettiest effects I have seen was a sheet of *Villarsia nymphæoides* belting round the margin of a lake near a woody recess, and it is too seldom seen in garden waters, being a pretty little water-plant, with its Nymphæa-like leaves and many yellow flowers.

Not rare—growing, in fact, in nearly all districts of Britain —is the Buckbean or Marsh Trefoil (*Menyanthes trifoliata*), with flowers elegantly fringed on the inside with white filaments, and the round unopened buds blushing on the top with a rosy red like that of an apple-blossom. In early summer, when seen trailing in the soft ground near the margin of a stream, this plant has more charms for me than any other marsh-plant. It will grow in a bog or any moist place, or by the margin of any water, and though a common native plant, it is not half enough grown in garden waters. For grace, few plants surpass *Equisetum Telmateia*, which, in deep soil, in shady moist places near water, often grows several feet high, the long, close-set, slender branches depending from each whorl in a singularly graceful manner.

For a bold and picturesque plant on the margin of water nothing surpasses the great Water Dock (*Rumex Hydrolapathum*), which is dispersed over the British Isles; it has leaves fine in aspect and size, becoming of a lurid red in the autumn. The *Typhas* must not be omitted, but they should not be allowed to run everywhere. The narrow-leaved one (*T. angustifolia*) is more graceful than the common one (*T. latifolia*). *Carex pendula* is excellent for the margins of water, its elegant drooping spikes being quite distinct in their way. It is rather common in England, more so than *Carex Pseudo-cyperus*, which grows well in a foot or two

The Great Water Dock.

of water or on the margin of a muddy pond. *Carex paniculata* forms a strong and thick stem, sometimes three or four feet high, somewhat like a tree-fern, and with luxuriant masses of drooping leaves, and on that account is transferred to moist places in gardens, and cultivated by some, though generally these large specimens are difficult to remove and soon perish. *Scirpus lacustris* (the Bulrush) is too distinct a plant to be omitted, as its stems, sometimes attaining a height of more than seven and even eight feet, look distinct; and *Cyperus longus* is also a good plant, reminding one of the *Papyrus* when in flower; and it is found in some of the southern counties of England. *Cladium Mariscus* is also another distinct British water-side plant, which is worth a place.

If one chose to enumerate the plants that grow in British

and European waters, a very long list might be made, but the recommendation of those which possess no distinct character or no beauty of flower is what I wish to avoid, believing that it is only by a selection of the best kinds that planting of this kind can give satisfaction; therefore, omitting a host of inconspicuous water-weeds, I will endeavour to indicate all others of real worth.

Those who have seen the flowering Rush (*Butomus umbellatus*) in flower are not likely to omit it from a collection of water-plants, as it is pretty and distinct. Plant it not far from the margin, as it likes rich muddy soil. The common *Sagittaria*, very frequent in England and Ireland, but not in Scotland, might be associated with this; but there is a very much finer double exotic kind to be had here and there, which is really a handsome plant, its flowers being white, and resembling, but larger than, those of the old white double Rocket. *Calla palustris* is a beautiful bog-plant, and I know nothing that produces a prettier effect over rich mud ground. *Calla æthiopica*, the well-known and beautiful "Lily of the Nile," is hardy enough in some places if planted rather deep, and in nearly all it may be stood out for the summer; but except in quiet waters, in the South of England and Ireland, will not thrive. The pine-like Water Soldier (*Stratiotes aloides*) is so distinct that it is worthy of a place; there is a pond quite full of this plant at Tooting, and it is common in the fens. It is allied to the Frogbit (*Hydrocharis Morsus-ranæ*), which, like the species of Water Ranunculi and some other fast-growing and fast-disappearing families, I must not here particularise; they cannot be "established" permanently in one spot like the other plants mentioned. The tufted Loosestrife (*Lysimachia thyrsiflora*) flourishes on wet banks and ditches, and in a foot or two of water. It is curiously beautiful when in flower. *Pontederia cordata* is a stout and hardy water-herb, with distinct habit, and blue flowers. There is a small Sweet-flag (*Acorus gramineus*) which is worth a place, and has also a well-variegated variety, while the common Acorus, or Sweet-flag, will be associated with the Water Iris (*I. Pseud-acorus*), and the pretty *Alisma ranun-*

culoides, if it can be procured; it is not nearly so common as the Water Plantain. The pretty Star Damasonium of the southern and eastern counties of England, an annual, is not to be recommended to any but those who desire to make a full collection, and who could and would provide a special spot for the more minute and delicate kinds. The Water Lobelia does not seem to thrive away from the shallow parts of the northern lakes, getting choked by the numerous water weeds. The Cape Pond flower (*Aponogeton*), a native of the Cape of Good Hope, is a singularly pretty plant, which is nearly hardy enough for our climate generally, and, from its sweetness and curious beauty, a good plant to cultivate in a warm spot in the open air. It is largely grown in one or two places in the south, and it nearly covers the surface of the only bit of water in the Edinburgh Botanic Garden with its long green leaves, among which the sweet flowers float abundantly. In the open air, plant it rather deep in a clean spot and in good soil, and see that the long and soft leaves are not injured either by water-fowl or any other cause. *Orontium aquaticum* is a handsome water-herb, and as beautiful as any is the Water Violet (*Hottonia palustris*). The best example of it that I have seen was on an expanse of soft mud near Lea Bridge, in Essex. It covered the muddy surface with a sheet of dark fresh green, and must have looked better so than when in water, though the place was occasionally flooded. The Marsh Marigold (*Caltha palustris*), that " shines like fire in swamps and hollows grey," will burnish the margin with a glory of colour which no exotic flower could surpass. A suitable companion for this *Caltha* is the very large Water Buttercup (*Ranunculus Lingua*), a very handsome British water-side perennial. *Lythrum roseum superbum*, a variety of the common purple Loosestrife, and *Epilobium hirsutum*, are two large and fine plants for the water-side.

ALPINE GARDENING IN ADVERSE CONDITIONS.

Among the best cultivators of alpine flowers, and under conditions less favourable than what are usual in many parts of the country owing to heavy soil and heavy rainfall, is Mr Wolley-Dod; and his advice is so good for amateurs in similar conditions that it is here given from a paper read before the Horticultural Society. Among alpine and rock plants, embracing so much and such infinite variety, some variety of teaching is better than any one formula, however good. Many parts of the country about our coasts, and on the mountains of the cold north of England, are so favourable to alpine plants that little trouble gives us good results; but readers who live in quite different conditions, in the West Midlands and other districts, will like to know how difficulties are met in such conditions.

"There are some favoured gardens where natural rock exists, or where the conditions of the soil with regard to quality or drainage are such that choice and delicate mountain plants may be grown on the ground level in ordinary borders. Such gardens exist in several districts in England, and are common in Scotland and Wales; few rules are necessary there, where plants have only to be planted and kept clear of weeds in order to thrive. But most of us who wish to grow choice alpine plants in our gardens have to make the best of conditions naturally unfavourable, and in doing this we can be helped by the experience of those who have made it their special study. We need not say much of climate and atmospheric conditions, because they are beyond our control. It may be remarked, however, that high elevation above the sea-level is a great advantage in the neighbourhood of towns, because the impurities in the air are more readily dispersed, and do not collect or settle as in lowland valleys. Good natural drainage is also a great advantage, because although we can drain the spot in which our alpines grow, and even our whole garden, still if the soil of the district is wet and retentive, the local damp seems to affect mountain plants unfavourably. Local differences of climate caused by soil and

PART OF ROCK GARDEN, FRIAR PARK. *June* 1910.

evaporation are no doubt important factors in the growth of plants, but it would be waste of time to dwell upon the endless particulars which make it impossible that the conditions which prevail on the Alps can be imitated in the valley of the Thames.

"The first necessity for growing choice alpines is to secure perfect drainage for the soil in which they grow. This may seem strange to those who have seen them growing on the mountains, often apparently in perpetual wet; but there the soil is never water-logged, or charged with stagnant moisture, but the wet is always in rapid motion and changing. Supposing that no part of a garden naturally gives the conditions in which alpines will thrive, we must make these conditions by artificial means. Those who wish to grow them on flat borders or retentive wet soils may do so on the ground-level by digging out the soil to a depth of 3 feet, and draining the bottom of the bed to the nearest outfall, and filling up to the surface with soil mixed with two-thirds of broken stone, either in small or large pieces. But in heavy soils, where large stones are easily obtained, still better beds for alpines may be made by enclosing the space with large blocks to a height of 2 feet or 3 feet, and filling up as before directed. The sides of these stone blocks can be covered with many ornamental plants in addition to those which are grown on the raised surface. But the commonest way of cultivating alpines is upon what are called rockeries, or loose rough stones laid together in different forms and methods.

"The forms in which the rockery, usually so called, can be constructed may be divided into three: (1) The barrow-shaped rockery, (2) the facing rockery, and (3) the sunk rockery. The first may be raised anywhere; the other two depend partly upon the configuration of the ground. No wood or tree roots should be used to supplement any of them; they must be all stone. The kind of stone is seldom a matter of choice; everyone will use what is most handy. The rougher and more unshapely the blocks the better. The size should vary from 40 or 50 lbs. to 3 or 4 cwts. No mortar or cement for fixing them together must ever be employed; they must be firmly wedged and interlocked, and depend upon one another, and not upon the soil between them, to keep them in their places. This rule is of the utmost importance; if it is neglected, a long frost or an excessive rainfall may cause the whole structure to collapse.

"Each successive part of the stone skeleton must be put together before the soil is added.

"THE BARROW-SHAPED ROCKERY.

"The most convenient size for the barrow-shaped rockery is about 4 feet high, and 6 feet or 7 feet through at the base. The length is immaterial. If the long sides face north-east and south-west, it will afford perhaps the best variety of aspect; but the amount of sunshine each plant gets will depend on the arrangement of each stone as much as upon the main structure. There cannot be too many projections, and care must be taken to leave no channels between the stones by which the soil can be washed down to the base. Overhanging brows, beneath which plants can be inserted, are very useful; large surfaces of stone may here and there be left exposed, and irregularity of form is far better than symmetry. A formal arrangement of flat pockets or nests offends the eye without helping the cultivator, as the tastes of alpines as regards slope of surface and moisture at their roots are very various. As for the degree of slope from the base to the summit of the barrow, it will not be uniform. In some places there will be an irregular square yard of level on the top, bounded by large cross keystones, for which the largest stones should be reserved. In other parts the sides will slope evenly to the ridge; or the upper half may be perpendicular, leaving only wide crevices to suit the taste of certain plants. If the blocks are very irregular in form, and their points of contact as few as possible, providing only for secure interlocking, there will be plenty of room for soil to nourish the plants. Everchanging variety of stone surface, both above and below the soil, is the object to be aimed at, and any sort of symmetry must be avoided. The second form, or

"FACING ROCKERY,

is dependent upon the natural shape of ground surface. Wherever there is a steep bank facing south or east, it may be utilised for the growth of alpines. The stones, as before advised, should be large and unshapely, and be buried to two-thirds of their bulk, and form a very uneven surface, all being interlocked from top to bottom as described. Rockeries of this form are less liable to suffer from drought; if the surface covered is large, access to all parts should be provided by convenient stepping-stones, because, although every stone in the structure ought to be capable of bearing the weight of a heavy man without danger of displacement, it is better not to have to tread upon the plants.

"THE SUNK ROCKERY.

"This is perhaps the best of all, but entails rather more labour in construction. Where subsoil drainage is perfect, a sunk walk may be made, not less than 10 feet or 12 feet wide, with sloping sides. The sides may be faced with stones, as described in the second form of rockery, and all or part of the excavated soil may be made into a raised mound, continuing the slopes of the excavated banks above the ground-level, and thus combining the facing rockery and the barrow-rockery. If the outer line of this portion above the ground be varied by small bays, every possible aspect and slope may be provided to suit the taste of every plant. However, unless drainage is perfect, a sunk walk, rising to the ground-level at each end, would not be feasible. But a broad walk, excavated into the side of a hill and sloping all one way, could be adapted to a structure nearly similar to that described, or the ground may be dug out in the form of an amphitheatre to suit the taste or circumstances. But whatever the form of rockery adopted, let the situation be away from the influence of trees, beyond suspicion of the reach of their roots below, or their drip, or even their shade, above.

Trees which only shelter from high winds are so far serviceable, and so are walls and high banks. There are few alpines for which a storm-swept surface is good, but trees are objectionable where they lessen the light, which is an important element in the welfare of most mountain plants. The shade and shelter afforded by the stones and form of the structure itself are the best kind of shade and shelter.

"SOIL.

"We now come to the subject of soil, which is very important, though I attach less importance to it than others do who have written on the subject. I hold that where atmospheric and mechanical conditions are favourable, the chemical combination of the soil is of secondary consideration. It is true that in Nature we find that the flora of a limestone mountain differs in many particulars from that of a granite mountain, and on the same mountain some plants will thrive in heavy retentive soil, whilst others will be found exclusively in peat or sand. But for one who is beginning to cultivate alpines to have to divide them into lime-lovers and lime-haters, lovers of sand and lovers of stiff soil, is an unnecessary aggravation of difficulties. So large a proportion of ornamental plants is contented with the soil which most cultivators provide for all alike—even though in Nature they seem to have predilections—that where an amateur has only one rockery, it would be too perplexing to study the partiality of every plant, and to remember every spot where lime-lovers or their opposites had been growing. While saying this, I confess that I have some rockeries where both soil and rock are adapted exclusively for lime plants; others from which lime is kept away, and where both soil and rock are granitic; but the great majority of plants thrive equally well on both. I know few better collections of alpine plants than one which I recently saw at Guildford, growing on a bank of almost pure chalk. I cannot say that I noticed any inveterate lime-haters there; but conditions of drainage and atmosphere were the chief cause of success. With regard to soil then, we must take care that it does not retain stagnant moisture, and yet it must not dry up too readily. Plants must be able to penetrate it easily with their roots, the lengths of some of which must be seen to be believed. Good loam, with a little humus in the form of leaf-mould or peat, and half or three quarters of the bulk composed of stone riddlings from the nearest stone quarry, and varying in size from that of rapeseed to that of horse beans, make up a soil with which most alpines are quite contented. The red alluvial clay of Cheshire, burnt hard in a kiln, and broken up or riddled to the above size, is an excellent material mixed with a little soil and a little hard stone. Where you are convinced that lime is useful, it may be added as pure lime, not planting in it till thoroughly slaked by mixture with the soil. Rough surface-dressing is a thing in which all alpines delight, as it keeps the top of the soil sweet and moist, and prevents their leaves being fouled. Use for this purpose the same riddled stone as described above, which is better than gravel, as round pebbles are easily washed off the slope by rain or in watering.

"PLANTING.

"It is better not to be in a hurry to see the stones covered. It would be easy to cover them with growth in a single season, but it would be demoralising to the cultivator. We must not degrade choice alpines by putting them to keep company with Periwinkles, Woodruff, large St John's Wort, dead Nettles, Creeping Jenny, fast-running Sedums, and Saxifrages, which do duty for alpines on raised structures of roots or stones in the shady, neglected corners of many a garden. Indeed, there are some plants, of which Coronilla Varia is one, which, when once established amongst large stones, cannot be eradicated by any means short of pulling the whole structure to pieces. Any plant which runs under a large stone and reappears on the other side should be treated with caution. As a rule, nothing should be planted which cannot be easily and entirely eradicated in a few minutes. If a rockery is large, there is no reason for limiting the area to be assigned to each plant, especially to such as are ornamental when in flower, and not unsightly at other seasons. If different rockeries or separate parts of the same can be assigned to rapid growers and to dwarf compact plants, it will be an advantage. There are many subjects which belong to the class of alpines which require to be displayed in a broad and high mass to do them full justice. Such things should make a train from the top of the rockery quite to the ground; Aubrietias, for example, and Veronica prostrata should look like purple or blue cataracts; others should be unlimited in breadth, like the dwarf mossy Phloxes and the brilliantly coloured Helianthemums. Such things do not like being cropped round to limit their growth, and if there is not enough room for them, they had better be omitted, though in stiff and cold soils they will not thrive in the mixed border. Whatever is grown, the small and delicate gems of the collection must run no danger of being smothered by overwhelming neighbours, and this requires both careful arrangement and constant watching. When first I began to cultivate alpines, I planted somewhat indiscriminately together things which I thought would make an ornamental combination, but the weaker soon became overwhelmed in the fight with the stronger, and there was nothing to be done but to build a new rockery and plant it more carefully. In this way I have now constructed at least a dozen rockeries, trying each time to benefit by past experiences and to exclude weedy plants. The first and second made still continue, and are still flowery wildernesses in Spring, but everything choice and delicate upon them has either long ago perished or been transferred to new quarters. But visitors to my garden in Spring, who are not connoisseurs in alpines, think these wild rockeries far more ornamental than the half bare stone heaps where my choicest plants are grown, and which they think will look very nice in a year or two when they are as well covered as the others. I have mentioned this to show that those who can appreciate the beauty of the smaller and more delicate alpines, and grow them for their own sake, must be contented to see their favourites surrounded in many instances by bare stones; but the stones, especially if they contain cracks, may often be clothed with plants without any danger of overcrowding. I have said little about choice of

"STONE FOR ROCKERIES,

though I have tried many kinds; and of all I have tried, I prefer the carboniferous limestone, common in North Wales, Derbyshire, and the north of Lancashire. The loose blocks of this which lie about the land are full of cracks and are varied in shape. I carefully avoid the furrowed and smooth-channelled surface slates of this stone often sold in London for rockwork, but most unsuitable for growing plants; I do not speak of these, but detached solid blocks, abounding in deep cracks and crevices. These crevices are the very place for some of the choicest alpines. Paronychia shows its true character in no other spot. Potentilla nitida flowers when fixed in them, and there only. They are excellent for Phyteuma comosum. The Spiderweb Houseleeks delight in them, and so do some of the smaller Saxifrages. These are only a few of a long list I might make, and things which grow in such tight quarters never encroach much. The little Arenaria balearica, which grows all over sandstone as close and in nearly as thin a coat as paint upon wood, does not grow well upon limestone; but this plant does encroach, spreading over the surface of small neighbours and smothering them. There are many things, however—some herbaceous, some shrubby and evergreen, —which do well only on condition of resting upon stone with their leaves and branches. It is so with Pentstemon Scouleri, and with that most charming dwarf shrub, Genista pilosa, which rises hardly an inch off the stone, though it may cover several square feet. I have said before that in planting, aspect must be carefully considered. The best aspect for alpines is east, and west is the worst; but there is not a spot on any rockery which may not be filled with a suitable tenant. Some of the most beautiful flowers abhor, in the atmosphere of my garden, even a glimpse of the sun. Ramondia pyrenaica is withered up by it in an hour; so is Cyananthus lobatus; and these must be shaded on every side but north. As a general rule, I find all Himalayan alpine plants impatient of sunshine; they may endure it in their own home, where they live in an atmosphere always saturated with wet. However, it is only the deep recesses of the rockery towards the north which get no sun at all, and plenty of things are quite contented on the north side of the slope. For instance, I must grow Lithospermum prostratum on stones or not at all. The white Erica carnea and several such dwarfs are included in the same number.

"As for bulbs, they may be ornamental enough at times, but I find they do as well or better elsewhere. Their leaves are untidy just at the time when the rockery ought to be most gay and neat; and watering in summer, which other plants require, is bad for them, so I have not included them in my list. While speaking of watering, I may say that rockeries such as I have described could not dispense with it in dry weather; it requires careful judgment; and I often prefer to water the soil holding the can close to the ground at the highest point of the stones, and letting the water run down the slope to get to the roots, rather than wet the plants themselves. Wet foliage and flowers often get burnt up by sunshine. Weeding, carefully done, is a necessity on rockeries, for weeds will come; but plants which seed about freely are to be avoided, as they greatly multiply the labour of weeding, and some of them are hard to eradicate from among the stones. The Harebells, and alpine Poppies, pretty as they are, must be excluded on this account; so must that weedy little plant, Saxifraga Cymbalaria, which can be grown on any wall. The fewer weeds there are, the more likely are seedlings of choice and rare plants to assert themselves. For instance, Geranium argenteum grows in crevices into which the seeds are shot when ripe, and where plants could not be inserted, and keeps up the supply of this elegant alpine.

CULTURAL

"RAISING ALPINE PLANTS FROM SEED.

"A few words may be in place here about raising alpines from seed; for constant succession is necessary, the duration of their life in cultivation being, for many obvious reasons which need not be discussed here, far shorter than in their native home. Reproduction from seed, where seed can be obtained, ensures the healthiest and finest growth, and there is no better way of getting seed than in saving it yourself. In several cases the first hint I have had that a plant has ripened fertile seed has been the recognition of a seedling near the parent, and this experience has taught me always to look carefully for seed after the flowering of rare specimens. I need not say, therefore, that I disapprove of the practice of cutting off flower heads as soon as they wither; in some cases the seed-head is nearly as ornamental as the flower, but I have before said that discretion must be used, even in this, as seedlings of some things are troublesome from their number. When ripe seed is gathered I recommend its being sown at once. It is then more likely to come up quickly, and as all such plants as we grow on rockeries are better sown in pans, there is seldom difficulty in keeping small seedlings through the winter. The greatest enemy we have in the process is the growth of Lichen, the worst being the Marchantia or Liverwort fungus, which completely chokes tender growth. A coating of finely sifted burnt earth on the surface, and a piece of flat glass laid over the pan, especially if no water is used for them unless it has been boiled, reduces this trouble to a minimum. But sowings of choice and rare seed should be carefully watched, and the fungus picked off at the first appearance. Many alpines seem never to form seed in cultivation, and must be reproduced by division or cuttings. The skill required to do this varies greatly with different subjects; where a shoot can seldom be found more than half an inch long, as in the case of two or three hybrid alpine Pinks, the striking needs delicate manipulation. Other things grow very slowly, though not long-lived, and a constant succession from cuttings must be ensured. Some of the terrestrial Orchids, such as Bee, and Fly, and Spider, we must be contented to keep as long as they choose to live, as they seem never to increase in cultivation at all, though they may flower well year after year. But there are not a few plants which refuse to be tamed, and from the time they are planted in our gardens, seem always to go from bad to worse, and are never presentable in appearance for two seasons together. Of these I may instance Gentiana bavarica and Eritrichium nanum, which I believe no skill has ever yet kept in cultivation without constant renewal, and which perhaps are never likely to repay the trouble of trying to keep them alive. In all alpine gardening there will be, even where equal skill is exerted, different degrees of success, according to the surrounding conditions; and it must not be expected that the same soil and treatment which keep a hundred rare alpines in perfect health at Edinburgh will be equally fortunate at Kew.

"FRAMES FOR ALPINES.

"Where the area of rockery is considerable, a cold frame should be assigned for keeping up the supply of plants for it—cuttings and seedlings—in pots. I think all attempts to imitate natural conditions, such as snow and long rest, by unnatural means are mistakes. During warm winters mountain plants will grow, and must be allowed to grow, and to keep them unnaturally dark or dry when growing is fatal to their health. Even in severe frosts, air must be given abundantly in the daytime, and the frames must not be muffled up. Stagnant air, whether damp or dry, is their worst enemy; but if the weather is warm enough to set them growing, they may easily die for want of moisture. I will not say more than this, for experience is the best guide, and every one thinks he can manage his frames better than his neighbour; but of the use of frames for flowering alpines in pots I must add a few words. There are certain very early-flowering alpines upon which a mixture of admiration and lamentation is bestowed at the end of every winter. Their flowers are often beautiful in a treacherous fortnight at the beginning of February, and are suddenly destroyed by a return of winter in its severest form. I may mention, amongst others, Saxifraga Burseriana and sancta, and their near relatives and hybrids, Primula marginata and intermedia, Androsace carnea, Chamæjasme and Laggeri, several dwarf species of Alyssum and Iberis, and there are a good many more. Pots or pans containing these may be grouped together in an open sunny spot, and plunged in sand or coal-ashes in a rough frame made for them, so that the lights may be not more than 3 inches or 4 inches above the pots. These lights should be removed in the daytime when the weather is fine, and air should be admitted, according to the temperature, at night. Such a sheet of elegant beauty, lasting, if well arranged, through February, March, and April, may be obtained in this way, that I often wonder why amateurs attempt to flower early alpines in any other fashion. With me April is the earliest month in which I can expect to have anything gay on the open rockery without disappointment. I am obliged to disfigure the slopes with sheets of glass and handlights to preserve through winter at all Omphalodes Luciliæ, Onosma tauricum, Androsace sarmentosa, and others which cannot endure winter wet, and the real pleasure of the rockery begins when the frame alpines are waning."

ALPINE FLOWERS IN PANS OR BASKETS.

So long as the exaggerated ideas of the difficulties of growing alpine flowers were prevalent, it was the custom, even in good gardens, to grow most of these plants in pots in frames, while at flower exhibitions we often see them now shown, and, bearing that in mind, it is important that they should be well grown in that way. Occasionally, too, we see them, as in the Alpine House at Kew, shown for their beauty in the Spring, in cool houses. Where there is the least difficulty as regards climate, such as the smoke of the town, having them slightly protected in pots will often gain a point or two, and in cold districts there is some reason why the early habit of flowering of so many beautiful kinds should not be taken advantage of, and by growing some of these early kinds in pots or pans, or shallow baskets, we might, when they were about to flower, transfer them to a very cool greenhouse, or to frames, to a pit with some path in it, or better still when in bloom to the cooler windows of the house, and so enjoy their beauty and save them from the vicissitudes of our often wretched Springs. In the case of the easily-grown kinds, such as our rosy native Rochfoil, Omphalodes, and Alpine Primrose, it would be easy to secure well-grown plants, of which pretty use might be made by many who do not exhibit alpine plants, while some such plan is essential for those who do.

I do not advocate their culture in pots at all where there is an opportunity of making a rock-garden; but there are cases in which they cannot be well grown in any other way. It is often well to keep rare kinds in pots till sufficiently plentiful.

Prizes are frequently offered at our flower shows for these plants, but the exhibitors rarely deserve a prize, for their plants are often ill selected, badly grown, and such as ought never to appear on an exhibition stage. In almost every other class the first thing the exhibitor does is to select appropriate kinds—distinct and beautiful—and then he makes some preparation beforehand for exhibiting them; but in the case of hardy plants, anybody who happens to have a rough lot of miscellaneous rubbish exhibits them. Yet such plants as the tiny shrubs *Cassiope*, *Menziesia*, and *Gaultheria procumbens*, the Alpine Phlox, and many others, might be found pretty enough to satisfy even the most fastidious growers of New Holland plants.

The very grass is not more easily grown than plants like Iberises and Aubrietias, yet to ensure their being worthy of a place, they ought to be at least a year in pots, so as to secure well-furnished plants. Such vigorous plants, to merit the character of being well-grown, should fall luxuriantly over the edge of the pots or baskets, the spreading habit of many of this class of plants making this a matter of no difficulty. In some cases it would be desirable to put a number of cuttings or young rooted plants into pots or pans, so as to form good plants quickly.

To descend from the type that seems to present to the cultivator the greatest number of neat and attractive flowering plants, we have the dwarf race of hardy succulents, and the numerous minute alpine plants that associate with them in size—a class rich in merit and strong in numbers. These should, as a rule, be grown and shown in pans: they are often so pretty and singular in aspect, as in the case of the silvery Rockfoils, that they are interesting when out of flower. All these little plants are of the readiest culture in pans, with good drainage and light soil.

Some few alpine plants are somewhat delicate or difficult to grow; and amongst the most beautiful and interesting of these are the Gentians, and certain of the alpine Primroses. In a general way, it would be better to avoid, at first, such difficult subjects. I believe that a more liberal culture than is generally pursued is what is wanted for these more difficult kinds. The plants are often obtained in a delicate and small state; then they are, perhaps, kept in some out-of-the-way frame, or put where they receive but chance attention; or, perhaps, they die off from some vicissitude, or fall victims to slugs, or, if a little unhealthy about the roots, are injured by earth-worms, whose casts serve to clog up the drainage, and thus render the pot uninhabitable. With strong and healthy young plants to begin with, good and more liberal culture, and plunging in the open air in beds of coal-ashes through the greater part of the year, the majority of those supposed to be difficult would thrive. I have taken species of Primula, usually seen in a very weakly and poor state, divided them, keeping safe all the young roots, put one sucker in the centre, and five or six round the sides of a 32-sized pot, and in a year made good "specimens" of them, with a greater profusion of bloom than if I had depended on one plant only. Annual division is an excellent plan to pursue with many of these plants, which in a wild state run each year a little farther into the deposit of decaying herbage which surrounds them, or, it may be, into the sand and grit which are continually being carried down by natural agencies. In our long summer, some of the Primulas will make a tall growth and protrude rootlets on the stem—a state for which dividing and replanting them firmly, nearly as deep down as the collar, is a remedy.

There are many plants which demand to be permanently established, and with which an entirely different course must be pursued, *Spigelia marilandica*, *Gentiana verna*, *G. bavarica*, and *Cypripedium spectabile*, for example. The Gentians are rarely well grown, and yet I am convinced that few will fail to grow them if they procure in the first instance good plants; pot them carefully and firmly in good sandy loam, well drained, using bits of grit or gravel in the soil; plunge the pots in sand or coal-ashes to the rim, in a position fully exposed to the sun; and give them abundance of water during the spring and summer months, taking precautions against worms, slugs, and weeds. And such will be found to be the case with many other rare and fine alpine plants. The best position in which to grow the plants would be in some open spot, where they could be plunged in coal-ashes, and be under the cultivator's eye. And, as they should show the public what the beauty of hardy plants really is, so should they be grown entirely in the open air in spring and summer. To save the pots and pans from cracking with frost, it would in many cases be desirable to plunge them in shallow cold frames, or cradles, with a northern exposure in winter; but, in the case of the kinds that die down in winter, a few inches of some light covering thrown over the pots, when the tops of the plants have perished, would form a sufficient protection.

ALPINE FLOWERS IN POTS.

Alpine and herbaceous plants in pots, and kept in the open air all the winter, are best plunged in a porous material on a porous bottom, and on the north side of a hedge or wall, where they would be less exposed to changes of temperature, and less liable to be excited into growth at that season.

The most suitable kind of pots for alpine flowers that I

have yet seen were those used by Mr G. Maw, in his gardens at Benthall Hall. These pots are of a peculiar size—8 inches broad by 4 inches deep. They seem peculiarly well suited to the wants of alpine plants, securing, as they do, a good body of soil, not so liable to rapid changes as that in a small vessel; while in stature, being only 4 inches high, they are exactly what is wanted for these dwarf plants. The common garden pan suits some alpine plants well, but is not so well suited to the stature of alpine plants, or the wants of their roots, as a pot of this pattern.

For growing the Androsaces and some rare Rockfoils, a modification of the common pot may be employed with a good result. This is effected by cutting a piece out of the side

Pot for Androsaces, etc.

Alpine Plant growing between stones in a pot.

of the pot, 1½ or 2 inches deep. The head of the plant potted in this way is placed outside of the pot, leaning over the edge of the oblong opening, its roots within in the ordinary way, among sand, grit, stones, etc. Thus water cannot lie about the necks of the plants to their destruction. This method, which I first saw in use in M. Boissier's garden, near Lausanne, is a good one for fragile plants. The pots used there were taller proportionately than those we commonly use, so that there was plenty of room for the roots after the rather deep cutting had been made in the side of the pot.

An even better mode is that of raising the collar of the plant somewhat above the level of the earth in the ordinary pot by means of half-buried stones.

In this way we not only raise the collar of the plant so that it is less liable to suffer from moisture, but, by preventing evaporation, preserve conditions congenial to alpine plants, and keep the roots firm in the ground; the small plants looking more at home springing from tiny rocks. It should, however, be understood that such attention is required only for the rarer of the higher alpine plants.

No matter in what way these plants may be grown in gardens, it is often well to keep the duplicates and young stock in small pots plunged in sand or fine coal-ashes, so that they may be easily removed at any time. The best way of doing this is shown in the wood-cut, which represents a four-foot bed in which young alpine plants are plunged in sand, the bed being edged with half-buried bricks. In bottoms of beds of this kind there should be half a dozen inches of coal-ashes, so as to prevent worms getting into the pots. Sand, or grit, or fine gravel, from its cleanliness and the ease with which the plants may be plunged in it, is to be preferred, but finely sifted coal-ashes will do if sand be not at hand.

Bed of small Alpine Plants in pots plunged in sand.

Such beds should always be in an open situation, near to a good supply of water, and, if several are made, should be separated by gravelled alleys of about 2 feet wide. The watering is important, and in a large collection it should be laid on. This certainly is the most convenient and economical way. Over some of the beds in Mr Backhouse's Nursery at York, may be seen an ingenious way of giving a constant supply of water to Primulas, Gentians, and other plants. Two perforated half-inch copper pipes are laid just above the plants in the beds, as shown in the cut. From the perforations in every 2 feet or so of the pipe, drops continually trickle down in summer,

Bed kept saturated by perforated pipes.

saturating the beds of sand, and the porous pots and their contents. In winter or very wet weather the water can be readily turned off.

ANNUALS FOR THE ROCK-GARDEN.

Although we do not connect annuals much with the alpine flora of Europe, yet in other mountainous countries, as in Mexico and California, annuals are less rare, and they need not be entirely omitted from our view even in the rock-garden, and some interesting rock plants are biennial or annual, like the little Sun Rose of the Channel Islands. Apart from the value of such plants, we have often to face bare spaces in the rock-garden, owing to clearances, deaths, or other causes, and it is not always easy to get perennial rock plants enough to cover them as they ought to be covered. The plan of dots of green on bare earth is one which it will take a long time to eradicate from the gardening mind, but those who care to fight against it may find annuals help us much in covering freely open spaces, until such times as we can afford to get plants of a more permanent character.

A choice must be made of the most elegant and dwarf kinds, and these that, by their stature or other characteristics, are fitted for the rock-garden.

Where there are plenty of means, these plants may be raised in the elaborate ways usually recommended in books, but those are not at all necessary. Raising them in the ground work, where we want them, is so simple that a child could do it. Choosing the last week in April or first week in May in cool districts, we have only to provide ourselves with the seeds in packets, and to make level and rake over the different surfaces which we want to cover, and sow the seed broadcast over all the surface destined for each group. Cover the soil lightly, putting a mere sprinkling over the fine seeds. It will be found that in that way the choicer annual flowers will come very well, and be a useful aid, until our stock of the perennial alpine plants are ready.

SOME DWARF AND MORE REFINED ANNUALS FOR THE ROCK-GARDEN.

We must be very careful in this case to avoid coarse plants, or very vigorous ones, however good; also, and particularly, colours of more popular than refined quality. In cases of large genera being named below, as Campanula, it is the dwarfer annual kinds that are meant, individual taste here, as elsewhere, having its own sway.

Abronia	Gypsophila	Nycterinia
Adonis	Heliophila	Omphalodes
Alyssum	Iberis	Phacelia
Anagallis	Ionopsidium	Phlox
Argemone	Isotoma	Platystemon
Brachycome	Jasione	Portulaca
Calandrinia	Kaulfussia	Sabbatia
Campanula	Koniga	Saponaria
Clintonia	Leptosiphon	Schizopetalon
Collinsia	Linaria	Silene
Erinus	Linum	Sphenogyne
Eutoca	Malcomia	Vesicaria
Felicia	Nemesia	Viscaria
Gilia	Nemophila	Whitlavia
Grammanthes	Nolana	

RAISING ALPINE PLANTS FROM SEED.

The difficulty of getting plants in sufficient numbers to give us broad effects has to be faced, and there is no royal route to avoid it—no one way of getting all we seek. We must get our plants where we can, and in various ways, not trusting wholly to Nurserymen, who rarely offer the plants liberally by the dozen or hundred, so as to allow of effective grouping. And it is not only this that often stops us, but the fact that even easily raised plants are often sent out in such a feeble state, that they are useless to clothe our miniature mountain. So it will often be wise to raise kinds from seed; as in that way we get numbers to carpet our fairy fields; the greater vigour of the seedling plant, and the chances of novelty or variety that seedlings sometimes gives us. Moreover, there is some reason to believe that plants we often fail with in cultivation, as some of the beautiful wild Columbines, are not truly perennial, but in their native countries renewed from seed. And, apart from the seeds offered in seed-lists, travellers may often gather the seeds of alpine plants, and send them easily by post—often where it would be difficult or impossible to send living plants. So that all who have rock-gardens to clothe with good plants, must not forget the seed-bed.

ALPINE PLANTS FROM SEED IN THE OPEN GROUND.

Many alpine plants may be raised from seed, and in every place where there is a good collection, it is well to sow the seeds of as many rare or new kinds as are worth raising. A good deal will depend on the appliances of the garden as to the precise way in which they are to be raised; but whether there be greenhouses on the premises or not even a glass hand-light, alpine plants and choice perennials may be raised there in abundance. Supposing we are supplied with a good selection of seeds in early spring, and have room to spare in frames and pits, some time might be gained by sowing in pans or pots, and by placing them in those frames, or by making a very gentle hotbed in a frame or pit, covering it with 4 inches or so of very light earth, and sowing the seeds on that. If this mode be adopted, they may be sown in March; and, thus treated, many will flower the first year.

In gardens without any glass, they may be raised in the open air. The best time to sow is in April, choosing mild open weather, when the ground is more likely to be in the rather dry and friable condition so desirable for seed-sowing. But it should be borne in mind that they may be sown at any convenient time from April till August, as it is not till the year after they are sown that they display their full beauty, or perhaps flower at all; and, therefore, should a packet or more of choice seed come to hand during the summer months, it is always better to sow it at once than to keep it till the following spring, as thereby nearly a whole season is lost. Those who already possess a collection of good hardy flowers may find a choice perennial—say, for instance, an evergreen *Iberis*, a *Campanula* or a *Delphinium*—ripening a crop of seed in May, June, or July. Well, suppose we want to propagate and make the most of it, the true way is to sow it at once, instead of keeping it over the winter, as is usually done. By winter, the seedlings will be strong enough to take care of themselves, and be ready to plant out for flowering wherever it may be desired to place them.

As to the immediate subject of raising them in spring, we will suppose the seeds provided, and the month of April to have arrived. If not already done, a border or bed should be prepared for them in an open but sheltered and warm position, and where the soil is light and fine. It would be as well to prepare and devote two or three, or more, little beds to this purpose of raising rock plants and hardy flowers. They would form a most useful nursery reserve ground, from which plants could be taken at any time to fill up vacancies, to exchange with those having collections, or to give away to friends; for assuredly it is one of the pleasures of gardening, to be able to share with friends who admire one of our "good things"; and by raising them from seed we can be more liberal. If the ground happen not to be naturally fine, light, and open, make it so by adding plenty of sand and leaf-mould, and then surface it with a few inches of fine soil from the compost-yard or potting-shed. The sifted refuse of the potting-bench will do well. Then level the beds, and form little shallow drills in them for the reception of the seed. Let the beds be about 4 feet wide, with a little footway or alley between each about 15 inches wide, and let them run from the back to the front of the border, not along it. Make the little drills across the beds, and, instead of making these drills with a hoe or anything of the kind, simply take a rake handle, a measuring rod, or any rod perfectly straight that happens to be at hand, and, laying it across the little bed, press it gently down till it leaves a smooth impression about 1 inch deep. Do this at intervals of about 6 inches, and then the little nursery bed is ready for the seed. From these smooth and level drills the seeds will spring up evenly and regularly.

Before opening the seed packets, it is necessary to have clearly written wooden labels at hand on which to write the name of each species, so that there may be no confusion when the plants come up. These labels should be about 8 or 9 inches long, and an inch wide, and the name should be written as near the upper end as possible, so that it may not be soon obliterated by contact with the moist earth. Now, this labelling process is usually done at the time of sowing the seeds, but a speedier and better way is to lay out all the seeds on a table some wet day, when out-of-door work cannot be done, and there and then arrange them in the order of sowing. Write a label for each kind, tie the packet of seeds up with a piece of bast, and then, when a fine day arrives for sowing them, it can be done in a very short time. In sowing, put in at the end of the first little drill the label of the kind to be sown first, then sow the seed, inserting the label for the following kind at the spot to which the seed of the first has reached, and so on. Thus there can be no doubt as to the name of a species when the same plan is pursued throughout. Near at hand, during the sowing, should be placed a barrow of finely-sifted earth; with this the seeds should be covered according to size, and then watered from a very fine rose. Minute seed, like that of Campanula, will require but a mere dust of the sifted earth to cover it.

Once sown, the rest may be left to Nature, save the keeping down of weeds, the seeds of which abound in the earth in all places, and will be sure to come up among the young plants. But these being in drills, we can easily tell the plant from the weed, and nothing is required but a persevering weeding. In these little beds the finest rock plants will come up beautifully, and may be left exactly where sown till the time arrives for transplanting them. This is a better way than sowing in pots, where they are liable to vicissitude, and from which they require to be "potted off." Of course, in the case of a very rare kind, the seedlings might be thinned a little, and the thinnings dibbled into a nursery bed, but, by sowing rather thinly, the plants will be quite at home where first sown till the time arrives for planting them out finally.

I am convinced that in finely pulverised earth, with, if convenient, an inch or so of cocoa-fibre and sand between the drills to prevent the ground getting hard and dry, much better results will be obtained than by sowing in pots. In the open air they come up much more vigorously, and never suffer from transplantation or change of temperature afterwards.

ALPINE PLANTS RAISED FROM SEED IN POTS.

Nevertheless, as few will venture the very finest and rarest kinds of seed in the open air, how to treat them in frames is of some importance, and the following notes on this matter are by the late Mr Niven, of the Hull Botanic Garden, in the *Gardener's Chronicle.*

"Presuming that the selection of the seeds is made, and that the seeds themselves are in the hands of the purchaser, sowing should take place as early as may be in March. First of all, the requisite number of 5- or 6-inch pots should be obtained, so that each seed-packet can have a separate pot for itself. Some nice light soil, mixed with a fair amount of sand and leaf-mould, should be prepared, and passed through a coarse sieve, keeping a sharp eye after worms, and at once removing them; the rough part which remains in the sieve should be placed above the drainage in the bottom of the pots to the extent of two-thirds of the depth, filling the remaining third with the fine soil; the whole should then be well pressed down, so that the surface for the reception of the seeds may be half an inch below the brim of the pot, and tolerably even. Each packet of seed should then be sown, and covered with a sprinkling of fine soil, which should be pressed down by means of a flat piece of wood, or, what will be perhaps more readily available, by the bottom of a flower-pot.

"The best guide as to the thickness of covering required is

to arrange so that no seeds shall be seen on the surface after the operation. If the seeds are minute, a very small quantity will be required to attain this end; if they are large, more will be requisite. This completed, and each pot duly labelled with the name of the plant and height of growth, the pots should then be placed in a cold frame tolerably near the glass, taking care that each pot is set level or as nearly so as practicable.

"In preparing the frame for their reception, it is desirable to have a good thickness of lime-rubbish in the bottom, say from 9 to 12 inches deep, as a protection against worms.

"Many seeds come up a long time after others; in fact, seed-pots are often thrown away in the supposition that the seeds are dead, when they are perfectly sound; and some will come up a year or so after being sown. All that is necessary with the seeds that do not come up during the spring is to give them occasional watering, and to guard against the growth of the Marchantia. This is frequently a great pest in damp localities, and is only to be kept in check by carefully removing it on its first appearance, for if allowed to make too much headway, any attempt at removal carries away the surface soil, and with it the seeds. In the month of October each pot should be surfaced with a sprinkling of fine soil, well pressed down; in fact, the process before described after sowing should be repeated. The pots may remain in the frame till the spring, nor should they be despaired of altogether till May or June, or in some instances later.

"To those who may not have the advantage of a cold frame to carry out the foregoing instructions, I would still recommend the use of flower-pots rather than sowing in the open ground; but under these circumstances I would say—sow one month later; place the pots in a warm sunny corner, and arrange some simple contrivance so that you can shade with mats during hot sunshine, and also cover up at night, in order to keep off heavy rains; the same care in watering should be observed, and the same watchful eye after snails, wood-lice, and other depredators, should be maintained.

"So much for the seeds in their seed-pots. Now a word or two as to the treatment of the plants afterwards. My practice is to pot off, as soon as they are sufficiently strong to handle, as many as are required, in 3- or 4-inch pots, say three in each pot. In these they will grow well during the summer, and become thoroughly rooted, ready for consigning to their final habitat, be it rock-garden or border, in the early part of spring, after the borders have been roughly raked over; thus giving them ample time to establish themselves before autumn arrives, and their enemy, the spade, is likely to come in their way. Failing a supply of pots sufficient for all, some of the stronger-growing ones may be planted in a sheltered bed of light soil, care being taken to shade them for a few days after being planted; or a few old boxes, 5 or 6 inches deep, may be used with even greater advantage for the same purpose, as they may readily be moved from the shady side of a wall to a more sunny locality after they have sufficiently recovered the process of transplanting; and, finally, they may receive the shelter of a cold frame as soon as winter sets in. This recommendation must not be considered as indicative of their inability to stand the cold weather, but as a preventive of the mechanical action of frost, which, in some soils especially, is apt to loosen their root-hold, and force the young plants, roots and all, to the surface.

"In the case of the smaller-growing alpines, such as the Drabas, Arabises, etc., I generally find that they stand the first winter best in pots of the smallest size, and in this form they may be the more readily inserted in interstices of rocks, where they will permanently establish themselves."

WATERING ALPINE AND ROCK PLANTS.

The notion that alpine plants want shade arises from the fact that those placed in the shade do not perish so soon from drought as those in the sun. The reason that alpine plants perish so soon on bare flower-borders, the surface of which may be saturated with rain one day and be as dry as snuff the next, at least to the depths to which the roots of a small or young alpine plant would penetrate, is therefore easily accounted for. Matted through a soft carpet of short grass in their native hills, or rooted deeply between stones, they can stand many degrees more heat than they ever endure in this country. As a rule, it is difficult to water them too freely if the drainage be good, which of course it will be in a well-formed rock-garden. To have the water laid on and applied thoroughly with a fine hose, is the best plan in districts not naturally moist. Some lay small copper pipes through the masses and to the highest points of the rock, allowing the water to gently trickle from these, but, except in special cases, the plan is not so good as the hose. Whatever way be adopted, the rule should be: Never water unless you saturate the soil, say with from $1\frac{1}{2}$ to 2 inches deep of water over the whole surface. As a rule, pretentious, wall-like, erect masses of "rock-work" require half a dozen times as much water as those made with plenty of soil, so arranged that it is easily saturated by the rains. Indeed, nothing but ceaseless watering could preserve plants in a healthy state on the "rock-work" commonly made. As regards the time of watering, it is a matter of very little importance, though, for convenience' sake, it is better not done in the heat of the day.

PLANTING ROCK AND ALPINE FLOWERS.

There is a mischievous way of planting almost every kind of small plant, which is particularly injurious in the case of the hardy orchids (whose roots are easily injured), and of all rare alpine plants. I refer to the practice of making a hole for the plant, and, after a little soil has been shaken over the roots, pressing heavily with the fingers over the roots and near the neck of the plant. What is meant will be understood from fig. 2, if the reader assumes that there is a little soil between the fingers and the roots. Where the roots are not all broken off in this way, many of them are mutilated, and often plants perish from this cause. The right way, after preparing the ground, is to make it firm and level, and then make a little cut or trench. The side of this trench should be firm and

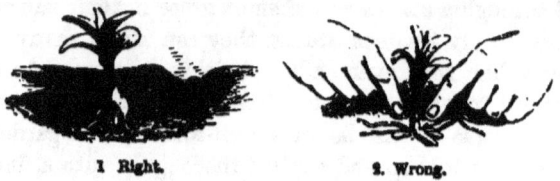

1 Right.   2. Wrong.

smooth, and the plant placed against it, the roots spread out, and the neck of the plant set just at the proper level, as in fig. 1. Then the fine earth of the little trench is to be thrown against the roots, and as much *side* pressure applied as may be necessary to make the whole quite firm. In this way not a fibre of the most fragile plant need be injured.

THINGS TO AVOID.

It is essential to keep clear of the UGLY, unhappily strewn too freely about the garden world. In man's attempt at rock-gardening, many hideous things have been made—even in public gardens, and illustrations of them printed for our guidance in books. Even now, in the public gardens of London, the most hideous and wasteful things are done in the shape of ignoble masses of spoiled brick, as in Waterlow and Dulwich Parks. It is brickyard waste, valueless for any purpose save a bottom for roads, and its use in public gardens is hardly to be explained, except as jobbery or gross ignorance. The mere cost of carting it to Dulwich Park, if rightly applied, might have given us a true rock-garden, formed of some of the natural sandstone, found south of London, that might have been a lesson in beauty. We have not only to avoid these brutalities in material because of their ugliness, or of their bad effect on our plants, but because every cobbler who rushes from his last to write a book on garden design will assume that the ugly way is the only way, and so do his little best against truth and beauty.

WHAT TO AVOID.

In the selection of a few illustrations showing with what deplorable results rockwork is generally made, my first intention was to have had them all engraved from drawings taken in various gardens, public and private; but as this course might have proved an invidious one, I have preferred to take most of them from our best books on Horticulture—the works of authorities like Loudon, Macintosh, and others, and that, if such ridiculous objects occur in books of repute, they must be yet more absurd in many gardens.

The first example is copied from the frontispiece of a small book on alpine plants, published not many years ago. Growing naturally on the high mountains, unveiled from the sun by wood or copse, alpine plants are grouped here beneath a weeping tree—a position in which they could not attain anything like their native vigour, or do otherwise than lead a sickly existence.

What to Avoid.
Frontispiece of a book on Alpine Plants.

One form which "rockwork" is made to assume is that of a rustic arch; and the following illustration, from Loudon, is less hideous than many that may be seen about London. Frequently they are formed of spoiled clinkers, but even if composed of good stone, they are useless for the growth of plants. How many rock Pinks or Primroses would find a home on such a structure, set in a part of the Alps favourable to vegetation? Probably not one, and should a few establish themselves on

What to Avoid.
Rustic Arch (after Loudon).

its lower flanks, they would in all probability perish from heat and drought if their roots had not a free course to the earth beneath. Even persons with some experience of plant life may be seen sticking plants over such objects as these, as if they were bits of metal, able to bear as many vicissitudes. The fact that plants push their roots far into old walls is no justification for the rustic arch as a home for alpine flowers. If the cement, burrs, and clinkers permitted them even to enter it, they have nothing of any kind into which to descend. There is rarely an excuse for constructing such arches; where they occur, they should be clothed with Ivy or other vigorous climbers.

The sketch, made at Hammersmith, shows something of the harsh, bare, and unnatural effect of structures of this sort.

What to Avoid.
Rockwork in Villa at Hammersmith.

The next scene is one in which a miniature representation of various mountains is attempted. Efforts of this kind usually end in the ridiculous. Let us succeed with a few square yards of stony mountain turf and flowers before we attempt to build the mountains.

The next illustration shows a rockwork and fountain in what we may call the true mixed style—huge shells, "cascades," and "rockwork." How any such object can be conceived to be in any sense ornamental is not

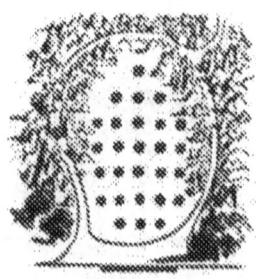

What to Avoid.

easily explained, but it has been taken from a work of authority.

Mrs Loudon's design, while not so repulsive as some of the

What to Avoid.
Fountain and Rockwork.
(after Loudon).

What to Avoid.
Rockwork (after Mrs Loudon).

others, shows in its elevated nodding head the tendency to make such arrangements useless by raising them too high, and by so placing the stones that the rain cannot nourish the plants. Like the arches, such structures as this should in all cases be covered with Ivy, or some kindly veil of vegetation, or broken up to make the bottom of a road or path. It should be noted that when rocks or stones are properly placed in the rock-garden, they do not require any cementing, but are surrounded by and placed on moist stony earth or grit, inviting to every fibre of the root that descends. From this we may deduce the rule—Rockwork consisting of stones cemented together is bad in all respects.

A "rockery" is occasionally seen bordering drives, often with large stones arranged in porcupine-quill fashion, and showing a dentate ridge of rocks springing up close from each side of the drive for a considerable length near the entrance gate—a style dangerous for coachmen on dark nights. Such a position is the last that should be chosen for the rock-garden.

Without alluding to even half the varieties of the ridiculous rockwork tribe, I have the pleasure of here presenting a plan of some recently constructed on the margin of a stream in a great London park. It shows exactly what *not* to do with any rocks introduced near the margin of water. So far from these illustrating exaggerated or extreme instances, I should have no difficulty in finding many, even uglier and more unsuitable, in a few hours' walk near London. That such blemishes are not confined to obscure places, where the light of modern

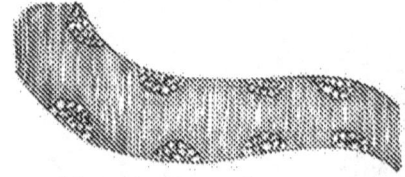

What to Avoid.
Ground-plan of "Rockworks" recently made in a London Park.

progress in these matters has not yet shone, is evident, as one of the most absurd was sketched in one of our greatest parks, and another in one of the most popular of London public gardens.

No public garden should show anything in the way of rock or alpine garden that is ugly or useless for its purpose. And

What to Avoid.
Sketched at Kew in 1872.

this rule should particularly apply to botanic gardens. Better far content ourselves with the good effects which we can get from trees and shrubs, and flowers on the level ground, than add to the hideous piles of rubbish that go by the name of "rockwork" all over the country. And where these excrescences do occur in public gardens, the right thing to do is to convey the offensive pile to the rubbish-yard some time when the ground is hard in winter.

Lastly, among the illustrations of how *not* to do it, is the rockwork figured on this page, which occurred in the Botanic Gardens, Regent's Park.

What to Avoid.
Sketched in the Botanic Gardens, Regent's Park, 1872.

What a check to progress in this direction are such "rockworks" as these! And yet there is no way in which our public gardens would do more good than by growing well, in the open air, the numerous brilliant flowers of the mountains of our own and other cold and temperate regions.

ON THE GEOLOGICAL ASPECTS OF ROCKWORK.

When rockwork has to be erected in a garden, it may be found that success will be attained in the proportion in which some broad principles, based on a study of Nature's own work, have been followed.

Every lover of Nature must have envied her power of adorning rough stony nooks by means of a few of the commonest plants; a fern or two and a little moss convert a few weather-beaten rocks into objects of beauty. And success *is* attainable in almost every case, if sufficient attention be only paid to the rules, which, it will be seen, are as sacred to the physical agents which model our scenery as they ought to be to every gardener. It is a trite observation to say that what pleases us in Nature is the perfect fitness of things which pervades all her belongings. The most rugged, abrupt, and even grotesque rock masses, when untouched by man, never repel us by a sense of incongruity;

they may be pleasing or awful, as the case may be, but they do not strike us as being out of place. Who, on the other hand, has not seen a lovely view marred by some unintelligent human hand, whether its work took the form of a quarry, a statue, or a vase? A secret of the difference lies in the words *weather-beaten*: rain, the chief rock-sculptor, working uniformly, slowly and gently, leaves to each stone which it is fashioning its proper character, models it according to its peculiarities of composition and structure—in short, uses it fitly; while men, with the most

Granite tor.

artistic pretensions, and armed with ruthless tools, too often misuse their materials.

The first great rule which it behoves constructors of rock-gardens to look to is one easily followed but constantly broken—it is that the work should be characteristic of the part of the country in which it stands. That is to say, use chalk at Brighton and sandstone at Tunbridge, granite on Dartmoor, and trap near Edinburgh; but the experience of every one must include cases in which this is ignored. Some artists have even carried their Philistinism in this respect so far, that the more they have succeeded in giving to their rockwork the appearance of a miscellaneous collection of mineralogical specimens from all parts of the world, the better they have been pleased. The familiar burnt brick of the South of England, and the slag and painted coke of the northern coal districts, are better than these.

It is needless to point out in detail what rocks are suitable for alpine gardens in the different parts of Britain; a walk in the country will show the rocks, and a glance at any geological map will tell their names.

The second rule not to be departed from is one not so easy to adhere to, but quite as important as the last, viz.:—The form of your rocks should be that which in Nature is assumed by the particular kind of rock of which it is composed. In order to appreciate the amount of observation which this rule renders necessary, we must consider what are the various agencies which together bring about on rocks the result which geologists know by the name of "weathering." Nature's mode of making her rocks weather-beaten requires such an amount of time, that we cannot attempt to imitate her in that respect; but if we cannot use her means, we can copy her results. Now, the weathering of a rock depends, before all things, on the structure of that rock, on its composition, and on the manner in which it is exposed to sun, rain, frost, wind, and the atmosphere itself, which are the great weathering and rock-carving agents. On many rocks water acts mechanically only; or, to be more accurate, its power of dissolving some rocks, such as quartz, is so limited, even when, as is almost always the case, it is charged with carbonic acid, that it is inappreciable, and may for practical purposes be left out of the reckoning. On a great mass of quartzite rock, for instance, the effect of rain would be of this kind. It could scarcely dissolve any of it away; but it would insinuate itself into every crevice and fissure and crack with which such hard rocks abound near the surface, and thence, by the help of frost, it blasts to shivers, winter after winter, layer after layer of this tough rock, just in the same manner as it bursts the water-pipes of our houses. By observation it is found that every rock affects a more or less peculiar kind of fracture; so that in bursting splinters from them, as has just been shown, the lines of fissure are not arbitrary and accidental. They, like everything else in Nature, form part of a plan. Hence a particular class of form is the result for each rock of this purely mechanical action of the weathering agent. In the case of quartzite, for instance, the fracture is "conchoidal," or shell-shaped, concave and wavy; this, on a large scale, gives rise to peaks with somewhat hollow sides and ridged with sharp serrated edges.

This may serve as an example of simple weathering on a homogeneous, hard, and practically insoluble rock. Let us see what takes place with more complex rocks, of which granite may serve as a representative. This rock is made up essentially of three minerals—quartz, felspar, and mica in various proportions. Now here the water with its carbonic acid will act not only mechanically, as in the case of quartzite, but as a powerful solvent and disintegrator. The fissures in granite are large and continuous, taking the form of immense joints, which cross and recross each other, often, but not always, in a regular manner; but besides these larger lines of weakness, which affect the whole rock, there are those minute lines which separate the constituent minerals from one another. Into all these the water trickles, decomposing the granite along the joints and cracks, "widening them, and rounding off the angles of their intersections, and ultimately only the harder masses, or the hearts of the blocks defined by the joints, remain as solid crystalline granite; some—though little—of the quartz is dissolved away by the water; the iron," which is usually present in small quantities in granite, "becomes oxidized and weakens the rock; but it is chiefly the felspar that is decomposed by the action of carbonic acid, its alkalies are removed, and its residue is washed away in the form of fine china clay. . . . The quartz crystals remain as sand; the mica remains, but is less observable, and is partially decomposed." (Professor Rupert Jones.) It is by processes such as that described, that the many fantastic shapes assumed by granite rocks have been arrived at, whether they be those of the curious balanced "Logging" stones of Cornwall or Brittany, the bare rounded tors of Devon, or the grey sterile mountain-tops of Aberdeenshire. All felspathic rocks of eruptive origin, such as porphyries, are moulded into the shapes which they now exhibit in the same way as granite, and such also is the case with those sedimentary rocks which consist to a considerable extent of felspar, such as many of our gritstones. In these, however, a great uniformity of weathering is caused by the regular lines of bedding which take the place of the horizontal joints of the former class of rocks. The vertical joints are similar in both. In igneous rocks, such as

Limestone.

basalt and greenstone, the jointing and fissuring is often of such a kind as to give rise to very striking effects, very various in their appearance, though probably closely allied in their origin. Thus, from the simple dark brown, or black, trap, without apparent structure, forming shapeless masses of a rounded, somewhat unpicturesque, outline, there is but one step to the bold semi-columnar escarpments of trap, which are so conspicuous in Northumberland and in many parts of Scotland; from these to the wonderful assemblages of rigid geometrical pillars of Staffa and the Giant's Causeway, with all their suggestiveness to rock-builders, the transition is shorter still; whilst in many parts of the three countries, we have examples of trap weathering into a mass of many-coated spheres of every size, decomposing layer by layer, with only a small core of the untouched rock in the centre of each ball. It is a noteworthy fact that basalt in this spheroidal condition weathers and decomposes much more rapidly than it does in the prismatic or columnar state. Rocks such as those we have been considering (with the exception of the grits and quartzite) have all been thrown up in a molten or pasty condition, which precluded their being subject to many of the rules which water-deposited rocks are bound by. Their structure is in a great measure the result of cooling; and although they frequently have a bedded appearance, they are not under the rigid sway of dip and strike, which in other rocks is all-powerful in producing, or rather in preparing, the structure of a country. Indeed, in the great majority of cases, it is the advent of the eruptive rocks which has given the sedimentary deposits their present positions, or what is technically called their "lie." Few of the latter, whether sandstones, limestones, shales, clays, or sands, are now lying in the horizontal positions in which they were formed, especially in much-disturbed and dislocated Britain. Great geological operations have taken place since then, and have squeezed, tilted up, and broken these beds of rock into every shape. And it will be obvious to all that had it not been for these great changes, the edges of these rocks could never have been brought under the influence of rivers and glaciers to carve them on the large scale into hill and dale, and of rain more delicately to "weather" and ornament them. It is therefore very necessary to observe the dip, or general mode of lying of the beds of any district which it is desired to make use of for rockwork purposes. The writer has seen a large rock-garden in the north of England which was laid out with great care and at vast expense, which is spoilt by one apparently small but fatal oversight—the dip of the beautifully arranged rockery-blocks is westerly and strongly-marked, while the dip of the real "live" rock immediately beneath is due east. Now this seems a small thing to find fault with; and it is true that an uneducated eye might be well pleased, in ignorance of the defect. But consider that this easterly dip in that part of the country is the *raison d'être* of the shape of the hills and valleys which make its beauty; without it the fine slope on which this garden stands would not be in existence—the entire district would be altered, to say nothing of the fact that, were it not for this dip, and the vast industries which it fosters, the wealth which built the rock-garden would have been elsewhere. "Follow Nature in all things," is the only safe motto for the landscape gardener. It would be tedious and perhaps not very useful to enumerate

Old Red Sandstone.

the different kinds of water-bedded rock which can in Britain be used for rock-gardens. A glance at the chief members will suffice.

Of the grits we have already spoken, and their mode of weathering is that of the entire class of sandstones, coarse and fine-grained, massive and flaggy. With regard to the latter, it may be allowable to point out, for special reprobation, a mode of rock-building which seems to be gaining favour in many districts. It consists in placing a number of broken flagstones on end, and in every position relatively to one another ; the result is peculiarly hideous, and resembles no possible combination of Nature's art, since the flags, at whatever angle they may be dipping, must be always parallel among themselves, except in the case of the arrangement known as "false bedding," which is one not likely to be successfully imitated. Sandstones are, as a rule, peculiarly adapted for rock-gardens by the forms they assume on weathering, by their great frequency, and by the great variety of their colours. From dark brown to bright red, from red to yellow, from yellow to white, thence through every tint of grey to blue and purple, the choice of colouring is great indeed in these rocks. They are found everywhere— as hard grits in the old Silurian and Cambrian districts, as great rugged crags throughout the Carboniferous regions, forming the well-known Old Red and New Red sandstones, more sparsely distributed among the Oolites, but forming occasional bands of striking character among the sands and clays of the Wealden (witness the "Greys" of the Lover's Seat and other marked natural rocks in the neighbourhood of Hastings and Tunbridge Wells), and in the much more recent tertiaries appearing occasionally, as in the sand of Brussels, as lines of grotesque fistulous masses running through incoherent sand, very much as flints lie in our Upper chalk.

Many sandstones and grits pass gradually into more or less coarse conglomerates, that is to say, rocks formed of rolled pebbles and blocks of stones derived from other pre-existing formations. Of such conglomerates there are many examples in Britain, and they are often very suitable for rockwork, owing to the uneven weathered surface which is the result of the different sizes of the pebbles, and occasionally of their different hardness, and which causes them to be dislodged unequally. The Permian conglomerates, in many places of Central England, are great additions to the natural beauty of the scenery, and should be taken advantage of for the formation of rock-gardens.

Stybarrow Crag, Ullswater.

LIMESTONES.

Under the name of limestone must be included a very large number of rocks different in texture, hardness, and general aspect, but having this in common—that they are chiefly composed of carbonate of lime. The result of this composition is that more than any other rocks they are liable to the solvent, as distinguished from the disintegrating, action of water charged, as rain-water always is to some extent, with carbonic acid. This action we see displayed on a large scale in the great stalactite-lined caverns in the Carboniferous limestone of the North of England, or in the sand-pipes running deep into the chalk of the South country. On a smaller scale, the effects of this dissolving power are marked on every exposed face of limestone of every age, and help to make them everywhere worthy of the attention of the rock-gardener. In some instances thin beds of hard limestone are weathered into a curious honeycombed state, the exposed parts being of a lighter colour than the inner stone; in others the faces of the beds present the appearance of a clumsy balustrade of the Louis XIV. style, the interstices having been gradually eaten away by the water running down the original lines of upright joints. Sometimes the most peculiar forms are assumed in this manner by limestones, and each kind has its own special characteristic shape, to be known only by constant observation; but perhaps no rock equals the great Magnesian limestone of Durham in the eccentricity or in the multiplicity of its disguises. This limestone is of a yellowish colour, and its structure is wonderfully diversified, sometimes hard and compact, sometimes friable, often concretionary and botryoidal, occurring as a mass of radiated concentric spheres of all sizes, generally crystalline, often as a distinct breccia or agglomeration of angular fragments held together by a cement of similar material. A walk along the coast of Durham, from South Shields to Roker, will show to what vagaries of weathering and denudation this extraordinary variety of conformation has given rise. The high cliffs are in places worn into deep caves, in others slender pillars of rough rock have been separated from the main mass, and stand solitary on the beach, while larger islands of rock stand out at sea, through which arches of every size and shape have been excavated. No rock can be better suited for rock-gardens if used rightly, and it is moreover known that its chemical composition is such as to be very beneficial to rock-plants. These magnesian limestones are called Dolomites, and it is notable that their fantastic shapes are by no means confined to England, since no mountain range is so remarkable for abruptness and startling variety of configuration as that in the Italian Tyrol, known as the Great Dolomites. Besides the hard old stony limestones of which we have spoken, there are in England a number of other kinds, from the oolitic limestones to the chalk, which can occasionally serve the landscape gardener's purpose. Their appearance is too well known to need description here. In the newer geological series there are frequently beds of a light porous limestone, very similar in appearance to the sinter which is deposited by petrifying springs. In many places this is called "ragstone," and it is extremely well adapted for our purpose; their distribution is, however, very local in Britain, so that, according to our theory as to æsthetics of rock-gardens, they cannot be very widely used. Abroad, in Tertiary districts, they are far more common, especially on the shores of the Mediterranean, both on the European and on the African side.

SCHISTS AND SHALES.

These may, for the purposes of rock-building, be considered together; the former being simply the hardened and altered form of the latter. Their weathered appearance, where exposed, varies very much with the angle of their dip and with the degree of crystallisation to which they have attained. Some schists are quite as crystalline as granite, and they then weather in the same manner, with this proviso, that the lines of foliation, or lamination, direct the operation. Where such beds are highly inclined, as on the south-west coast or in Brittany, a curious appearance is often seen, which may be called the "Artichoke form," as it exactly resembles the mode of arrangement of the Artichoke leaves. At lower inclinations, schists and the harder shales do not form striking features; but, by offering slight rocky elevations, above a more or less level ground, with distinct "craig and tail shapes," they can be made highly effective in rock-gardening where they occur naturally. This has been done with the greatest success in the Central Park, New York. The softer shales may be dismissed as rockery materials, except for the purpose of forming the lower of the two beds of rock essential to the construction of a good waterfall or of an overhanging crag. While on the subject of waterfalls, it may be as well to remind the landscape gardener that, with very few exceptions, the rocks forming waterfalls in Nature dip *up-stream*, and this holds good for great and small falls alike. The clays and sands need not detain us; where these unrocky materials prevail, the rock-maker is clearly entitled to do the best he can to try and imitate the rock-masses of more favoured districts. But even then he should be bound by what we will call our third rule, which flows naturally from our other two, enounced above: "In no case should the rock-garden be constructed in a manner contrary to the broad geological laws to which all rocks are subject in their natural state."

In this brief survey of a large and interesting subject, it has only been intended to suggest some points for the consideration of rock-builders, and to show that success in their art, as in every other, is to be attained only by careful observation and study of Nature's own models.   G. A. LEBOUR, F.G.S.

SOME NOTES OF A JOURNEY IN THE ALPS OF EUROPE AND THE ROCKY MOUNTAINS OF N.W. AMERICA.

"The best image which the world can give of Paradise is in the slope of the meadows, orchards, and corn-fields on the sides of a great Alp, with its purple rocks and eternal snows above; this excellence not being in any wise a matter referable to feeling or individual preferences, but demonstrable by calm enumeration of the number of lovely colours on the rocks, the varied grouping of the trees, and quantity of noble incidents in stream, crag, or cloud, presented to the eye at any given moment."—*Ruskin*.

As many lovers of alpine plants have no opportunity of seeing them in a wild state, I have thought it well to include a few notes of my first short excursion in an alpine country, which may serve to give some notion of such regions to those who have no better means of knowing anything of it. They relate no exciting accounts of attempts to mount any peaks, but only deal, in passing, with one of the many texts that may be read in the great book of the Alps.

The first day's work was devoted to the ascent of the Grande Salève, which, though not a great mountain, and with green meadows instead of snow at its top, is nearly 5000 feet high, and is a way of commencing training for more serious work.

The limestone chain, to the highest point of which we have to walk, is situated a little to the south of Geneva, and has vast escarpments looking toward that town, with many alpine

flowers, and a noble view of the mountains around, of Lake Leman, and valleys, hills, and far-off Alps, all aglow with the sun of a June morning. A few miles' drive through the fragrant, sparkling air brings us from the margin of the lake to the foot of the mountain before six o'clock, and then we begin the ascent, through the last patches of meadow land, for the most part very like English meadow land, but fuller of Pinks, Harebells, Sages, and Peaflowers, making the land gay with colour. Soon we pass the cultivated land, and enter on the hem of an immense belt of hazel and copse wood, with numerous little green and bushless carpets of grass here and there, which cuts off vine, and corn, and meadow, from the slopes of the mountains. Here, at six in the morning, the nightingale is singing, while white-headed eagles float aloft, now over the lake, and now over plain and hill, sometimes on motionless wing, and silently gliding along on the look-out for prey. From floating bird in glowing air fragrant with Lily-of-the-Valley, the white bells of which may be seen leaning out of its leaves at the base of the bushes, to the flower-clad heaps of stone, and in every peep which the eye obtains through the bush and wood to the villa-dotted margins of the lake, the scene is one of beauty and abounding life.

Some gorges and precipices are reached, every crevice having some plant in it, and all the ledges being clothed with the greenest grass or bushes, but as yet few of such as are generally termed alpine plants are seen. Many of the most delicate and minute of these would grow well in such spots, but the long grass and low wood would soon overrun them. The copse-wood gets a home on the shattered flanks of the mountain. Among it we find numbers of beautiful flowers that may be termed sub-alpine, and occasionally plants that are found of diminutive size near the top of a mountain, are here met with larger in size. The plants that occur in such places should have an interest for all who love gardens, because they flourish under conditions like those of the greater part of our islands. Every copse, shrubbery, thin wood, or semi-wild spot in pleasure-grounds, throughout the length and breadth of the land, may grow scores of these copse-herbaceous plants, that now rarely find a home in our gardens.

That fine rock-plant, *Genista sagittalis*, in bushy masses of yellow flowers, forms the very turf in some spots. Dwarf neat bushes of *Cytisus sessilifolius* become very common; and soon I gather my first wild Cyclamen. The Lily-of-the-Valley forms a carpet all under the brushwood. The Martagon Lily shoots up here and there among the common Orchids and Grass, and Hawthorn Bush is in flower here later than on the plain. The Laburnum is mostly past; but on high precipices we see it in flower. The great yellow Gentian begins to be plentiful, and *Globularia cordifolia* is in dense dwarf sheets here and there, showing its latest flowers. *Anthericum Liliago* is very plentiful and pretty; and we see all this by the side of well-beaten paths, from which many flowers have been gathered. Trifolium, Dianthus, Anthyllis, and Euphorbia struggle for the mastery wherever a little grass has a chance to spread out, and every chink in the rocks where a little decomposed mould has gathered, supports some plant.

After a walk of three hours we reach the top, having often stopped to admire the varied views. From the bottom the visitor might have expected a barren mountain-top, with stunted vegetation; but it is an immense plateau, miles in length, and covered with the freshest verdure. The best meadows of Britain could not vie with it in these points, while the grass is gay with flowers to which they are strangers, and here and there young plants of the great yellow Gentian, with their large leaves, form the fine-leaved plants of the region. Trees there are none; but occasionally the Hazel, Cotoneaster, and other shrubs form a little group of mountain shrubs, enclosing some spot, so that the cattle that are driven up here in the summer months cannot eat down the flowers so easily. The mountain is of limestone, but now and then we meet with a great block of solid granite, a remembrancer of the days when glaciers from the far-off Mont Blanc range stretched to this. In several places there is a large expanse of well-worn rock, a level well-denuded mass, with cracks in it, in which Ferns grow luxuriantly. The surface is indented with roundish hollows, as if great lizards had left their impress on it; these have in the course of ages become filled with a few inches of mould from decomposed moss, etc., and in them grow Vacciniums, Rockfoils and Stonecrops and Ferns, quite as well as if the "most perfect drainage" were secured.

I was very glad to meet with my first silvery Rockfoils in a wild state, having long held that these so often kept in pots, even in Botanic Gardens, require no such attention, and may be grown everywhere in the open air. The plants grew in many positions: at the bottom of small narrow chasms; under the shade of the bushes; in little thimble-holes on the surface of the rocks in a tiny and sometimes flaccid condition from the drought; and here and there among short grass and fern, where the gathered soil was a little deeper.

The vernal Gentian is known as the type of much that is

charming in alpine vegetation: its vivid colour and peerless beauty stamp themselves on the mind of the traveller that crosses the Alps as deeply as the wastes of snow, the silvery waterfalls, or the dark plumy ridges of Pines, though it be but a diminutive plant. It is there a little gem of life in the midst of death, buried under the deep all-shrouding snow for six or even eight months out of the twelve, and blooming during the summer days near the margin of the wide glaciers, and within the sound of the little snow cataracts that tumble off the high Alps in summer. But it is not confined to such awful spots; it descends to the crests of low mountains like this, where the sun's heat has power to drive away all the snow in spring, and where the snow is quickly replaced with boundless meadows of the richest grass, that form a setting for innumerable flowers. Among these the "blue Gentian" occurs, and blooms abundantly late in spring, while acres of the same kind lie deep and dormant, under the cold snow, on the slope of the high neighbouring alp for months afterwards. This brilliant Gentian is very plentiful in the pastures here, but it is already passed out of flower, and the seed vessels, full and strong, are seen among the taller herbage. Alpine travellers, botanists, and writers say that this lovely plant and its fellows cannot be cultivated, and Dean Close echoed this in describing in *Good Words* his passage over the Simplon—an idea quite erroneous, as the plant is of easy culture, even on the level ground.

On one side we have the Jura range, and the wide sunny valley, cultivated in every spot below the town of Geneva, and, between the Jura and our position, the lower part of the Lake of Geneva, scarcely fluttered by the breeze, the countless pleasant spots along its shores, and issuing from it the blue waters of the Rhone. Many green and well-pastured mountains lie beyond, with dark clouds of Pinewoods on their sides and summits. Others still higher, and with the verdure less visible, are behind, and, above all, a great, bony, steep-scarped, dark range, stretching all across the view.

The variety and beauty of the country traversed on descending the other side of the Salève, and the margins of calm celestial-looking Lake Leman, with vast ranges of snowy mountains beyond its broad expanse, give the traveller a rose-coloured impression of the Alps, which forty-eight hours' journey from Geneva was quite sufficient to modify in my case. The country has every conceivable variety of attractive pastoral scenery, and, better still, the human beings in it seem to partake of the felicity which appears to be here the lot of all animated nature. Their cottages and houses, nestling in nooks in flowery fields, and carved out of the abundant wood of the region, snug gardens, vine-clad slopes, numerous flocks, and high ridges of mountain-lawn, with noble groups of Pines, in vast natural parks, form pictures of which the eye never wearies.

THE SAAS VALLEY.

Compared to the shores of the lake I had passed the day before, the Saas valley, with its deeply-worn river-bed and vast sides of gloomy rock, looked anything but a cheerful pass to the Monte Rosa district; but, fortunately, I had other resources than those of the landscape or the sky, and as yet the weather permitted of enjoying them, for here were countless tufts of the Cobweb Houseleek. It was the first time I had ever met with it in a wild state, and cushioned in tufts, over the bare rocks, in the spaces between the stones that here and there had been built up to support the side of the pathway, and in almost every chink there were thousands of it. Although some of the House-leeks are among the most singular of dwarf plants, they are the succulent plants of the Alps: they are among the hardiest of all plants, enduring any weather, and living even in smoky towns.

Next, an old friend, the Hepatica, came in sight, peeping here and there under the brushwood, but rarely in such strong tufts as one sees it make in our gardens. In a wild state it has, like everything else, to fight for existence, and is none the worse for it. To meet this in its wild home would have rewarded one for a day's hard walking in these solitudes, and it had many interesting companions; not the least welcome being the Swiss Club Moss, which mantled over the rocks in many places, pushing up little fruiting stems from its green branchlets.

The scenery now began to get very bold and striking, and, after a walk of nearly two hours, we reached a village with a very poor inn, where we had some black bread and wine. By this time a slight misty rain had begun to fall, and bearing in mind the long valley we had to traverse before reaching a place where we could rest for the night, we resolved to move on as rapidly as possible, and shut our eyes to all the interesting

An Alpine Village.

objects around us. A soaking rain helped us to carry out this part of the plan. With rapid pace and eyes fixed on the stony footway, on we went, the valley becoming narrower as we progressed, and in some parts dangerous from almost perpendicular walls of loose stone. Presently a little rough weather-beaten wooden cross was passed beside the footway.

"Why a cross here?"

"That great stone or rock you see killed, on its way down,

a man returning with his marketings from the valley," the guide replies! He must have formed but a small obstacle to that ponderous mass—big as a small cottage—which fell from its bed and leaped from point to point, at last right over the torrent-bed, resting on a little lawn of rich grass and bright flowers on the other side.

An Alpine Waterfall.

Ten minutes afterwards we came to a group of three more rough wooden crosses, and loosely fixed in the stones at its sides. They marked the spot where two women and a man had been buried by an avalanche. "And how," said I, "do you recover people's bodies who are thus overwhelmed?" "We wait till the snow melts in spring, and then find and bury them." In many places along this valley these wooden crosses, marking the scene of deaths from like causes, occurred so thickly as to remind one of a cemetery.

In the wide valleys and level land about the lakes life is as easy as need be; but where man creeps up to occupy the last tufts of verdure that are spread out where the Alps defy him with forts of rock and fields of ice and snow, it is very hard. Even the procuring of the necessaries of life makes him liable to dangers of which in our own country we have no experience; almost every commodity used has to be dragged up these valleys on the backs of men or mules from the villages and towns in the Rhone valley; while in their dwellings, made of stems of the Pine, and usually placed on spots likely to be free from danger from avalanches, they are sometimes buried alive.

Soon the rain began to be mingled with flakes of snow, and soon it became a heavy fall; and, as we gradually ascended, every surface was covered with it, except that of the torrent beneath, which roared away with as much noise as if the waters of a world, and not those of one hollow in a great range, were being dashed down its picturesque bed—sometimes cutting its way through walls of solid rock of great depth, at others dashing over wastes of worn and huge stones, carried down and ground by its action. Often we crossed it on small rough bridges of Pinewood, fragile-looking, and heavily laden with fresh-fallen snow. The hissing splash of many cascades accompanied the tumult of the river-bed—many of these born of the melting snow and previous heavy rain, the main ones much swollen by it, the air full of large flakes of snow, the Pines on the white mountain side began to look quite sharp-coned from the pressure of its weight.

We had by this time got into a region abounding with flowers, as every one of the caves was lined with the little yellow *Viola biflora*. Every cranny was golden with its flowers. On entering one of these caves, I saw some crimson blooms peeping from under the snow about the roof or brow. They were those of the first Alpine Rhododendron I had ever seen wild. Occasionally, pressed by the snow, the handsome flowers of a crimson Pedicularis might be seen; and in almost every place where a little soil was seated on the top of a rock or stone, so straight-sided that the snow only rested on the top, the beautiful, soft, crimson, white-eyed flowers of *Primula viscosa* were to be seen. It grows in all sorts of positions—wherever, in fact, decomposed moss forms a little soil. In dry places it is smaller than in wet ones, and is usually particularly luxuriant on ledges where a gradual or annual addition of moss or soil takes place, so that the tendency of the stems to throw out rootlets is encouraged.

Several hours in falling snow, feet saturated with deep snow-water, and beginning to chill, notwithstanding the hard walking, make Saas, and Saas only, the one object to attain To gain it, we pass through one or two small hamlets, the inhabitants of which were as much surprised as ourselves at the sudden fall of snow early in June, and we reached Saas just as night was falling. By this time nearly a foot of snow had fallen on the corn, already far advanced in the ear.

As the country for miles around was covered with a dense bed of snow, my hopes of seeing the plants of the high Alps in this region were over; and rather than return by the same long and dreary valley, I determined to cross the Alps and descend into the sunny valleys of Piedmont, where we should, at all events, probably see some traces of vegetable life.

An Alpine Stream

Next day we set out for Mattmark, nearly 9 miles from Saas, more than 7,000 feet higher than the sea-level, and above the level of the Pine or any exalted vegetation. Only a few spots under ledges, etc., were bare, but we found many well-known plants, as well as the rare *Ranunculus glacialis* in full beauty, some of the flowers measuring nearly an inch and a half across. Near where we found this, a great sea-green arch shows the end of a large glacier, apparently a wide and deep river of ice beneath a field of snow, except where in places it is riven into glass-green crevasses. We have to skirt this field of ice to reach Mattmark, where there is a lake, the overflow from which passes right under the glacier.

Lloydia serotina we met with in great abundance in the

region of the glacial Ranunculus, and also *Androsaces* and the alpine Forget-me-not. By scraping off the snow here and there, we could see the very pretty little *Pyrethrum alpinum*, reminding one of a Daisy with its petals down in bad weather. Several not common Rockfoils and a few *Geum*, *Linaria alpina*, very dwarf, but with the flowers much larger than usual; *Gentiana verna*, abundant; a pink Linum, *Polygala Chamæbuxus*, *Loiseleuria procumbens*, *Senecio uniflorus*, with deep orange flowers, and the most silvery of leaves an inch or so high ; and the beautiful *Eritrichium nanum*, from half an inch to an inch high, and with cushions of sky-blue flowers—were among those not hidden from us by the snow.

Next morning we were up early to cross the Pass of Monte Moro into Italy; the snow was very deep, and we were the first strangers who had crossed during the year. The snow was 18 inches thick even in the lower parts of our three hours' walk, so that it was impossible to gather any plants; and this was unfortunate, as the neighbourhood of the little lake of Mattmark, between two glaciers, is said to be very rich in plants. However, there was quite enough to do to ascend Monte Moro, with its deep coating of snow. Arrived at the cross which marks the top, a magnificent prospect bursts upon us—the white clouds lie in three thin layers along the sides of Monte Rosa, but permit us to see its crest, while the great mountains whose snowy heads tower around it are here seen in all their beauty. On the Swiss side nothing but snow is seen on peak or in hollow; on the Italian, a deep valley has wormed its way among the mountain peaks, crested with sun-lit snow and dark crags, and guarded by vast ice rivers and unscaleable heights. We can gaze into this valley as easily as one does from a high building into the street below; and, crouched on the sunny side of a cliff, to gain a little shelter from the icy breeze that flowed over the pass, view its signs of life and green meadows, and, above their highest fringes, the vast funereal grove of Pines on every side, guarding, as it were, the green valley from the death-like wastes of snow above it. Its effect was much enhanced by the snow that had just fallen, and covered up thousands of acres of the higher ground. The contrast between the valley flushed with life and the great uplands of snow was very beautiful.

We had several miles to descend through the snow before a trace of vegetation could be seen, when fairy specimens of the nearly universal *Primula viscosa* began to show their rosy flowers here and there on ledges, where they were pressed down by the snow; and by clearing little spaces with the alpenstock, we found the ground nearly covered with them. Then the glacial Buttercup began to make its appearance in abundance. Another minute gem was here in quantity—the silvery *Androsace imbricata*, growing on the hollowed flanks of rocks—the tufts, not more than half an inch high, sending roots far into the narrow chinks. These having a downward direction, the water could reach the roots from above. One plant was gathered in the hollow recess of a cliff, with at least one hundred little rosettes and flowers, forming a tuft 3 inches in diameter, all nourished by one little stem as thick as a small rush, and which was bare for a distance of 2 or 3 inches from the margin of the chink from which it issued. The tuft, bloom, and minute silvery leaves suspended by this were, in all probability, as old as any of the great larches in the valley below.

The *Androsaces*, with very few exceptions, have not until quite recently often been successfully cultivated. Their silvery rosettes are more delicately chiselled than the prettiest encrusted Saxifrage; their flowers have the purity of the Snowdrop, and occasionally the blushes of the alpine Primroses. They are the smallest of beautiful flowering plants, and they grow on the very highest spots on the Alps where vegetation exists, carpeting the earth with loveliness wherever the sun has sufficient power to lay bare for a few weeks in summer a square yard of wet rock-dust.

The icicle-fringed cliffs, on the concave sunny faces of which the only traces of vegetation seen about here were found, and the rocky precipices seen from the spot, make all this diminutive flower-life the more interesting.

A very pretty dwarf Phyteuma, with blue heads, was found on the rocks here, and as we got down the mountain, *Geum montanum*, with its large yellow flowers, gilded the grass somewhat after the fashion of our Buttercups, and the fine *Saxifraga Cotyledon* was also coming on; one plant found had a rosette of leaves 8 inches across. *Pyrethrum alpinum* here takes the place of the Daisy, and is full of flower. The *Arnica* is in great abundance, and very luxuriant, looking like a small single Sunflower. *Silene acaulis* is everywhere, and no description can convey an idea of the dense way in which its flowers are produced. Starved between chinks, its cushions are as

A Glacier.

smooth as velvet, 1 inch high—though perhaps a hundred years of age—so firm that they resist the pressure of the finger, and so densely covered with bright rosy flowers that the green is totally eclipsed in many specimens. These flowers barely rise above the level of the diminutive leaves.

Soon we reached the meadow-land towards the bottom of the warm valley, and found this Piedmontese meadow almost blue with Forget-me-nots and strange Harebells, enlivened by Orchids, and jewelled here and there with St Bruno's Lily. The flower is nearly 2 inches long, of as pure a white as the snows on the top of Monte Rosa, each petal having a small green tip, like the spring Snowflake, but purer, and golden stamens. The pleasure of finding so many beautiful plants, rare in cultivation, growing in the long grass under conditions very similar to those enjoyed in our meadows, was greater than that of meeting with the more diminutive forms on the high Alp, verifying, as they did, the conviction that no flowers grow in those mountain meadows that cannot be grown equally well in the rough grassy parts of many British pleasure-grounds and copses.

Alpine Larch-wood.

Coming over the pass of Monte Moro, *Primula viscosa* was in perfect condition and full bloom, and yet so small that a shilling would cover the entire plant, while in lower spots on the opposite side of the valley single leaves of it were nearly 3 inches across and 5 inches long. This will help to show the fallacy of supposing that, because a plant is found in almost inaccessible places and hard chinks of cold alpine rock, we must attempt the nearly impossible task of imitating such conditions, or give up the culture of such an interesting class of plants.

The cliffs here rise in some parts like a vast wall to a height of 8000 feet—stupendous and beautiful towers of rock and sunlit snow, perfectly lifeless, but reverberating now and then with tumbling avalanches of the recently fallen snow. Above the village of Macugnaga, as in many other parts of the Alps, some of the Larchwoods are beautiful from the evidences of the struggle for life. Once the breath of summer has passed over the earth, the dwarf herbage is all freshness and life—the smallness and feebleness of the minute vegetation preventing us from seeing the stamp of the destroyer. The winter snow weighs down the little stems, and then when in spring their successors come up in crowds, the earth is covered with a carpet, as if winter would never come again. But not so with the trees. Many lay prostrate, dead, barked, and bleached nearly white among the flowers that crowded up around them. Others were in the same condition, but leaning half erect amidst their green companions: others were dashed bodily over the faces of cliffs: others had their heads and trunks swept over the cliffs by the fierce mountain storms, but holding on by their roots, and, in the most contorted shapes, endeavoured to lift their living tops above the rocky scarp from which in their pride of youth they had been cast. I never in any wood saw anything so wildly and grimly beautiful as this.

WOOD PLANTS.

We next resolved to descend into the plains of Lombardy, cross the lakes of North Italy, go as far as Lecco on the Lake of Como, ascend Monte Campione, and find *Silene Elisabethæ*, a plant as rare as beautiful, and any good plants which that region might afford. The long and ever-varying Val Anzasca, which runs from the foot of Monte Rosa to the great road from the Simplon, is unsurpassed for the beauty and variety of its scenery. We started from the Hotel Monte Moro at half-past three in the morning, when several of the highest peaks were illumined by a ruddy light, and all the lower ones were in the dull grey of daybreak. The Orange Lily in the meadows was not growing higher than the grass, and in single plants, not tufts; the effect was not what we are accustomed to see in Lilies. But by looking over a ledge now and then, those small alpine meadows, apparently stolen from the vast wilderness, were thinly studded with large fully-expanded Lily blooms, the

Cascade in a high wood.

flowers relieved by the fresh grass. *Asplenium septentrionale* was extremely abundant. Of flowers we saw but few, for the taller tree vegetation cuts off the view and runs up, and clothes the secondary mountains to the very summits, except where grass that is like velvet spreads out, as if to show the small silvery streams, which soon hide in the woods, and by-and-by are seen in the form of cascades falling over wide precipices, to be again lost in deep, wet, tortuous, stony beds, and presently forming larger cascades. Then lower down they break and shoot perhaps for 300 feet, till they join the main stream of the valley below, which has cut itself an ever-winding, diving and foaming bed between terraces, and cliffs, and gullies of rock, affording scenes of infinite beauty and variety.

We walked 12 miles down the valley before breakfast, and every step revealed a new charm. Before us, a great succession of blue mountains; on each side, mountain slopes green to the line of blue sky; behind, all the glory of the Monte Rosa group, in some places flat-topped and of the purest white, like vast unsculptured wedding-cakes—in others, dark, scarred, and pointed to the sky, like some of the aged Pines on their lower slopes, standing firmly but with branch and bark seared off by the fierce alpine blast. Lower down, the valley begins to show signs of human life, with well-built and clean-looking houses; the slopes of the hills are frequently terraced, to give the necessary level for pursuing a little cultivation. Vines begin to appear, and for the most part are trained on a high loose trellis from 5 to 7 feet above the surface of the ground, so as to permit of the cultivation of a crop underneath. The trellises are frequently held up by flat thin pillars of rough stone, which support branches tied here and there with willows. It seems a good plan for countries with a superabundance of light and sun.

From nearly every rock and cliff along the valley spring the pretty rosettes and foxbrush-like panicles of flowers of the great silvery Rockfoil. But the charm of the valley is its ever-varying and magnificent scenery—a foreground of Italian valley vegetation—the deep-cut river-bed below, the ascending well-clothed mountains to the right and left, and then up the valley the higher Pine-clad slopes, all again crowned by the majestic mountain of the rosy crest. The most passionate and unreasoning love of country would be excusable in the inhabitants of these happy spots, enriched with the vine and other products of the south, sheltered by evergreen woods, and walled in by arctic hills.

We will hasten by the streams that feed Lake Maggiore, and stop for a while near the islands on its fair expanse. Mountains with dense green woods creeping to their very tops are reflected in the transparent water, in which they seem to be rooted, so near do they rise from its margin, and only showing their stony ribs here and there, where a deep scar or scarp occurs, too precipitous for vegetation.

The isles look pretty, but not beautiful, because of the rather extensive and decidedly ugly buildings and terraces upon them; but they are only specks in a great natural garden. Brockenden is quite right when he says of one of the islands: "It is worthy only of a rich man's misplaced extravagance, and of the taste of a confectioner." The Maidenhair Fern is abundant on the islands. The vegetation here and on the margins of the lake is often of an interesting character, quite sub-tropical in some places; but as our business is with alpine and rock plants only, we must pass all this by, and hasten on to the shores of Como. When approaching Isola Madre, the first thing that struck my attention was a plant like a greyish heath, covered with light rosy flowers, growing out of the top of a wall. It proved to be an old friend, the Cat Thyme, and in beautiful condition; as grown in England, nobody would ever suspect it to be capable of yielding such a bright show of flowers. *Trachelium cœruleum* grows very commonly on the walls, and so does the Caper, a noble plant when seen springing from a wall, bearing numbers of its large blooms.

MONTE CAMPIONE.

Arrived at Lecco to hunt for the handsome Catchfly on the crest of this mountain, we start at three o'clock in the morning, as it is our aim to get up a little out of the warm valleys before the dew had fled. Soon we find ourselves on the spur of a mountain, on which Cyclamens peep forth here and there from among the shattered stones—sometimes handsome tufts, where the position has favoured them, and now and then springing in a miniature condition from some chink, where there was very little "soil." Lower down we met with the neat *Tunica* on the tops of walls, and it continued to appear for some distance higher up, rarely looking so pretty as when well cultivated. The Maiden-hair Fern does not ascend up the mountain sides, nor even find a home in the villages up the valley, though in the town of Lecco it adorns the mill-wheels and moist walls near watercourses, with abundance of small plants adhering closely to the wall, and dwarf from existing on moisture or very little more. As we ascend, the fine flowers of *Geranium sanguineum* are everywhere seen; while Aconites, Lilies, are here and there. The Orange Lily is a great ornament hereabouts—one on most inaccessible cliffs of the mountain, with its bold flowers like a ball of fire in the starved wiry grass. The Martagon Lily is also abundant. Dwarf Cytisuses are great ornaments to the rocks, and here and there the leaves of Hepatica are mingled with those of Cyclamen, suggesting bright pictures of spring. The Cyclamens are deliciously sweet, and the great spread of the alpine Forest Heath, seen in all parts, must afford a lovely show of colour in spring.

We think we have taken leave of all the meadow-land, when the hills again begin to break into small pastures, where Orchises, Phyteumas, Arnica, Inula, Harebells, and a host of meadow plants, struggle for the mastery. Soon we come to great isolated masses of erect rock, whose surface is quite shattered and decayed in every part; and, after half-an-hour among these, see, far up, rosettes of the blue flowers of *Phyteuma comosum*, projecting about two inches from the rock. The rosettes are as wide as the plant is high, and much larger than the leaves, which are of a light glaucous colour. We ascend far above these rocks, and find the mountain-side has broken into wide gentle slopes, park-like, with birch and other indigenous trees here and there, but, for the most part, a great spread of meadow-land, adorned in every part with a lovely carpet of flowers. Conspicuously beautiful was the St Bruno's Lily, growing just high enough to show its long and snow-white bells above the grass. It should be called the Lady of the Meadows, for assuredly no sweeter or more graceful flower embellishes them. In every part where a slight depression occurred, so as to expose a little slope or fall of earth on which the long grass could not well grow, or along by a pathway, *Primula integrifolia* was found in thousands, long passed out of flower.

In wandering leisurely over the grass, an exquisite Gentian, of a brilliant deep and iridescent blue, came in sight. At first we thought it was the fine *Gentiana verna*, but on taking up some plants, it proved to be an annual kind, quite as beautiful and brilliant as either *G. bavarica* or *G. verna*. Whereever a boulder or mass of rock showed itself, *Primula Auricula* was seen, often in the grass and always on the high rocks and cliffs. A showy Epilobium and Dentaria are also seen among the taller vegetation, while the compact little blue Globularia creeps from the surrounding earth over every rock. As we mount, the mist of the higher points begins to envelop us, and hide the lovely and ever-varying scenery below and on all sides, except now and then when the breeze clears the vapours away.

As the upper lawns are reached, the extraordinary nature of the mountain begins to be seen through the increasing mist. Lower down, and indeed in all parts, erect, isolated masses of rock are met with; but towards the great straight-sided mass that forms the central and higher peak, huge *aiguilles* are gathered together so thickly that, dimly seen through the mist, they seem like the ghosts of tall old castles and towers creeping one after the other up the mountain-side. Lower down, cliffs of the same nature and great height form one side of the mountain, their giant and weird appearance being much heightened by the mist which completely hid the valley and made them seem as if poised in the air.

Hereabouts we came upon some little tufts of the most diminutive and pretty *Saxifraga cæsia*. In little indentations in rocks it sometimes looked a mere stain of silvery grey, like a Lichen; on the ground, it spread into dwarf silvery cushions, from 1 to 3 or 4 inches wide. It seemed quite indifferent as to position, sometimes growing freely along, and even in, a channel, the sides and bed of which are a mass of shattered rocks, and which is in winter a stream and a torrent after heavy rains and thaws. Some plants were as large as a dessert plate, a mass of Liliputian silvery rosettes, each about the eighth of an inch across, and formed of from fifteen to twenty-five diminutive leaves, and hundreds of rosettes going to form a tuft about an inch high.

This is one of the gems in the large Saxifrage family, which affords a greater number of distinct plants worthy of cultivation in the rock-garden than any other. These plants grow upon the mountain tops, far above the abodes of our ordinary vegetation, not only because the cool, pure air and moisture are congenial to their tastes, but because taller and less hardy vegetation dares not venture there to overrun and finally extinguish them. But though they dwell so high in alpine regions, they are the most tractable of all plants in British gardens, and grow as freely as our native lowland weeds in gardens where Gentian and alpine Primula and precious mountain Forget-me-not require all our care. They are evergreen, and more beautiful to look upon in winter than in summer, so far as the foliage is concerned, and their foliage is beautiful, while, unlike many other plants which have attractive leafage, or a peculiar form and habit, they flower freely in the early summer.

One would think that coming from habitats so far removed from all that is common to our monotonous skies, it would be impossible to keep these little stars of the earth in a living state; but our climate suits them well, and they are the chief stay of the cultivator of alpine plants. In autumn, when most plants quail before the approach of darkness, winter, and frost, and casting off their soiled robes, the Rockfoils glisten with silver and emerald when the rotting leaves are hurrying by before the stiff, wet breeze.

The Lion's-paw Cudweed is very abundant on Monte Campione. Daphne and Rhododendron in small quantities, and the pretty little *Polygala Chamæbuxus*, often crop out

The limit of the Pines.

less beautiful than when in cultivation. A blue Linum, probably *L. alpinum*, is very common; the rare *Allium Victoriale* we found sparsely on high rocks; and *Dryas Octopetala* abundantly in flower, with *Anemone alpina* in a very dwarf state; while pale flowers of the common *Gentiana acaulis* looked up singly here and there. In the higher and barer parts of the meadows, *Aster alpinus* was charming, not in tufts or masses, but dotted singly over the turf. Having climbed so high for the chief object of our ascent, we failed to find it there after a long search, and, disappointed, were descending the mountain down a long and rocky chasm formed of a vast bed with banks of shattered rock, when, much to our pleasure, a little plant with a few leaves was discerned growing from a chink on a low mass of rock. By carefully breaking away portions of this, we succeeded in getting the plant, roots and all, out intact, and by very diligent searching, found a few more specimens of it. It was not yet in flower, but pushing up the stem preparatory to it. Then a long trudge down mountain, valley, and hilly road brought us home to our quarters at half-past nine, after a day of nearly twenty hours' walking.

With a few words on the vegetation of some parts of the Simplon great range, these notes will end. The chief feature of the smaller vegetation alongside the great Simplon Road is the foxbrush-like flowering pyramids of the great *Saxifraga Cotyledon*, and on the highest parts of the road, wherever the ground near it softens into anything like turf, the fine blue of the vernal Gentian sparkles amongst yellow Potentillas and Ranunculi. It is pleasant to meet with it in flower weeks after one has left it in full flower in England in April, and seen it bear seed on mountains about 5000 feet high. About the end of June it was in fresh and perfect condition here, and likely to remain so for some time to come. Observe the capabilities of the plant, and the changes that it endures without losing health in any case. In perfect health in England, without a covering of snow through the winter, and flowering strongly in early spring, it flowers here in the month of June, and higher up in July.

Let us ascend one of the highest mountains of the range a little way, climb upwards for two hours, passing the limits of the Pines, till we get at the base of the bed of an enormous glacier, a vast high field of snow apparently, which fills the upper portion of a wide gap between two mountains. The wide expanse of ground which we are traversing is simply a mighty bed of shattered rock, which at a remote day was carried down by this colossal, ever-levelling machine, and it is now covered with a scanty vegetation of alpine Rhododendron and high mountain plants.

Everywhere, and very pretty, is the mountain form of the Wood Forget-me-not, but no trace of the true *Myosotis alpestris*.

Everywhere the large white flowers of the mountain Avens are covering the surface; but as we are in such rich ground, we had better confine ourselves to plants not British, and—climb. That exertion is above all things necessary; the vast slopes of shattered rock seem interminable—an hour's hard work only brings us to a point that we thought we could reach in five minutes, and this point, instead of proving the resting-place and exploring-ground we had expected it to be, merely shows us that still the wide and mighty mass of shattered rock creeps higher and higher, far beyond our powers of approach, until at last the wall of ice, "durable as iron, sets death-like its white teeth against us." On a great ridge beneath it are some scattered fragments of vegetation rooting deeply among the stones, and gaining a scanty subsistence from the sandy grit which results from the decomposition of the fields of brittle rock. The Crimson Rockfoil is a mass of flower; we cannot see anything but flowers on its dense cushions, beautiful in this awful solitude. Here and there a large yellow flower is seen, which proves to be *Geum reptans*, a fine plant, from 3 to 6 inches high. Presently, while admiring the great beauty of the crimson Saxifrage here, within a few feet of wide beds of snow, that lie on each side of the ridge on which we stand, what appears a giant plant comes in sight; the flowers are much larger, so that instead of little cushions made up of a multitude of blooms, we see the individual cup-like blooms standing boldly up, of much deeper hue, and the leaves also grown large and distinct. It is the noble *Saxifraga biflora*. It is a pleasure to gather this plant here, and also *Linaria alpina*, more familiar to me, and so beautiful here. Some alpine plants are prettier in cultivation than in a wild state. Not so *Linaria alpina*, which grows and flowers well in sandy soils and moist places at home, and gets so strong that its glaucous leaves form quite a strong tuft, but which here shows its rich orange and purple flowers, gathered in dense tiny tufts here and there among the stones, without any leaves being seen, and it is more lovely here than in cultivation, though its beauty in either case is of a high order. The very dwarf and pretty little *Campanula cenisia* was abundant among the higher plants, its tufts of light green among the *débris*. One solitary tuft of *Ranunculus alpestris* was met with by the side of a little rivulet; a plant about 6 inches in diameter, and quite pretty where "specimens" are rare, and where one thing struggles with another in the grass.

Descending, the ground, becoming more level, begins to form an undulating basin between two ranges, and here the short grass is jewelled with dwarf alpine plants and flowers. The silky-leaved and very dwarf *Senecio incanus* occurs in thousands; the Cudweeds, too, are abundant, while a few inches above the dense silvery turf formed by such plants, the large and beautiful purple flowers of *Viola calcarata* form, not quite a sheet of colour—for the flowers occur singly, and are separated one from the other by bits of green and silvery turf—but sometimes the eye is brought nearly level with the surface of a bank dotted with blossoms, and the effect is lovely. It is not the effect of "massing" flowers, but that of "shot" silk. The flowers of this Violet were generally very large—I measured several an inch and a half across, while the plants from which they sprang were almost inconspicuous, and generally I had to use the flower stem as a guide to the minute rosette of leaves in the grass. A still more beautiful effect, and perhaps more so than I have seen either in garden or wild, was observed when tufts of *Gentiana verna* occurred pretty freely amongst this Violet, the vivid blue of the Gentian in patches amongst the groundwork of the Violet. In quite a valley of Gentians—a little lawn at an elevation of about 7000 feet—were some growing in a watery hollow, of a vivid and exquisite blue; they were large tufts of *Gentiana bavarica*. The little Box-like leaves were in compact tufts, and the flowers were larger, of a deeper blue than *G. verna*, which is saying a great deal.

There were spots near at hand where *G. verna* formed a

The Home of the Purple Saxifrage.

turf of its own, and yet it was not so beautiful as *G. bavarica*, which was growing exactly in positions that would suit the Bog Bean and the Marsh Marigold. Attempts to cultivate *G. bavarica* in England have hitherto been a failure. It is very rarely seen with us even in Botanic Gardens, and, when it is seen, is usually in poor health. A few words, then, about the position in which I found it in such perfection, may prove useful. A little mountain streamlet diverges from its channel and spreads over the surface of the ground for 20 or 30 yards across, not destroying the grass, but simply showing itself in trickling patches here and there. On the little hillocks of grassy earth that stood a few inches above the water, I found the plant in very good condition, the roots certainly in the water, and the "collar" of each plant very little above it. Somewhat lower down, the waters gathered together again, leaving the sides of that marshy spot and the intermediate ground perfectly green, but very wet, and here and there dotted with clusters of blue stars, to which in brilliancy the choicest gems were but dull and earthy. In walking on this green spot the water hissed and bubbled up around. Here the plants were very fine, the pretty little close-growing tufts of light green leaves clearing spots for themselves in the longish grass. The slightest impression made here immediately became a small pool, and in no place did I find the plant but where the hand, if pressed into the grass, was at once surrounded by water. A few steps away, and *Gentiana verna* was everywhere in full beauty on dry banks; but in no case did either species manifest a tendency to invade the ground of the other. In fact, proof was there that *G. bavarica* is a true bog-plant. And what a beautiful companion for the Wind Gentian, the Water Violet, the peat-loving *Spigelia marilandica*, *Rhexia virginica*, the little creeping Bell-flower, and like plants!

Scene in the Rocky Mountains.

MOUNTAIN VEGETATION IN AMERICA.

THE passage of the great American desert which is crossed on the way from New York to San Francisco is, perhaps, the best preparation one could have for the startling verdure and giant tree-life of the Sierras. Dust, dreariness, alkali—the earth looking as if sprinkled with salt; here and there a few tufts of brown grass in favoured places; but generally nothing better than starved wormwood, that seems afraid to put forth more than a few small, grey leaves, represents the vegetable kingdom in the plains of the desert region. Where the arid hills—showing horizontal lines worn by the waves of long-dried seas—are visible, a few thin tufts of alders and poplars mark their hollows; while willows fringe the streams of undrinkable water which course through the valleys. A better idea of the country can scarcely be had than by imagining an ash-pit several hundred miles across, in which a few light-grey weeds, scarcely distinguishable from the parched earth, have sprung up.

As the train ascends the Sierra, it passes through dark-ribbed tunnels of long covered sheds, which guard it from the snow in winter. Dawn broke upon us as we were passing through these; and, looking out, dust, alkali, dreariness, harshness of arid rock and hopelessness of barren soil, are seen no more, but near at hand a giant Pine rushes up like a huge mast, while all around and in the distance are great Pines grouped in stately armies, filling the valleys and cresting all the wave-like hills, till these are lost in the distant blue.

Isolated Rocks in Rocky Mountains.

To the western slopes of the great chain of the Sierras one must go to see the noblest trees and the richest verdure. There every one of thousands of mountain gorges, and the pleasant and varied flanks of every vale, and every one of the innumerable hills, are densely populated with noble Pines and glossy Evergreens, like an ocean of huge land-waves, over which the spirit of tree-life has passed. The autumn days I spent among these trees were among the happiest one could desire—every day glorious sunshine, and the breeze as gentle as if it feared to overthrow the dead trees standing here and there leafless, and often perhaps, barkless, but still pointing as proudly to the zenith as their living brothers. Wander away from the little rough dusty roads, crossing, perhaps, a few long and straight banks of grass and loose earth—the stems of dead monarchs of the wood now given back to the dust from which they once gathered so much beauty and strength—and fancy willingly reminds us of the mast-groves of the Brobdingnags. There is little animal life visible, with the exception of a variety of squirrels, ranging from the size of a mouse to that of a cat, the graceful Californian quail, and occasionally a hare or a skunk. Everywhere vegetation is supreme, and in some parts finer effects are seen than in the most carefully-planted park. This results not more from the stately Pines (not often crowded together as in the Eastern States, and often near the crest of a knoll, standing so that each tall tree comes out clear against the sky) than from the rich undergrowth of evergreens with larger leaves that form a smaller forest beneath the tall trees. Grand as are the Pines and Cedars (Libocedrus), one is glad they do not monopolise the wood; the Evergreen Oaks are so glossy, and form such handsome trees. One with large shining leaves, yellowish beneath, and long acorns in thick cups, covered with a dense and brilliant fringe of fur, was the most beautiful Oak I ever saw; but most of the Evergreen Oaks of California, whether of the plain or hills, are handsome trees. One day, in a deep valley, darkened by the shade of giant specimens of the Libocedrus, I was astonished to see an Arbutus, about 60 feet high, quite a forest tree. This is Menzies' Arbutus, commonly known by the old Mexican name of the "Madrona"; and a handsome tree it is, with a cinnamon-red stem and branches. Here and there, too, the Californian Laurel (Oreodaphne) forms laurel-like bushes, and tends to give a glossy, evergreen

Mountain Woods of California.

character to the vegetation. Shrubs abound, the Manzanita (Arctostaphylos glauca) and the Ceanothuses being usually predominant; while beneath these and all over the bare ground are the dried stems of the numerous handsome bulbs and brilliant annual flowers, that make the now dry earth a living carpet of stars and bells of brilliant hues in spring.

On the very summit of the Sierra Nevada the vegetation is not luxuriant; there, as elsewhere on high mountain chains, is the frost that burns and the wind that shears. A solitary Pine that has been bold enough to plant itself among the rocks of the high summits, it is usually so contorted that it looks as if inhabited by demons; while here and there one has succumbed to the enemy, and a few blanched branches stick from a great, dead, barkless base, lapped over the earthless granite. But go a little lower down the mountain, and most probably you will find a noble group of Piceas, startling from the size and height of their trunks, though looking much tortured about the head by the winds that surge across these summits—the mast-heads of the continent.

Snow falls early and deep on the Sierras, and the stems of the higher trees are often covered with it to a depth of from 6 to 25 feet. Near the railway and near frequented places, thick stumps of Pines, 6 to 15 feet high, may be noticed; these are the trees which have been cut down when the snow was high and thick and firm about the lower part of their stems. But if the nights are bitterly cold, the sun is strong in the blue sky far into the winter months, so that the snow is melted off the tree-tops, and the leaves of the Pines live in light, throughout the winter. All the Pines that grow near the summit must resist intense cold.

The golden light of the sky and the blue of its depth, and the purity of the fresh mantle of snow, are not more lovely in their way than the robe of rich yellow Lichen with which the stems and branches of the Pines are clothed. Imagine a dense coat of golden fur, 3 inches deep, clothing the bole of a noble tree for a length of 100 feet, and then running out over all the branches, even to the small dead twigs, and smothering them in deep fringes of gold, and some idea may be formed of the glorious effect of this Lichen (Evernia). It is the ornament of the mountain trees only; in the valleys and on the foot-hills it is not seen.

It is a mistake to suppose the Sequoia (Wellingtonia) is such a giant among the trees here; several others grow nearly or quite as high, and it is very likely that in such a climate many Pines would attain extraordinary dimensions. There was a small saw-mill near where I stopped for some days, and several yokes of oxen were constantly occupied in dragging Pine logs to it. The owner never thought of bringing anything smaller to this than a log 3 or 4 feet in diameter in its smallest part, and usually left 100 feet or so of the portion of the tree above this on the ground where it fell, as useless. What is it that causes the tree-growth to be so noble there? Soil has very little to do with it. I often saw the trees luxuriating where there was not a particle of what we call soil, and, indeed, in places where 25 feet or so of the whole surface of the earth had been washed away by the gold-miners. A bright sun for nearly the whole year, and an abundance of moisture from the Pacific Ocean, explains the matter. This should draw our attention to the fact that, in planting, and especially in the planting of coniferous trees, we pay far too much attention to supplying them with rich soil, and far too little consideration to the climate in which we have to plant.

ALPINE FLOWERS ON THE ROCKY MOUNTAINS.

There is a foot or two of snow in some places on November 15, 1870; but the time for very deep snow has not yet come, and we are fortunately in time to see a patch of alpine plants here and there before they are tucked in under their wintry shroud. What are these brown tufts like withered moss among the rocks and boulders on exposed spots, some of them cushioned low and flat; others looking as if moss had assumed a shrubby habit, and died full of years, at 3 inches high perhaps, on a gouty stem nearly as thick as the finger? These are little Phloxes, withered almost beyond hope by the heats of summer; but pull up one, and the old roots are seen sending out a mass of fragile feeders in the snow-moistened earth, and in the very centre of each juniper-like truss of prickly leaves may be discerned a small speck of green. When the 20 feet of snow melts in spring, and the sun warms the saturated earth, these mites of Phloxes will be to the now arid solitudes as blossoms to the crabbed apple-tree. The dead moss will change to bright, shining green, and presently this will be obscured by as fair a host of flowers as ever fretted over the small herbs on Tyrolese Alp. The alpine Phloxes of the Rocky Mountains are as indispensable to the choice collection of alpine plants as Gentians or Primroses.

Everywhere on bare places there are tufts of dwarf, bushlike Pentstemons, from 2 to 5 inches high, and bearing nearly the same relation to the tall Pentstemon of our gardens as the alpine Phloxes do to the border Phloxes. The Pentstemons are among the most beautiful of rock-plants, their colours being of a more refined and delicate character than those of the tall varieties, good as these are. Indeed, no flowers possess such iridescent blues and purples as these. Like the little Phloxes, many of these have woody stems, probably as old as some of the Pines near at hand, and have embellished these lonely heights for ages unadmired, unless the "grizzlies" or the woodpeckers delight in such objects.

It might perhaps be thought that, however well the alpine plants thrive among rocks and boulders, the giant Pines would require good soil, or, at all events, level ground of some kind, to start from. It is not so. A seedling Pine springs up in some shallow chink or narrow crack in a mass of great stones; patiently it throws out long feeders on one side, which find their way down the steep faces of the rocks or run through any moist or narrow channels into the feeding ground beyond; it soon gathers strength enough to build a great trunk above the narrow chink from which it sprang, lapping its base over the close-embracing rocks much as a fungus would. I have seen trunks measuring 18 feet in circumference springing from masses of raised rocks, where one would not think a wiry juniper bush could live.

On looking at some compact brownish tufts of leaves, a few yellow Coronas are seen; these are somewhat "everlasting" in character, and have only faded with the snow-water. They belong to quite a distinct plant of the Buckwheat family—Eriogonum. The family we know is nearly all composed of weeds, and the genus, which has many members in America, is seldom in the least attractive; but this one is quite a gem of a rock-plant—handsome umbels of primrose-yellow springing abundantly from dull brownish tufts of leaves 2 inches high, making it as pretty as it is distinct. Far away, on a bare, gravelly hillside, vivid red tufts are seen; these prove to be another equally beautiful kind of Eriogonum, the leaves of which assume a deep, shining blood colour.

Here and there the withered stems of Lilies may be seen; Washington's Lily—a tall, noble, and fragrant kind—and several other Lilies occur abundantly. The stems of some which I found in little ravines were quite 8 feet high. The Soap-plant—a bulbous perennial—is abundant on all the lower mountains and on the coast hills. Numerous bulbs of a high order of beauty occur on the mountains and plains of California, but they mostly bloom in spring and we only see their withered stems.

Another very beautiful rock-shrub, quite distinct from anything we have in our European Alps, is the Bryanthus. After trudging for hours over snow and rock in quest of this, I had given it up, when a spray, with a withered truss of bloom, was seen, and soon I had dug a few score plants of it from beneath a couple of feet of snow. This Bryanthus may be roughly described as having the leaf of a heath, with handsome crimson flowers, like those of a small rhododendron, and forming bushes from 4 to 10 inches high. Another rock-shrub, quite distinct from all others, is a creeping Ceanothus, which runs along the ground as closely as Twitch. On the lower hills, where it grows more freely, the shoots march in parallel lines over the ground, covering it with a rigid carpet of dark green leaves.

One of my objects in coming here was to see the Californian Pitcher-plant (*Darlingtonia*) in a wild state. This plant resembles the Sarracenias of the eastern side of the continent, the chief difference being that it has a cleft appendage to the margin of the orifice of the pitcher, each lobe being from 1 to 2 inches long. I came upon the Darlingtonia, greatly to my pleasure, on the north side of a hill, at an elevation of about 4000 feet, growing among Ledum bushes, and here and there in sphagnum, and presenting at a little distance the appearance of a great number of Jargonelle pears, with their larger ends uppermost, at a distance of from 10 to 24 inches above the ground. This resulted from the pitchers being quite turned over at the top, so as to form a full rounded dome, and the uppermost part of the pitcher being of a ripe pear yellow. The plants grow in small bogs, from springs on the hillside; the soil peat resting on a quartz gravel. The plant is quite a strong grower. I found one large colony growing so well among common rushes that Darlingtonia seemed to be quite beating them in the struggle. I was too late for seeds, but saw sundry stems 3 feet or more high, bearing empty seed vessels as large as large walnuts. All the pitchers have a spiral twist, which is much more marked towards the apex, and in the large specimens. But perhaps the most remarkable feature of the plant is its efficiency as a "fly-catcher." In the houses about here the pitchers are regularly used in summer for catching flies. Each of the developed pitchers that I cut off had from 3 to 5 inches of various forms of insect life, dead and closely packed in the lower part of its chamber. Pass a sharp knife through a lot of brown pitchers withering round an old plant, and the stumps resemble a number of tubes densely packed with the remains of insects. What attracts them is not so very clear, as the orifice is half hidden in the turned-over head, and by its two-lobed appendage. But, by raising the pitcher above the eye, and looking up into its dome, often 3 inches through in fair specimens, it seems a curvilinear roof of miniature panes set in a golden network. This is in consequence of the greater portion of the upper part of the pitcher being transparent in all the space between the veins, though no one transparent spot is more than a line or two across. Within the pitcher the surface is smooth for a little way down; then isolated hairs appear; and soon the chamber becomes densely lined with needle-like hairs, all pointing down, so decidedly indeed that they almost lie against the surface from which they spring. These hairs are very slender, transparent, and about a quarter of an inch long, but have a needle-like solidity, and are colourless. The poor flies, moths, and ladybirds travel down these conveniently arranged stubbles, but none seem to turn back. The pitcher, which may be a couple of inches wide at the top, narrows very gradually, and at its base is about a line in diameter. Here, and for some little distance above this point, the vegetable needles of course all converge, and the unhappy fly goes on till he finds his head against the firm thick bottom of the cell, and his retreat cut off by myriads of bayonets; and in that position he dies. Very small creatures fill up the narrow base, and above them larger ones densely pack themselves to death. When held with the top upwards, sometimes a reddish juice, with an exceedingly offensive odour, drops from the pitchers. The plant throws out runners rather freely, by which means it increases. As to its culture, there can be no doubt about that—a soil of peat, or peat and chopped sphagnum, kept wet —not merely moist—the pots or pans to be placed on a moist bottom. Frame or cool house treatment is best in winter; warm greenhouse or temperate stove in summer. It is hardy in the south of England and Ireland.

Armeria, but the plants form branching, cushion-like tufts; the leaves rigid and spiny. They are dwarf evergreen rock-garden plants, but, coming from eastern regions not now of easy access, are not easy to introduce, and for this and other reasons make slow progress in gardens. They are beautiful plants, flowering usually in July and August, when many of the early flowers are past. Slow in growth and difficult to increase as regards their general propagation, and

Acantholimon venustum (*Prickly Thrift*).

where large plants of the rare kinds exist, it is a good plan to work some cocoa-nut fibre and sand in equal parts into the tufts in early autumn. Before working in this material, some of the shoots should be gently torn, so as to half sever them at a heel or junction; then gently work in more material around, and water to settle the soil. Many of the growths thus treated will root by spring. Cuttings made in the ordinary way are by no means certain, but when this method is adopted, August or September is the best time. All cuttings so-called should be torn off with a heel and inserted without further ado.

Acantholimon glumaceum is the best known as the most vigorous grower, forming cushions of narrow dark green leaves, spiny at the point, and spikes of rose-coloured flowers from June to August. At Tooting, many years ago, this species formed an edging a foot or more wide, and about 150 feet long, and when in flower was a pretty sight.

A. venustum.—A delightful plant when seen in good condition. I lost the finest specimen I have ever seen during the great frost of 1895. The plant, unfortunately, had been left fully exposed with other alpines in pots. This lovely species in the summer of 1894 produced some forty spikes of its pink blossoms. The tufts are dark green, with a slightly greyish or glaucous tint overlying the same. This species is of much slower growth than *A. glumaceum*, and requires some good sandy loam, with leaf-soil and broken brick rubbish mixed freely with the soil. It bears its rose-pink flowers in July, on one-sided, slightly arching spikes, and is certainly one of the most charming of midsummer alpines. Firm planting, a rather sheltered spot, and a deep soil, well-drained, should be given. Cilicia.

A. androsaceum.— This species is distinguished by the more dense tufts which it forms when established, as also by the rosettes being less spiny. This is not so much due to the spines as to the pliant nature of the leaves. It is of easy culture, spreads somewhat freely over a ledge of rock, and bears pink blossoms on sprays 4 inches high.

A. acerosum.—The dense character of this species and the grey glaucous hue of the leaves at a short distance, remind one of *Dianthus cæsius*. A closer inspection, or even an unwary placing of the hand upon the spines, will quickly dispel any such idea, since the short, greyish glaucous leaves are the most spiny of all.

PART II.

ALPINE FLOWERS FOR GARDENS

A SELECTION OF ALPINE FLOWERS ALPHABETICALLY ARRANGED, WITH INSTRUCTIONS FOR THEIR CULTURE AND POSITION IN GARDENS.

ACÆNA (*Tufted Bur*).—Dwarf tufted and spreading plants of secondary value only for the garden, but often useful for dry banks or poor places in borders where we seek a little repose in the shape of a carpet of soft green or grey. They are of easy culture in the common soil, increase rapidly by division, and though mostly South American, the cultivated kinds are quite hardy. There are, perhaps, twenty kinds in cultivation in Europe, but a few only are worth having, where effect is sought.

Acæna microphylla (*Rosy-spined A.*).—A minute trailer from New Zealand, curious from its small round head of inconspicuous flowers furnished with long crimson spines. The plant spreads into dense tufts, and in summer and autumn is thickly bestrewn with the showy globes of spines. It is easily increased by division, is hardy, grows in ordinary soil, but thrives much the best in that of a fine sandy nature. Its home is on bare level parts of the rock-garden, usually beneath the eye, and it is also good as a border, or even an edging plant. Occasionally it may be used with a good effect as a carpet beneath larger plants not thickly placed. Syn., *A. novæ Zealandiæ*.

Acæna Argentea is stronger growing, the leaves always larger and very glaucous. It is nearly related to

A. pulchella, which, owing to its trailing habit and abundance of bronzy leaves, is more useful. The graceful branches of this, when hanging over large stones or old walls, have a pretty effect, and it is hardy and evergreen.

A. Buchanani.—In this, the foliage is what may be called "Pea-green," although this fails to convey any idea of the prevailing hues of green which make up the colour of the finely divided foliage, thickly set with pretty red spikes of bloom. Although of free growth, it does not seem to have the encroaching habit of some of the New Zealand Burs, and it should on this account be more valued for the choicer parts of the rock-garden.

A. ovalifolia.—This has bright green foliage, and being of vigorous growth, will be found very useful for draping large stones in the rock-garden.

A. millefolia, A. myriophylla, and A. sanguisorbæ are also useful trailers. The flowers, bright green foliage, and long graceful stems entitle them to a place.

ACANTHOLIMON (*Prickly Thrift*).—Dwarf mountain plants, extending from the east of Greece to Thibet. The flowers resemble those of *Statice* and

The flowers are pink, on stems nearly 6 inches high. Asia Minor.

These, I believe, are all the species at present in cultivation. The following information has been gathered from the dried specimens at Kew :—

Acantholimon kotschyi is about 4 inches high, with distinctly broad leaves, being spiny and freely flowered, blossoms white.

A. armenum has pink blossoms on sprays nearly 6 inches high.

A. cephalotes has rosy pink flowers in globose heads, while the spiny leaves are less numerous in the rosettes than in most kinds. This comes from Kurdistan.

A. laxiflorum is the tallest species, growing about 9 inches high, the leaves long and narrow.

A. libanoticum is exceedingly woody and dense in growth. It is a Syrian species, with flowers of pink hue.

A. pinardi also has pink blossoms, the specimens varying in stature, possibly on account of age.

So far as could be determined by dried specimens, many of these not now in cultivation are very beautiful, and, from the general scarcity of good midsummer alpine plants in the rock-garden, would be greatly prized. E. J.

ACHILLEA (*Yarrow*).—Herbaceous and alpine plants numerous through N. Asia, S. Europe, and Asia Minor, varying in height from 2 inches to 4 feet; their flowers pale lemon, yellow, and white, rarely pink or rose. Many of the cultivated kinds are too rampant for grouping with alpine garden plants. The dwarfer kinds, on the other hand, come in for groups for the rock-garden or the margins of rock borders, and as edging plants, most of them growing freely and being easy of increase; some of the higher alpine kinds are not very enduring in our open winters, and often in our gardens get "staggy" after a few years' growth, requiring division and replanting.

Achillea Ageratifolia.—A silvery-leaved plant from the sub-alpine districts of Northern Greece, 4 to 7 inches high, with white flowers resembling Daisies; early in summer. The leaves are narrow, tongue-shaped, crimped, and covered with white down. This is a very neat and distinct plant, and easy of cultivation in light soil.

A. aurea (*Golden Yarrow*).—One of the showiest kinds, about 12 inches high; leaves finely cut, flowers bright yellow; freely on upright stalks. Caucasus.

A. Ægyptiaca (*Egyptian Yarrow*).—A silvery plant in all its parts, with finely cut leaves, and handsome heads of yellow flowers, with something of the grace of a fern in its leaves. A native of Egypt and the East, it is not hardy in all soils and positions, but it survives on well-drained sunny spots, flowers in summer, and is easily multiplied by division.

A. clavenae (*White Alpine Yarrow*).—A dwarf kind, covered with a short, silky down, which makes the plant almost of a silvery white; flowers in summer of a good white. It likes a light, free, loamy soil. Alps of Austria; increased by division of the roots, and also by seed.

A. Huteri (*Huter's Yarrow*), with bright green foliage, and pure white flowers. It likes a sunny part of the rock-garden, and grows well in common soil. Exempt from the struggle for life in the alpine turf, this, like so many spreading plants in our gardens, is best divided and replanted every second year.

A. Tomentosa (*Downy Yarrow*).—One of the tufted plants that help to form the carpets of silver, whereon large Violets and Gentians display their charms on the Alps, itself sending up flat corymbs of bright yellow flowers. On such ground it is dwarf, but in rich soil in gardens it is taller, 12 inches high. It is a good plant for the margins of mixed borders, and also for the rock-garden. European Alps, thriving in ordinary soil.

A. rupestris (*Rock Yarrow*).—A pretty and early-flowering kind from Calabria, thriving in poor soil and on warm banks; *A. nana*, *moschata*, and *umbellata*, a Greek plant, have like value to the above-named for the rock-garden.

ACIS.—A small genus of bulbous plants, natives of South Europe, of which few species are in cultivation.

Acis autumnalis (*Autumnal A.*).—A like slender-leaved little bulbous mountain plant, with stems 3 or 4 inches high, bearing flowers, resembling delicate pink snowdrops, drooping elegantly on short reddish footstalks, and blooming in autumn before the leaves appear. It is a true gem for the rock-garden, where

Acis Autumnalis.

it should be planted in a warm soil and sunny position, sheltered with a few stones, and on which it would look very well springing from a carpet of delicate, feeble-rooting *Sedum* or other dwarf plant. I have never seen it in nurseries except about Edinburgh. Where the soil is of a fine sandy nature, it will thrive as a border plant, but is as yet rare. Europe.

The other kinds are **Acis trichophylla, rosea,** and **hyemalis**, all of which will thrive where the soil is of a fine sandy nature, but are yet so rare as to be worthy of the best position and care. Mr Elwes doubts if any of these plants will thrive in the open air in England. Syn., *Leucojum*.

ADONIS (*Ox-Eye*). — Handsome plants of the Buttercup order; dwarf in stature, with finely divided leaves, and red, yellow, or straw-coloured flowers. There are about fifteen or sixteen species, most of which are annuals, and, with the exception of two or three fine kinds, they are not suitable for the rock-garden, but the kinds named are excellent for it.

Adonis vernalis (*Ox-Eye*).—A handsome alpine perennial, forming dense tufts, 8 inches to 15 inches high of finely divided leaves in whorls along the stems. It flowers in spring, when the tufts are covered with large, yellow, Anemone-like flowers, 3 inches in diameter, a single flower at the end of each stem.

Of **A. vernalis** there are several varieties, the chief being *A. v. sibirica*, which differs in having larger flowers. *A. apennina* is a later blooming form.

A. pyrenaica is a closely allied kind from the Eastern Pyrenees, with large yellow flowers like *A. vernalis*, but with broader petals, flowering in April and May. It may be grown in free, sandy moist loam, and not often disturbed, robbed, or shaded by coarser plants.

A. amurensis.—Like the *A. pyrenean* in habit, this flowers with the snowdrops, and is of easy culture—save that, until plentiful, it should be grown on the rock-garden, in moist, sandy loam, well drained.

ÆTHIONEMA (*Silvery Cress*).— Elegant greyish rock plants, found on the sunny mountains near the Mediterranean. The little plants grow freely in borders of well-drained sandy loam, but their home is the rock-garden. As the stems are prostrate, a good effect will come from planting them where the roots may descend into deep earth, and the shoots fall over the face of rocks at about the level of the eye. Easily raised from seed, and thrive in sandy loam. There are many species, but few are in gardens, owing to their inhabiting countries often under the rule of the Turk, and for that and other reasons not so easy to introduce as the plants of the Alps and Pyrenees. All the cultivated kinds are dwarf, and may be well grouped with rock plants on the warmer slopes of the rock-garden. Among the most charming of plants for gardens, let us hope the future will see many of the kinds introduced and grown. The following is an abstract of a paper on them in the *Garden*, by Mr W. B. Hemsley, of Kew.

The geographical range of the genus is from the Pyrenees to the Western Himalaya. There are, perhaps, half-a-dozen in Europe, including the beautiful *Æ. cepeæfolium*, better known as *Hutchinsia rotundifolia* and *cepeæfolia*. One only reaches India, where it is found at an elevation of from 12,000 to 16,000 feet, and the remainder are natives of the countries indicated above. Nearly all the species are natives of alpine regions, and grow naturally in stony or rocky places, and many of them are reported from chalky districts. The perennial species will, therefore, require to be kept tolerably dry at the root; a light soil in a well-drained border, or a place in the rock-garden, will best suit them. Old plants should be replaced by young ones as often as convenient. These may be raised from seed or cuttings, which is better done in a cool frame or pit. The annual species, excepting *Æ. Buxbaumii*, are not, so far as we know, in cultivation. In habit and foliage *Æthionemas*, especially the half shrubby species, have very much the aspect of some of the woody Candytufts, but the petals are all equal in size. The flower-spikes are usually very dense, and the seed-vessel relatively large, and very much crowded, so that in some species, as *Æ. Buxbaumii*, they bear some resemblance to the catkins of the common Hop. The flowers are usually some tint of red or lilac, or combination of the two. A few species have yellow flowers, and there are white-flowered varieties of several species. About fifty species are known, all natives of the mountains of Europe, Asia Minor, Syria, and Persia.

Æthionema cepeæfolium (*Iberidella; Hutchinsia rotundifolia*, Hort. Kew).—A densely-tufted, more or less glaucous-green, glabrous barb, with a long perennial tap root, that burrows deeply amongst stones. Stems, 3 to 6 inches long, ascending; leaves, mostly opposite, small fleshy, one-third to three-quarters of an inch long, those from the root broadly obovate or almost orbicular, quite entire, or obscurely toothed, those on the stem sessile, obtuse, or auricled at the base; flowers, half an inch in diameter, in cylindrical, crowded, erect racemes, pale lilac with a yellow eye; pedicels, horizontal. A native of the Alps of Europe, where it is widely dispersed, and abundant in many parts of Switzerland.

Æ. trinervium.—Leaves, hard, more or less distinctly three-nerved, oblong or narrowly lanceolate, the lower ones narrowed at the base, upper ones obtusely heart-shaped and stem-clasping. Flowers, rather large, white, seed-vessel oblong linear, rounded or truncate at the top, crowned with the equally long style.

Mountains of Persia. There is a variety of this species, called *ovalifolium*, with broader ovate-oblong leaves. It is a native of Armenia.

Æthionema sagittatum.—Leaves, rigid, many-nerved, oblong, or lanceolate, deeply hastate at the base, with acute lobes; flowers, rather large, white; seed-vessel, oblong, narrowed at the base. Persia.

Æ. tenus heterophyllum and *cæspitosum* are dwarf, densely-tufted alpine species, with small white or pink flowers. The only Indian species (*Æ. Andersoni*) also belongs to this group. It is a diminutive plant, with white or pink flowers.

Æ. rubescens.—Leaves, alternate, obovate; flowers, large, rose: seed-vessel, elliptical, tapering at both ends. A native of the alpine summits of the Sicilian Taurus, etc., at an elevation of 11,800 feet. This is a very showy species.

Æ. bourgæi.—Leaves, opposite, obovate; flowers, large, rose; seed-vessel, oblong-elliptical, rounded at both ends. Found in stony places in the alpine region of Mount Akdagh, Syria. Differs chiefly from the last in its opposite leaves.

Æ. chloræfolium (*Iberis* of Sibthorp and Smith).—Leaves slightly papillose and scabrid at the margin; flowers rather large; petals, obovate, rose, much longer than the calyx. A native of Asia Minor.

Æ. rotundifolium.—Very near *Æ. oppositifolium*, differing chiefly in the shape of the seed-vessel, and the panicle being free instead of adnate to the seed. A native of stony places in the Western Caucasus. This is quite different from *Iberidella rotundifolia*.

Æ. thesiifolium.—Stems, tall, slender, and twiggy; leaves, long, narrow, lanceolate, upper ones, acute; flowers in an elongating raceme pink. A native of stony places in the mountains of Cappadocia. It grows about 18 inches high, has long narrow leaves, and large fleshcoloured flowers, elegantly marked with purple.

Æ. grandiflorum.—Branches, long, slender, simple, about 1 foot high; leaves, oblong-linear, rather obtuse; flowers, purple, as large as those of *Arabis alpina*; petals, four times as long as the sepals.

A native of Mount Elbrus in North Persia; discovered by Hohenacker in 1843, and subsequently collected by Haussknecht, in Kurdistan, at an elevation of 4000 feet in 1857.

Æthionema pulchellum (*Æ. coridifolium* of Botanic Gardens, not of De Candolle).—Similar to the last, of which it was formerly considered a variety; but it is a more diffuse plant, having smaller flowers, the petals being about two and a half times as long as the sepals. Armenia, Persia, and Kurdistan.

Æ. membranaceum.—Stems, erect, simple, about 6 inches high; leaves, oblong-linear, smaller than those of the two preceding. The seed-vessel of these three species is very broadly winged, and the wings are entire, or very slightly toothed, at the margin. A native of Persia; formerly figured in Sweet's "Flower Garden."

Æ. diastrophis (*Diastrophis cristata*).—In habit, foliage, and flower, this comes very near to *Æ. pulchellum*, but it differs from that and others of this sub-section in its very long fruiting racemes and small seed-vessel, with elegantly toothed wings. It is a native of Russian Armenia, and was in cultivation at Dorpat in 1841, and is now in cultivation at Exeter, Mr Veitch's.

Æ. armenum.—This, judging from dried specimens, although smaller-flowered than its immediate allies, must be a very pretty species when growing. It is of dwarfer (3 or 4 inches high), more diffuse habit, with more leafy stems and dense spikes of small purplish rose-flowers; seed-vessel, crenate. It inhabits the mountains of Armenia, and Cappadocia, growing in stony places.

Æ. coridifolium.—Stems, numerous, thick, only a few inches high; leaves, crowded, short, linear-oblong, or linearobtuse, or somewhat acute; flowers, large, but not equalling those of *Æ. grandiflorum*; seed-vessel, boat-shaped. This handsome species is a native of the chalky summits of the Lebanon and Taurus.

Æ. capitatum.—This species, of about the same stature as the last, but with longer stems and more scattered leaves, is remarkable for its short dense fruiting heads of boat-shaped seed-vessels with entire wings; the flowers are small and inconspicuous. Alpine region of Cappadocia.

Æthionema speciosum. — A densely-tufted species with ovate-oblong leaves, and rather large rose-pink flowers; seed-vessel elegantly toothed, and tinged with purple. It is described as one of the prettiest of the genus, growing in dense tufts 3 to 4 inches high, and producing a profusion of large flowers. *Æ. lignosum, sublulatum, stylosum, lacerum,* and *fimbriatum* belong to the same group. They have rather small flowers, but in all of them the seed-vessel is very elegant. Armenia.

Æ. cordiophyllum. — Stems, few, rigid, densely leafy; leaves, rigid, quite sessile, deltoid-cordate, the lobes embracing the stem, the lower ones opposite; flowers, rose-pink, of medium size; boat-shaped seed-vessel, toothed. This plant grows from 6 to 12 inches high. Armenia.

Æ. cordatum.—Stems, few, rigid, densely-leafy; leaves, sessile, deltoid-cordate, acute; flowers, rather large, sulphur-yellow. A native of dry, rocky places in the alpine region of Armenia and Syria. It is similar to the last, but differs in its larger yellow flowers, and less distinctly toothed seed-vessel.

Æ. moricandianum.—Stems, few, short, and leafy; leaves all opposite, nearly sessile, ovate, obtuse, the upper ones sometimes cordate at the base; flowers, large, yellow. A native of Mount Caira, where it was discovered by Cinard in 1843. This species comes very near to *Æ. cordatum,* differing in its obtuse leaves, which are all opposite and scarcely cordate, and in its flowers, which are twice as large.

Æ. græcum.—Stems, numerous, short; leaves, crowded, very small, ovate-oblong; flowers, rather large, similar to those of the European *Æ. saxatile,* but twice as large. A native of the chalky mountains of Greece.

AJUGA (*Bugle*).—Dwarf sage-like perennials of easy culture and increase; and though not of first value among rock-plants, useful, from their freedom and good colour.

Ajuga genevensis (*Geneva Bugle*).— This has violet-blue flowers, the stem being a cone of flowers for a length of 4 or 5 inches or more. Suitable for rock-garden, it will hardly be well to give it a place there, except by the margins of walks. The true plant, widely distributed on the continent, is not found in Britain, but the variety with the floral leaves large and longer than the flowers, and having a dense leafy spike (*A. pyramidalis*), is found in Scotland, and is sometimes grown in gardens.

The British Creeping Bugle (*A. reptans*) is grown in gardens under various names, for the sake of its dark browny-purple leaves, and a variegated variety of it is sometimes grown.

ALLIUM.—These plants are often given in large numbers in Dutch and other lists, and with slight reason, as their beauty is little, from the garden point of view. Beyond a few, they are hardly worth cultivation, and though some kinds are often seen among rock-plants, they are out of place with them, and the kinds worth growing are easily grown without the aid of the rock-garden.

ALLOSORUS CRISPUS (*Parsley Fern*).—A beautiful Fern, found in some mountainous districts, where it grows out of the crevices of the rocks; the fronds grow in dense tufts. It requires light, and should only be shaded from the hot sun. On the rock-garden it thrives, planted between stones, with broken stones about its roots, and just its fronds peeping out of the crevices. Growing in this way, it seems to be quite at home. It is well suited for planting in chinks on the rock-garden, and associates well with alpine plants. Careful division.

ALSINE (*see* **ARENARIA**).

ALYSSUM (*Madwort*).—Rock and alpine plants, numerous in alpine countries, the species much resembling each other, so that only few of the best are worthy of culture for the rock-garden, and these are of the easiest culture in almost any soil, and of rapid increase by cuttings, seed, and some by division. They are usually more fitted for borders and banks than for the select alpine garden.

Alyssum Alpestre (*Alpine A.*).— A pretty species, partaking of the brilliant colour and free-flowering properties of the well-known Rock Alyssum, and the neatness of habit and dwarfness of the Spiny or the Mountain, forming neat tufts of hoary leaves, the whole plant being covered with minute, shining, star-like hairs, and, not growing more than 3 inches high. A native of the Pyrenees and Alps, its home with us is in sunny spots on the rock-garden; the soil to be of poor, rather than of a rich, nature. Flowers in early summer, and is readily increased by seed or from cuttings.

The silvery A. (*A. argenteum*), a native of Corsica, is closely related to this species, but is taller and more robust, has small flowers, and is not so well worthy of culture.

A. montanum (*Mountain A.*).—A distinct species, spreading into compact tufts of glaucous green, 3 inches high, the plants studded with yellow, alpine wall-flower-like blooms, fragrant, flowering in early summer. The beautiful stellate hairs are large enough on this kind to be seen by the naked eye. It is a native of many parts of Europe, on hills and low mountain ranges, chiefly on calcareous soils, and to succeed, it is best to place it on the rock-garden in sandy soil, and, well grown, it will prove a beautiful ornament, especially when it grows into large cushions, on one side perhaps falling over a stone. Readily increased by division, cuttings, or seeds, though it does not often seed freely with us.

A. saxatile (*Rock A.*).—A popular plant, and one of the best of the yellow flowers of spring. Hardy in all parts of these islands, the profusion of its masses of showy yellow bloom, with its freedom of growth in any soil, have made it one of the most grown of rock plants. It is best for borders and walls or banks, and also for association with the evergreen Candytufts, and *Aubrietia*, and on wet ground it is better to plant in raised beds and in poor soil: it perishes in winter in some heavy rich clays. Very easily raised from seed, or by cuttings. Comes from Podolia in Southern Russia, and flowers with us in April or May.

There is a somewhat dwarfer variety, distinguished by the name of *A. saxatile compactum*, but it differs very little from the old plant and forms, differing slightly in colour (*citrinum*), but these are not so effective as the old plant.

Alyssum spinosum (*Spiny A.*).—The flowers of this are small and not pretty, but the plant forms a silvery and pretty little

Alyssum montanum.

bush on any kind of soil, that I think it has quite as good a right to be named here as many others valued for their flowers alone. Small plants quickly become Liliputian silvery bushes, 3 to 6 inches high; when fully exposed, almost as compact as moss. The leaves are covered with small stellate hairs, and form interesting objects under the microscope. On established plants the old branches become transformed into

ALPINE FLOWERS FOR GARDENS

spines: hence its name. It is distinct in appearance from anything else in cultivation, and merits a place on some not overvalued spot on the rock-garden. It is readily increased from cuttings. S. Europe.

Alyssum pyrenaicum is a neat rock-plant, with white fragrant flowers; a good rock-garden plant.

A. serpyllifolium is a grey-green leaved kind with yellow flowers. Small plants quickly become Liliputian bushes, 3 inches to 6 inches high; and, fully exposed, are almost as compact as moss.

A. maritimum (*Sweet Alyssum*), is a small annual with white flowers, growing on the tops of walls in the west country, and in sandy places. In these situations it is perennial, but in gardens is grown as an annual, sowing itself freely, and is for covering bare spaces as well worth a place as any.

The taller kinds of *Alyssum* are not well suited for the rock-garden.

A. podolicum is a small alpine plant from S. Russia. It has in early summer many small white blossoms, and is suited for the rock-garden or walls.

ANAGALIS (Pimpernel).—Pretty dwarf plants, chiefly half-hardy annuals, the best known of which is the Italian Pimpernel (*A. Monelli*), with large blossoms of a deep blue, shaded with rose. There being several varieties of this, they are among the annual flowers I should recommend where bare spaces occur in rock-garden, pending the coming of good perennial rock-plants.

Anagalis tenella is a native plant found in bogs, bearing slender stems with small round leaves, among which are tiny pink flowers. It may be grown easily in the bog- or rock-garden, or anywhere where the soil is moist and spongy, and the vegetation dwarf and fragile like itself.

ANDROMEDA.—Various bushy plants usually called *Andromeda* in gardens, belong strictly to several other genera. There is only one true species of *Andromeda* known—

A. polifolia. It is a pretty little grey bush, grouped in peat beds or in the bog garden.

For allied plants usually known as *Andromeda*, see *Cassandra, Cassiope, Leucothoe, Lyonia, Pieris,* and *Zenobia.*

ANDROSACE.—Tiny plants of the higher alps, often growing at elevations where the snow falls early in autumn, they flower as soon as it melts, growing on cliffs with a vertical face, or with portions of the face receding here and there into shallow recesses. Here they endure intense cold—cold which would destroy all shrub or tree life exposed to it. They are almost sure to perish in a smoky atmosphere; their small evergreen leaves, often downy, retain more dust and soot than larger-leaved evergreen alpine plants do. The Androsaces enjoy in cultivation small fissures between rocks or stones, firmly packed with pure sandy or gritty loam, not less than 15 inches deep. They should be so placed that no wet can gather or lie about them, and they should be so planted in between stones that, once well rooted into the deep earth —all the better if mingled with pieces of broken sandstone—they could never suffer from drought. A few kinds will do on level borders, such as *A. sarmentosa,* they are usually the jewels for the most carefully made and tended rock-gardens.

Androsace alpina.—This is a lovely little plant, but difficult to grow. It likes rather a moist place, and shaded from the hot sun, although it loves moisture at its roots. The plant must not be kept damp overhead; all the moisture must be directed to the roots, and so arranged as to allow of its free escape again. Syn., *ciliata.*

A. brigantica.—A handsome little kind, with pure white flowers, the foliage deep green. It loves to grow on sandy slopes, shaded from the melting sun.

Androsace carnea (*Rosy A.*).—One of the prettiest and most distinct coming from the summits of the Alps and Pyrenees, where it flowers in summer, when the snow melts. It is known from any of the other cultivated kinds by its small pointed leaves, not, as in them, gathered in tiny rosettes, but more regularly clothing a somewhat elongated stem, so as to remind one distantly of a small twig of Juniper, or of the Juniper Saxifrage. The flowers are pink or rose, with a yellow eye. It is not difficult to cultivate in a mixture of sandy loam and peat—the spot to be exposed, and the soil at least a foot deep, so that its roots may descend, and be less liable to suffer from vicissitudes. Thorough watering should be given during the dry season, particularly when the plant is young, and before it has taken deep root. Treated thus, it will form healthy tufts, and prove one of the most beautiful plants in the rock-garden in spring. Like most of the kinds, it may be raised from seed, sown in pans of sandy peat as soon as gathered. *A. Eximia* is a large form.

A. chamæjasme (*Rock Jasmine*).—This does not nestle into close moss-like cushions, like the Helvetian and other Androsaces, the foliage forming large rosettes of fringed leaves, the blooms borne on stout little stems, from 1 to 5 inches high. They are white at first, with a yellow eye, changing to crimson, the outer part becoming a delicate rose. It is one of the prettiest alpine plants, and one of the easiest to grow on an open spot on the rock-garden in well-drained light loam, the surface nearly covered with pieces of broken stone, with abundance of water in summer, exposed to the full sun, and not overrun by weeds or grazed down by slugs. A native of the Tyrolese and Swiss Alps, where it flowers later than in our gardens. In Britain it blooms in April, May, and June, earlier or later according to the season, is propagated by division, and may be grown very well in pots along with the rarer Rockfoils, plunged in sand or coal-ashes.

A. helvetica (*Swiss A.*).—This forms dense cushions, about half an inch high, of diminutive ciliated leaves, in little rosettes, each resting on the summit of a little column of old and dead, but hidden half-dried leaves. A white flower, with a yellowish eye, rises from every tiny rosette, each flower being almost twice as large as the rosette of leaves from which it has arisen, and resting on the little mass of glaucous green. Looked at from the height of a man, the leaves are not distinctly seen, the flowers quite so; and thus the effect is somewhat as if one were looking from a height down on some grey bush, with very large flowers and diminutive foliage. Requires some care in cultivation, full exposure to sun, and a well-drained spot, placed between and tightly pressed by stones about the size of the fist, which will guard it against danger from excessive moisture, and at the same time permit of the roots passing into the good soil in the crevices.

Androsace imbricata (*Silvery A.*).—This differs from the Pyrenean and Swiss Androsaces in having the rosettes of a beautiful silvery white colour. The pretty white flowers are without stalks, and rest so thickly on the rosettes as often to overlap each other. It will grow freely in loamy soil in free well-drained spots. Pyrenees, Alps, and is propagated by seeds and division. Syn., *A. argentea.*

A. lanuginosa. (*Himalayan A.*).—The European species of this diminutive family usually have their leaves in tufts as compact as the very Mosses and Lichens. This kind has spreading and, sometimes, long stems, branched, and bearing umbels of flowers of a delicate rose, with a small yellow eye; the leaves nearly an inch long, and covered with silky hairs.

A. l. leichtlini is a variety; flowers being larger and the colour deeper. It was grown for many years at the York Nurseries, under the name of *A. l. oculata*, which is the best name for it. Add a little limestone to the rock-garden light loam, and place the plant so that its shoots may fall over the edge of a low rock. Where the soil is free, and not wet in winter, it may be tried as a border plant. It is best propagated by cuttings, and flowers in summer and early autumn. Himalaya. In a district

ALPINE FLOWERS FOR GARDENS

where, from too heavy soil or other reasons, it does not thrive on the level

Androsace lanuginosa in the Rock-Garden at The Friars, Henley-on-Thames.

ground, I find it grows between the stones in a "dry" well.

Androsace obtusifolia (*Blunt-leaved A.*).—This has rather large rosettes of leaves, somewhat spoon-shaped, with stems clothed with short down, from 1 to 4 inches high, bearing sometimes one, but generally from two to five white or rose-coloured flowers, with yellow eyes. It seems to grow taller and more vigorously than *A. Chamæjasme*, and in a native state is often gathered by handfuls, and placed in vases, with Gentians and other alpine flowers. Widely distributed over the European Alps, and usually flowering in midsummer; but in this country opening in spring. The culture for *A. Chamæjasme* will suit this plant.

A. pubescens (*Downy A.*).—Allied to the Swiss and Pyrenean Androsaces in its rather large solitary white flowers, with pale yellow eyes, just rising above the densely packed, slightly hoary leaves, the surface of which is covered with stalked and star-like hairs. The unopened blooms look like small pearls set firmly in a tiny five-cleft cup, and are held on stems barely rising above the dwarf cushion formed by the plant. It may be distinguished from its fellows by a small swelling on the flower-stem close to the flower, and is an exquisite little plant, widely distributed over the Pyrenees, Alps, and other European ranges, flowering in July and August in its native state, and in our gardens in spring or early summer. It grows without difficulty on sunny fissures in deep sandy and gritty peat.

Androsace ciliata (*Fringed A.*).—Is by some considered a variety of the preceding, with the flower-stems twice as long as the leaves, which are glabrous on the surface and ciliated at the margin, the old leaves not forming a column beneath each rosette. It is, however, distinct. *A. cylindrica* is a variety with the stems rising to half an inch high, with persistent leaves, which form columns on the stems. It is by some considered a species, bears pure white flowers in spring, and should be treated like *A. pubescens*.

A. pyrenaica (*Pyrenean A.*).—This forms a dwarf, compact, and cushioned mass of tiny grey rosettes, something like the Swiss Androsace, but the paper-white flowers with yellowish eyes are not quite so well formed, and the flower, instead of being seated in the rosettes of leaves, rises on a stem from a quarter to half an inch high. The leaves are downy, and have a keel at the back, and, like those of *A. helvetica*, the old leaves are persistent, and remain in little columns below the living rosette. This plant was grown to great perfection by the late Mr James Backhouse, of York, in fissures between large rocks, with deep rifts of sandy peat and loam in them. It will also grow on a level exposed spot, but in that should be surrounded by half-buried stones.

A. villosa (*Shaggy A.*).—A very dwarf species, found on many parts of the Alps, with leaves, and thickly covered with soft white hair or down. The leaves are mostly covered with the silky hairs on the under side, united in a sub-globular rosette, and bear in umbels white or pale rosy flowers, with purplish or yellowish eyes, on stems from 2 to 4 inches high. It is more inclined to spread than any of the nearly allied sorts, as it throws out runners. It should be planted in fine sandy loam; it may be grown on level spots on borders.

Androsace villosa himalayica is a form of *villosa*, but much more vigorous, and flowers later in the spring. Pure white, with a very distinct red eye. In the early part of the season the foliage is not covered with the white silky hairs, but the foliage becomes pure white later in the season. It is also grown under the name of *A. Arachnoides*.

flower and foliage can be obtained. It also helps to keep the plant dry in winter.

Androsace sarmentosa Chumleyi differs in the stalks being shorter and stronger, and the flower much deeper in colour. It is a better plant, and is a gem for the rock-garden. This likes a sprinkling of limestone on the soil. If kept well top-dressed, it will send out young runners like a Strawberry plant, and root very freely from the same.

Androsace villosa.

A. sarmentosa.—This is a Himalayan species, growing at an elevation of over 11,000 feet. The flowers, borne in trusses of ten to twenty, at first sight resemble those of a rosy white-eyed Verbena. Like many other woolly-leaved alpines, this is difficult to keep alive through our damp winters. A piece of glass in a slanting position about 6 inches above the plant preserves it. Care should also be taken to put sandstone, broken fine, immediately under the rosettes of leaves and over the surface of the soil, to keep every part of the plant, except the roots, from contact with the soil. A dry calcareous loam is best. Where limestone can be had to mix with the soil, a much better display of

A. vitaliana (*Yellow A.*).—Rarely grows above an inch high, and produces, scarcely above the leaves, flowers large for so small a plant, and of a good yellow. On the Alps it reminded me of a Liliputian furze-bush, looked at through the wrong end of a telescope. It is lovely for association with the freer-growing Androsaces and dwarf Gentians, and it may even be grown on a border in a not too dry district where the soil is open and sandy. A dry soil or a heavy one it does not like, and when in suitable districts it is tried as a border plant on the level ground, it should be surrounded by stones, half plunged in the ground, to prevent evaporation, as well as to protect it. It is abundant on the Alps in various parts of Europe, and is increased by careful division or by seeds. Syn., *Aretia Vitaliana*.

Androsace laggeri.—This is one of the most distinct of the family, and is easily recognised by its tiny rosettes of sharp-pointed leaves. The flowers are of a bright pink, with a lighter centre.

A. foliosa is the handsomest species, the flowers borne in large bunches, rosy-red, and larger than in the others. This plant revels in good deep limestone soil. The stone should be broken into pieces about the size of a walnut, and add good heavy loam in full sun. Thus the plant will form bushes one foot across in one season.

A. wulfeniana.—This is a very distinct plant flowering later than *A. ciliata*, with much deeper blood-rose flowers, borne close to the foliage, the whole plant being very compact, and forms quite a cushion. It does much better when planted on the level, and makes a good companion for such as *A. carnea*, *A. C. eximia*, *A. ciliata*, *A. vitaliana*, *A. laggeri*, *A. chamæjasme*. The above all love the sandstone, and should be well looked to in the autumn and spring, and be well top-dressed with sand and leaf-mould.

ANEMONE (*Windflower*).—Beautiful alpine and meadow plants, to which is due much of the flower beauty of spring and early summer in northern and temperate countries. In early spring, or what is winter to us in Northern Europe, when the valleys of Southern Europe and all round the basin of the Mediterranean are beginning to glow with colour, we see the earliest Windflowers in all their loveliness. Those arid mountains that in the distance often look so barren, have on their sunny sides carpets of Windflowers in countless variety, often belonging to the old favourite in our gardens—the Poppy Anemone. Later on the Star Anemone troops in thousands over the terraces, meadows, and fields of the same regions. Climbing the mountains in April, one finds *A. Hepatica* nestling in nooks all over the bushy parts of the hills. Farther east, while the common Anemones are aflame along the Riviera valleys and terraces, the blue Greek Anemone is open on the hills of Greece; a little later the blue Apennine Anemone blossoms. Meanwhile our Wood Anemone adorns the woods throughout the northern world, and here and there through the brown Grass on the chalk hills comes the purple of the Pasque-flower. The grass has grown tall before the graceful Alpine Windflower blooms in all the natural meadows of the Alps; while later on bloom the high Alpine Windflowers, which are soon ready to sleep again for months in the snow. These are but a few examples of what is done for our northern world by these Windflowers, so precious for our gardens also.

With many handsome kinds, everyone is not worth growing, and so we make a choice of the best for the rock-garden. Whatever the difficulties in the growth of other alpine flowers, there are none with the Windflowers; free in most soils, and hardy. There are few groups of plants so precious for the garden, whether we look at the more strictly alpine kinds, the free-growing "florists'" kinds, such as the Poppy Anemone, or the autumn-blooming Japanese Anemones.

In the rock-garden alpine kinds are essential, and, although some are slow, they are not difficult to grow. As in the case of so many mountain plants which grow in soil composed of decayed rock, open or warm soils are usually best for the alpine kinds in our country. The Poppy Anemone is so free in such soils that many people raise it as an annual, and flower it within the year

It is somewhat too vigorous for the rock-garden, as are all the forms of the Japan Windflower.

Anemone alpina (*Alpine Windflower*). On nearly every great mountain range in northern and temperate climes, this is one of the most frequent plants. It may be seen in various stages on the same day, and on the lower terraces of the great mountains and on the green slopes of the valleys, it grows as tall as in our gardens. The interior of the flower is white, the outside tinted with pale purplish-blue. It flowers in its native country as the snow disappears, and in our gardens at the end of April or in the beginning of May. When plants are well established in good soil, they may be taken up and divided; it may be raised from seed. Sometimes the flowers are yellow, in which state the plant is known as *A. sulphurea*.

A. angulosa (*Great Hepatica*).—This is larger than the common *Hepatica*, with flowers of a fine sky-blue, as large as a crown piece, and with five-lobed leaves. It thrives in spaces between American plants and choice dwarf shrubs, as well as on the rock-garden. Where plentiful, it may be used as an edging to beds of spring-flowering shrubs, and for planting in open, rather bare, and unmown spots along the margins of wood walks, being more free in growth than the common *Hepatica*.

Anemone apennina (*Apennine Windflower*).—This has erect flowers of a fine blue, starlike, larger in size than a half-crown piece, paler on the outside than within, and thickly scattered over a low cushion of soft green leaves. Although figured in most of our works on British plants, and naturalised in various places, it is not a true native; but the hardiest of our native plants take not more kindly to our clime. It is one of the hardy

Alpine Windflower.

spring flowers, and, among the best plants that gem the Apennine hills, there is not one more worthy of being naturalised. It flowers in March and April, is readily increased by division, and grows from 6 inches to 10 inches high.

A. blanda (*Greek Windflower*).—A very lovely, dwarf, hardy plant, with flowers of a deep sapphire blue, opening in the dawn of spring, during mild open winters, and in warm districts showing as early as Christmas, flowering continuously too. From the harder and smoother texture of the leaves, it can stand exposure to cutting winds even better than the Apennine Anemone. It has every good quality of a hardy alpine plant; should be grown in every rock-garden, planted on bare banks that catch the early sun; when plentiful, may be naturalised on dry and bare banks. Increased by division and by seeds. Frequent on the hills of Greece.

Anemone coronaria (*Poppy Anemone*).—A showy handsome plant, grown in our gardens from the very earliest times, and of which there are a great number of varieties, both single and double. The single sorts may be readily grown from seed. These double varieties may be planted in autumn or in spring, or at intervals all through the year, to secure a succession of flowers; but the best bloom is secured by September or October planting, where the winters are not severe. The Poppy Anemone does best in a rich deep loam, but is not very fastidious. It flowers in April and May, and often through the winter, but though vigorous on many soils, is not quite hardy on heavy soils in cool districts. For the rock-garden choose the best single uni-coloured forms. The ordinary mixed kinds are for borders. Seed or division.

The Greek Anemone (*A. blanda*).

A. fulgens (*Scarlet Windflower*).—A brilliant, hardy, vigorous kind, the large scarlet flowers on stems about a foot high, springing from a dwarf mass of hard, deeply-lobed leaves. It does well as a border plant, thrives in the rock-garden, and I find it grows readily in Grass. The flowers, borne in April and May, are vivid scarlet. There are various forms of this. Division or by seeds.

A. halleri (*Haller's Windflower*).—This is one of the finest, as well as perhaps the rarest, of the alpine Pasque-flowers.

The deep lilac flowers grow singly on longish slender stems, and are larger than those of any of the same group. It does best in well-drained soil, rich, and not too heavy. It was first found by the gentleman whose name it bears, in the Valley of St Nicholas, in the Upper Valais, and since then, though sparingly, in the Eastern Pyrenees.

Anemone hepatica (*Hepatica*).—A beautiful mountain plant, long known in our gardens. It is hardy everywhere, is not fastidious as to soil, though it loves a warm loam, and presents a diversity of colour—single blue, double blue, single white, single red, double red, single pink (*Carnea*), single mauve-purple (*Barlowi*), crimson (*splendens*), and lilacina. Every variety of the *Hepatica* is worthy of care and culture, but I think the best of all is the wild plant with its lavender-blue flowers so free and so pretty, early in the year. The plant, a native of many hilly parts of Europe, is usually found in half shady positions, which will be found to suit it best in a cultivated state also. It is readily increased by division or by seed, the double kinds by division only.

A. nemorosa (*Wood Anemone*).—In spring this native plant adorns our woods, and also those of nearly all Europe and Asia. In heavy soils in the open fields it does not vary, but in woods, where the soil is gritty and free, it often varies much; so that we may now and then gather several varieties from the same place, and so large forms worthy of culture have been obtained. There is a large white form in cultivation, as well as the blue and purplish ones.

A. palmata.—Distinct, with leathery leaves and large handsome flowers in May and June, glossy, yellow, only opening to the sun. A native of N. Africa and other places on the shores of the Mediterranean, this fine plant should be grown in deep turfy peat, or light loam with leaf-mould, placed on level spots, where it can root deeply and grow into strong tufts. There is a double variety. Increased by division or seed.

A. patens (*Woolly Pasque-flower*).—This blooms early in March in England, and on this account it is worth growing. It somewhat resembles *A. pulsatilla*, but has larger flowers and leaves. Germany.

Anemone pulsatilla (*Pasque-flower*).— This fine plant is a true native one, and when it occurs on a bleak chalk down it is freely dotted over the turf. In the garden it forms handsome tufts, and flowers abundantly as a border or rock plant; it should be planted in various aspects to secure a longer season of bloom. There are several varieties, including red, lilac, and white kinds, but these are rare. It prefers well-drained and light but deep soil. Flowers in spring, purplish, on stems 5 inches to 12 inches high. Division or by seeds.

A. pratensis (*Meadow Pasque-flower*) is a native of most of the northern parts of Europe, and in some places grows abundantly in dry meadows, bearing small, drooping flowers of a deep purple colour, the leaves finely cut. Central Germany.

A. ranunculoides (*Yellow Windflower*). —Not unlike the common Wood Anemone in habit, this is distinct in its yellow flowers coming in March and April. It is S. European, and though usually less free on common soils than the Apennine Anemone, it is happy on light, open soil. On limestone soils it is best. It is charming for association with tufts of the Apennine or the Greek Windflower.

A. Robinsoniana (*Azure Windflower*). —A lovely plant; a large form of the Wood Anemone, or thought to be so. Whatever its origin, it is the most precious of all for its colour, hardiness, and use in all sorts of places. It is a vigorous plant, 6 inches to 10 inches high, with firm leaves, the flowers large and of a lilac-blue colour. The flower-bud is well formed and drooping, the flowers well opened out, always erect, and bearing in the centre a sheaf of yellow stamens. Nothing is more lovely than a patch of this in full bloom on a bright spring day, and it should form carpets on every rock-garden, on the sunny slopes, and also on the northern ones to prolong the bloom.

A. stellata (*Star Windflower*).—With star-like flowers, ruby, rosy purple, rosy, or whitish, usually having a large white eye at the base, contrasting with the delicate colouring of the rest of the petals, and the brown-violet of the stamens and styles of the flower. It is not so vigorous as the Poppy Anemone, and in Britain requires a warm position and a light, sandy, well-drained soil. In the rock-garden, where we may give this a raised and warm place, we may succeed with it, but generally it is not a hardy plant in Britain. Division and seeds. Syn., *A. hortensis*.

Anemone sylvestris (*Snowdrop Windflower*).—Distinct, with white flowers in spring as large as a crown piece, and beautiful buds,form a vigorous tufted plant, 12 to 15 inches high. A native of Central Europe, it is at home in Britain, but in some soils fails to flower. It is best in the lower part of the rock-garden or among the shrubs near it. Growing almost anywhere freely, it should not have the choicer places needed for the rarer alpine kinds. Division.

A. vernalis (*Shaggy Pasque-flower*).— One of the Pasque-flower division of the Anemones, but very dwarf, the flowers large and shaggy, and covered with brownish silky hairs. A native of Norway, and extreme northern countries, also of very elevated positions on the Alps and Pyrenees, and rarely seen in good condition in our gardens. It should be grown in some select spot on the rock-garden in well-drained and deep soil. The flowers, borne early in spring, are whitish inside.

The above-named Windflowers are the most beautiful. Some kinds are omitted which, if distinct as species, are too vigorous for our purpose, such as *A. rivularis*, and *A. narcissiflora*, and for the rock-gardener the best way is to make good use of the proved kinds. It is only where the aim is a botanical collection that every kind that comes will be sought.

ANTENNARIA (*Cat's-Ear*).—Small moor or mountain plants, the cultivated kinds of which are all perennial. They are of quite secondary use in the rock-garden. The Mountain Cat's-ears, *A. dioica* and *A. alpina*, and varieties *minima* and *tomentosa*, are neat-growing dwarf plants, with white downy

ALPINE FLOWERS FOR GARDENS

foliage, hence useful as carpeting plants. All are of the simplest culture in any ordinary soil. These are good rock-garden plants, and the pretty little rosy heads of one form of the Mountain Everlasting may be seen in the cottage gardens of Warwickshire. These last kinds only grow a few inches high, and are very easily increased by division.

A. tomentosa (Hort.) is a plant of a similar character that has been much used as a dwarf silvery plant in the flower garden. It is hardy, and of easy increase and culture in bare spots.

Anemone vernalis. (Engraved from a drawing by H. G. Moon.)

ANTHEMIS (*Camomile*).—Of the kinds of these in cultivation there are few worth a place on the rock-garden. *A. Aizoon* is a dwarf silvery rock-plant

—from 2 inches to 4 inches high, having small white Daisy-like flowers. Its chief beauty is in the leaves, which are covered with a white downy substance. It should be grown in the rock-garden in exposed places. Some handsome kinds are too vigorous for the rock-garden.

Anthemis Macedonica.

ANTHERICUM (*St Bruno's Lily*).—Graceful Lily-like alpine pasture-plants, among the most beautiful of hardy flowers. Though rather taller than most rock-plants, their alpine associations as well as their beauty should give them a place among the more vigorous plants or among the rock-garden shrubs.

Anthericum hookeri.—A showy plant, 1 foot to 20 inches high, flowering in early summer, bright yellow, nearly half an inch across, freely in racemes, 3 inches to 5 inches long. The leaves form dense tufts in ordinary soil, but the plant grows best in one that is moist and deep, or in peaty bog. New Zealand.

A. liliago (*the St Bernard's Lily*).—From 1 foot to 2 feet high, with flower-spikes that bear numerous pure white flowers in early summer. An easily grown plant, not so pretty as the St Bruno Lily.

A. ramosum has the flower-stems about 2 feet high, much branched, and bearing small white flowers; it has narrow Grass-like leaves, and the plant soon grows into large tufts.

Anthericum liliastrum (*St Bruno s Lily*).—A most graceful alpine meadow plant, in early summer throwing up spikes of white, Lily-like blossoms. The plants must be protected from slugs and caterpillars, from attacks of which they are liable to suffer. It thrives as a good colony or group in an open space between dwarf shrubs. Where plentiful, it would be an interesting subject to naturalise in a grassy place in cool soil. Syns., *Paradisea* and *Czackia*.

The major variety of the St Bruno's Lily has much larger flowers (2 inches across) than the wild plant, and has the peculiarity of sending up large single flowers from the root. These open before the flowers on the spike, and are larger, resembling the white blooms of a *Pancratium*. This habit of the plant points to it as distinct from the ordinary type of St Bruno's Lily. It grows 3 feet high in good soil, and is a fine plant, but though many think highly of it, the species is more elegant in form.

ANTHYLLIS (*Kidney Vetch*).—Dwarf mountain plants of the Pea family, of which there are some half a dozen species in cultivation. As far as now known, few are worth growing on the rock-garden.

Anthyllis montanus, the Mountain Kidney Vetch, is a very hardy rock-plant; dwarf, about 6 inches high, the leaves pinnate, and nearly white with down, the pinkish flowers in dense heads, rising little above the foliage, and forming with the hoary leaves pretty little tufts. I have never seen any alpine plant thrive better on the stiff clay of North London. Resisting any cold or moisture, it is among dwarf plants of the first order of merit as a rock-plant. Alps of Europe; division and seeds.

A. erinacea is a singular-looking, much-branched, tufted, spiny, almost leafless shrub, about 1 foot high, with purplish flowers.

A. Vulneraria (*Woundwort*). — A native plant, is pretty, and well worth growing on dry banks. There are varieties, white and red.

ANTIRRHINUM (*Snapdragon*).—Rock-plants and perennial herbs, mostly hardy and many of them from mountainous regions, but none so popular in gardens as the handsome Snapdragon (*A. majus*) which, like the Wallflowers, often grows on walls and stony places. Among the many species, some few are seen in cultivation from time to time, but they do not take a large place in gardens, among the best being *A. Asarina*, *A. rupestre*, *glutinosum*, and *sempervirens*, throwing in poor soil and dry spots. It is probable there are not a few of these plants of much beauty not yet in cultivation.

AQUILEGIA (*Columbine*).—Alpine or mountain copse perennials, often beautiful in habit, colour, and in form of flower, widely distributed over the northern and mountain regions of Europe, Asia, and America. Among them may be found great variety in colour—white, rose, buff, blue, and purple, and intermediate shades even in the same flower, the American kinds having yellow, scarlet, and delicate shades of blue. Though often taller than most of the plants strictly termed alpine, they are true children of the hills, and the alpine kinds, living in the high bushy places in the Alps and Pyrenees, and North Asian mountain chains, are among the fairest of all flowers. Climbing the sunny hills of the sierras in California, we meet with a large scarlet Columbine, that has almost the vigour of a Lily, and in the mountains of Utah, and on many others in the Rocky Mountain region, there is the blue Columbine (*A. cærulea*), with its long and slender spurs and lovely cool tints. Although many cottage gardens are alive with Columbines in early summer, there is some difficulty in cultivating the rarer alpine kinds. They require to be carefully planted in sandy or gritty though moist ground, and in well-drained ledges in the rock-garden, in half-shady positions or northern exposures. Most wild Columbines, however, fail to form enduring tufts in our gardens, and they must be raised from seed as frequently as good seed can be got. It is the alpine character of the home of many of the Columbines which makes the culture of some of the lovely kinds so difficult, and which causes them to thrive so well in the north of Scotland, while they fail in our ordinary dry garden borders. No plants are more capricious; take, for instance, the charming *A. glandulosa*, grown like a weed at Forres, in Scotland, and so short-lived in most gardens. Nor is this an exception; it is characteristic of other alpine kinds. The best soil for them is deep, well-drained, moist loam.

It is probable many of the species are biennial, and that it is well to raise them from seed frequently; and to avoid the results of crossing, it is better to get the seed, if we can, from the home of the species. The seeds should be sown early in spring, and the young plants pricked out into pans, or into an old garden-frame, as soon as they are fit to handle, removing them early in August to the borders; select a cloudy day for the work, and give them a little shading for a few days.

Mr Whittaker, of Mosely, near Derby, has been very successful with both *A. glandulosa* and the blue variety of *A. leptoceras*, and he grew them in a thoroughly drained, deep, rich, alluvial soil; the same were the conditions of Mr Grigor's success.

Mr Brockbank speaks hopefully of growing the finer kinds from seed. He says: "I attribute failures to plants sent by nurserymen in very small pots, and it will be found that you can never get up a good stock of *Aquilegias* by purchase. The proper way is to grow your own from seed. Sow in shallow wooden trays, or in pots, and grow the plants on carefully in a cold frame. When the seedlings are sufficiently large, prick them out into the places wherein you wish them to grow—some in pots and some in the garden—and plant them in various situations, here in the shade and there in the open, so as to have as many chances of success with them as possible. I always plant three plants in a triangle, 4 inches apart, so that any group can readily be taken up and potted if we wish it. Once planted, leave them alone ever afterwards, or, if you move them, take up a large ball of earth with them, so as not to loosen the soil about the roots more than can be helped. When the plants have flowered and the seed has ripened, my practice is to gather some for future sowing, and to scatter the rest around the plant, raking the soil lightly first, and shaking the seed out of the pods every three or four days. From the seed thus scattered young plants come up by hundreds, often as thick as a mat, and may be transplanted, when suitably grown, into proper situations. In this way, I have here abundance of Columbines, and amongst these plenty of *A. glandulosa* self-sown, and as strong and hardy as any."

The late Mr J. C. Niven, of the Hull Botanic Gardens, one who knew alpine and hardy plants so well, suggests that all the Columbines, except the common one, should be looked upon as biennials rather than good perennials. The seeds should be sown early in spring, and the young plants pricked out into pans or into an old garden-frame as soon as they are fit to handle, removing them early in August to their permanent positions; select a cloudy day for the work, and give them a little artificial shading for a few days. Carry out the same process year after year, the old plants being discarded after flowering. Any attempt at dividing the old roots usually fails. There are, however, instances, especially on light soils and hilly districts, where several of them remain good for years.

Aquilegia alpina (*Alpine Columbine*). —A pretty alpine plant, widely distributed over the higher parts of the Alps of Europe, the stems from 1 foot to 2 feet high, bearing showy blue flowers. There is a lovely variety with a white centre to the flower, which, from its colour, is certain to be preferred, and many will say they have not got the "true" plant if they possess only the variety with blue flowers. It does not require any very particular care in culture, but should have a place among the taller plants of the rock-garden, and be planted in a rather moist but not shady spot in deep loam, with leaf-soil.

A. californica (*Californian Columbine*). —One of the stoutest of the American kinds; the spurs are long, bright orange, attenuated. To appreciate the beauty of the flower, it must be turned up from its pendent position; then the beautiful shell-like arrangement of the petals is seen, the bright yellow marginal line gradually shading off into deep orange. The seeds of this kind should be saved, as having once blossomed, the old plant is apt to perish. I have never been disappointed with the seedlings diverging from their parent type in character. This plant thrives best on a deep loam and moist. Syns., *A. eximia*, *A. truncata*.

A. canadensis (*Canadian Columbine*).— The flowers of this are smaller than those of the Californian kinds; this, however, is compensated for by the brilliancy of the scarlet colour of the sepals and the bright yellow of the petals. It is a slender grower, about 1 foot in height, with sharply-notched leaves, and is easily raised from seed. There is a yellow form. Writing of this species, Mr W. Falconer says: "To see it at its best, you should see it among the rocks, where it grows in abundance in our woods, and always in high rocky places; there it springs from the narrowest chink, a little bush of leaves and flowers, or maybe in an earthy mat upon a rock you find a colony of Columbines, Virginian Saxifrages, and pale Corydalis; they usually grow together."

Aquilegia chrysantha. (*Golden Columbine*).—This plant was at first by persons who look at herbarium distinctions only, erroneously supposed to be a variety of the Blue Columbine, and named such by Torrey and Gray. After cultivating the plant, however, for several years, Dr Gray described it as a new species. The plant comes from a different geographical range, grows taller, flowers nearly a month later, and blooms for two months continuously. It has a very long and slender spur, often over 2 inches in length, is hardy, and thrives even on the stiff clay soils north of London, and enjoys wet, though it is none the less free in more happy situations. It comes true from seed, which is best raised under glass, the seedlings being pricked out carefully when young. Attaining a height of 4 feet under good culture, it is a fine plant for grouping among the shrubs of the rock-garden. It would be a pity if such a distinct, beautiful, hardy plant should degenerate in our gardens, by crossing with other kinds.

A. cærulea (*Blue Columbine*).—Beautiful and distinct, the spurs of the flower almost as slender as a thread, a couple of inches long, twisted, and with green tips. It is in the blue and white erect flower that the beauty lies, the effect being even better than in the blue and white form of the alpine Columbine. It is a hardy plant, blooming rather early in summer, and continuing a long time in flower. It grows from 12 inches to 15 inches high, and is worthy of the choicest position on the rock-garden. Unlike the Golden Columbine, it is not a true perennial on many soils, though a better report in this respect comes from the cool hill gardens. To get strong healthy plants that will flower freely, seeds of this kind should be sown annually, and treated after the manner of biennals, as it rarely does well after standing the second year, and in many cases dies out before that time. The flowers are, however, so lovely and so useful for cutting, that it is deserving of care to have it in good condition.

This is one of the plants which deserve a home in the nursery in a choice little bed to itself, from which its flowers could be gathered for the house. The seed is best sown as soon as may be after it is ripe, in cool frames near the glass, or in rough boxes in cool frames. With abundance of fresh seed, there will be no difficulty in raising it in fine beds of soil in the open air, protecting the beds from birds or slugs. The seed is usually too precious to risk in the open air.

What is supposed to be a white variety of this plant is sometimes called *A. leptoceras*, which was indeed the first name given to the plant.

"M.," writing from Utah, says: "Some plants of this species seen in Utah seem to belong to a distinct variety; their colour is not blue, or blue and white, but pure white or yellowish-white. They were flowering in great quantity 10,000 feet above the sea, wherever any tiny stream trickled down the mountain slopes, and the flowers at a little distance reminded one more of those of *Eucharis amazonica* than anything else. The plant grows in handsome tufts 2 feet or 3 feet high, the flowers large and broad, and the spurs very long (2 inches at least), with a rounded knob at the top."

Aquilegia fragrans (*Fragrant Columbine*).—This is very distinct, growing about 1 foot high, with downy, somewhat clammy leaves, and very free-flowering. The flowers are pale yellow or straw, with short hooked spurs. Himalayas.

A. glandulosa (*Altai Columbine*).—A beautiful species, with handsome blue and white flowers, and a tufted habit, flowering in early summer—a fine blue, with the tips of the petals creamy-white, the spur curved backwards towards the stalk, the sepals dark blue, large, and nearly oval, with a long footstalk. A native of the Altai Mountains, and one of the best kinds for the rock-garden, in well-drained, deep, sandy soil. Increased by seed and by very careful division of the fleshy roots, when the plant is in full leaf. Mr William Jennings informs me that, if divided sowing, and when full grown is impatient of removal, but if not transplanted when more than two years old, it continues to flower for at least five or six years, sometimes for more. Those who can get true seed of this fine plant will do well to raise it with care and plant out when very young into well-prepared beds of moist, deep peaty or sandy soil, putting some of the plants in a northern or cool position. It would be well, also, to sow

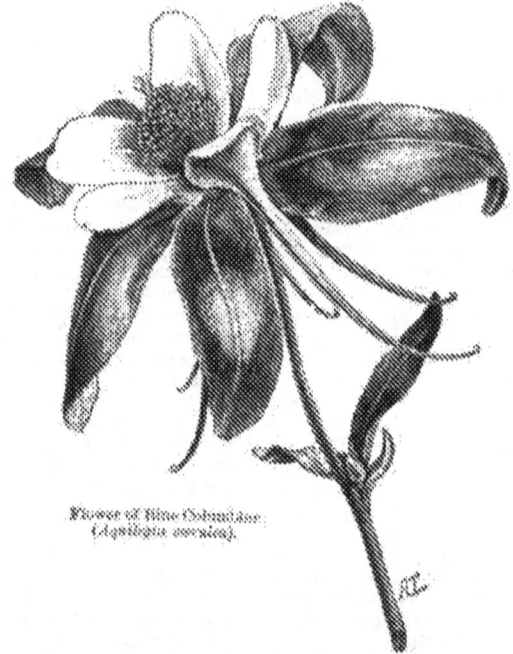

Flower of Blue Columbine (Aquilegia cœrulea).

when it is at rest, the roots are almost certain to perish—at least, on cold soils. The Forres Nurseries, in Morayshire, have long been famed for the successful growth of this plant; it has no special care there, and there is no secret about the culture, which is wholly in the open air. The soil is described as "a rich mellow earth, partaking a little of bog or peat earth, and rather cool and moist than otherwise." It flowers the year after some seeds where the plants are to remain, and in various other ways to try and overcome the difficulty which has hitherto generally attended the culture of this lovely plant. The seeds of other Columbines have a bright perisperm, while those of this species are unburnished, arising from little corrugated markings with which the microscope shows them to be covered.

Mr Brockbank writes: "I have referred

to the original specimen of *A. glandulosa*, sent by Prof. Regel, of the St Petersburg Botanic Gardens, from the Altai Mountains. It is a different plant from the *A. glandulosa jucunda*, being more than twice as tall, and in every way more robust. The specimen at Kew is nearly one and a half times the height of the large folio paper in which it is preserved, and the flower measures 4½ inches in diameter. The plants in Kew Gardens are not this variety—the true variety—of *A. glandulosa*, and, as far as I know, it is not to be found with any of our Nurserymen."

Aquilegia glauca (*Grey-leaved Columbine*).—A distinct and interesting plant, though not so showy as some of the other kinds. It grows from 18 inches to 2 feet high, with glaucous foliage, the spurs of the flowers being rather short and red, and shading into the pale yellow of the other parts of the flower.

A. Skinneri (*Skinner's Columbine*).—A distinct and beautiful kind, the flowers on slender pedicels, the sepals being greenish, the petals small and yellow; the spurs nearly 2 inches long, of a bright orange-red, and attenuated into a slightly-incurved club-shaped point, the leaves glaucous, their divisions sharply incised; the flower-stems 18 inches to 2 feet high. Though coming from so far south as Guatemala, owing to the fact that it is met with in the higher mountain districts, it is nearly, if not quite, hardy. Here, again, crossing steps in, and too frequently mars its beauty. While the name may be often seen, the plant is rare, nor are the conditions that insure its thriving well known, if they exist with us. It is a late bloomer.

A. Stuarti (*Stuart's Columbine*).—This, a cross between the true *A. glandulosa* and *A. Witmanni*, was raised by the late Dr Stuart, who tells us that it is, in his opinion, an improved form of *A. glandulosa*, refined in colouring, free flowering, very large and attractive. It is perfectly hardy, flowers three weeks before any other Columbine, and always comes true from seed. He recommends that a bed be trenched 2 feet deep, with plenty of manure in the bottom, sowing the seed in rows, and allowing the seedlings to flower where they are to stand. The plants may be thinned out to 8 inches apart, allowing 12 inches between the rows. In time the foliage will cover the entire bed, and the plants will produce an abundance of bloom. By top-dressing in the autumn the plants improve in vigour every season, a three-year-old bed being a mass of bloom.

Aquilegia viridiflora (*Green Columbine*).—A modest and pretty kind, with sage-green flowers. Out-of-doors in the border the plant may not be noticed, but if a flowering spray or two be cut and placed in a small glass, its beauty of form and colour too, may be seen. There is a variety of it, known as *A. atropurpurea*, of which the sepals are green, the petals deep chocolate. The plant is a strong grower, a native of Siberia, and is the same as Fischer's *A. dahurica*. It has a delicate fragrance, too. It is a rare plant in gardens. Seed.

A. vulgaris (*Common Columbine*).—The only native Columbine, and as beautiful, I think, as some of the rarer alpine kinds, and no one who has once seen it wild, will readily forget its beauty. It would be most desirable also to select and fix varieties of the Common Columbine of good distinct colours. Being a native of mountain woods and copses, this may be grouped with good effect in the shrubby part of the rock-garden. The best white form of this plant is a beautiful and stately Columbine, which sows itself freely in various positions when once brought into the garden, and looks well wherever it comes. The hybrid forms raised in gardens and much grown and talked of, are not so beautiful as this and other wild kinds.

ARABIS (*Rock Cress*).—Early and brave, these mountain plants have few of striking importance for the rock-garden, and these are of easy culture, and increase so free, indeed, that they are grown as edgings, and often fall over cottage garden banks and rough walls, giving pretty effects. In this family, it may be that, as the mountain world becomes better known, gems for the rock-garden may appear, but, so far, as already tried in our gardens, few of the kinds are attractive in colour.

Arabis albida (*White Rock Cress*).—Through long years of neglect of all sorts of dwarf hardy plants, this, the "white *Arabis*" of our gardens, has held its own, and is now seen in almost every garden. A native of the mountains of Greece, and many parts in adjacent regions, it is as much at home in Britain as is the Daisy, and will grow in any soil or situation, in cities as well as in the open country, where its profuse sheets of snowy bloom may expand unblemished under the earliest suns of spring. By seed, or cuttings, it is easily increased, and a valuable ornament of the border and the spring garden. On the rock-garden it is well fitted for falling over the brows of rocks; it may also be used as an edging. It is closely allied to the Alpine Rock Cress (*A. alpina*), so widely distributed on the Alps, and by some would be considered a sub-species of that plant, but it is sufficiently distinct, and by far the best kind.

A double form has recently been grown, and it is a good plant. There is a variegated variety in cultivation, known by the name of *Arabis albida variegata*, which is useful as an edging-plant, both in spring and summer flower-gardens. It is the dwarfest and whitest of the variegated Rock Cresses that are grown under the names of *A. albida variegata*. The yellower and stronger variety, frequently called *A. albida variegata*, and which is the best for general purposes, is a form of *Arabis crispata*, of which the ordinary green form is not worthy of cultivation.

A. blepharophylla (*Rosy Rock Cress*).—Like the white *Arabis* in its habit, size, and leaves, the flowers are of a rosy purple, and like a miniature Rocket, and thriving as freely as the old single plant, distinct from any flower of the same order in cultivation. It varies a good deal, and there is no difficulty in selecting a strain of the brightest rose, but it does not seem to take to our country, and is rare. It is best raised every year from seed, which it yields freely. In mild districts, and on light soils, plants should be tried out in winter. The brighter forms are effective a considerable distance off. A native of North America.

Among other kinds of *Arabis*, *A. procurrens* is a dwarf spreading kind, with shining leaves and small whitish flowers. There is a variegated form of it (*A. p. variegata*) which is worthy of a place among variegated hardy plants. The prettiest of the variegated Rock Cresses is *A. lucida variegata*. It forms very neat and effective edgings in winter, spring, and summer flower gardens, thrives best and is easiest to increase by division in open, sandy, and yet moist soil. The best time to divide it is early in autumn, April, or very early in May. *A. purpurea*, an interesting species for botanical, large, or curious collections, and bearing pale bluish and lilac flowers, is not worthy of general cultivation while we possess such brilliant plants as the *Aubrietias*. *A. arenosa*, from the south of Europe, is a pretty annual kind that may prove useful in the spring garden, and which might be naturalised on dry banks. *A. petræa* is a neat, sturdy little plant, with pure white flowers, a native of some of the higher Scotch mountains, and very rarely seen in cultivation, but when well developed in a moist yet well-exposed spot, is pretty. *A. aubrietiodes* is a pretty soft rosy kind, not yet much known.

ARCTOSTAPHYLOS (*Bearberry*).—Trailing mountain shrubs, usually evergreen, of good habit and hardy, and useful among the dwarf shrubs of the rock-garden. The berries of some kinds are a favourite food of game.

All are interesting little shrubs, thriving in peaty loam. Seeds offer the readiest means of increase, though all may be increased by layer. The two native kinds are excellent rock-plants.

Arctostaphylos alpina (*Black Bearberry*).—A plant very rarely seen in cultivation, a native of high alpine or arctic regions, and of the northern Highlands of Scotland, distinguished by its thin, toothed leaves, which are not evergreen, but wither away at the end of the season, and by its bluish-black berries.

Arctostaphylos alpina (*The Black Bearberry*).—The badge of the Clan Ross is rare as a native plant, being confined to dry, barren Scotch mountains from Perth and Forfar northwards, and ascending to elevations of nearly 3000 feet above sea level. It forms compact, woody patches, with stout, leafy branches, and scaly bark. The deciduous leaves, wrinkled above, have ciliated margins, and are narrowed into a short stalk. They vary in length from $\frac{1}{2}$ inch to $1\frac{1}{2}$ inches, and are coarsely toothed above the middle; the white blossoms are produced in twos or threes, and appear with the young leaves. The berry is black, and measures $\frac{1}{4}$ inch in diameter.

A. uva-ursi (*Bearberry*).—A small and prostrate creeping mountain shrub, with leathery leaves, and their under side netted with veins, and with the sepals at the base and not at the crown of the berry. The flowers are of a rose-colour in clusters at the apex of the branches; the berries of a brilliant red. It is a native of dry heaths and barren places in hilly countries, and is easier to cultivate than almost any other small mountain or bog shrub, thriving well in common garden soil. It is a useful plant in the rock-garden, when its shining evergreen masses of leaves fall over rocks, and also on the margins of beds of shrubs.

Another kind widely different from all the foregoing, is one cultivated in the Edinburgh Botanic Gardens under the name of *A. californica;* this is a very vigorous, trailing, evergreen shrub, with spathulate, leathery, entire leaves. *A. pungens* is a much branched, erect-growing shrub, with leathery-pointed leaves, from 1 inch to $1\frac{1}{2}$ inches long, downy when young and smooth, when old, the blossoms white, tinged with rose.

A. Manzanita.—A native of California, where it gets to be a good-sized shrub, and bears abundantly large drupe-like fruits of a pleasant taste, which are much used as food by the Indians of that region, but it is not of proved hardiness in our islands.

ARENARIA (*Sandwort*).—Mountain and heath plants of great variety, of dwarf and sometimes mossy habit, and some bearing pretty flowers. They are easy to cultivate, quite hardy, and though not alpine plants of the highest importance, they are, nevertheless, of value in the rock-garden, grow freely in almost any garden soil, and are of facile increase by division or seed.

Arenaria balearica (*Stone Sandwort*).— A tiny self-nourishing plant, coating the face of stones with a close, Thyme-like verdure—as with Moss—and then scattering over the green mantle countless little starry flowers. I write this sitting on a rock, to which it clings closer than Moss. It has crept over the edge of some rocks which slope to water, and dropped its little mantle of green down to within 18 inches of the water, but all the flowers look up from the shade to the light. Right and left there are boulders in various positions, on every face of which it may be seen, as every tiny joint roots against the earthless face of the stones. To establish it on stones, plant in any soil near on the cool side, and it will soon begin to clothe them. It flowers in spring, is readily increased by division, and quite easy to grow on most soils, and even on the face of walls (north side), and on stones and rocks in the sunnier districts on the cool sides. Easily naturalised in rocky places.

A. Huteri (*Huter's Sandwort*), is a charming alpine form, growing freely in sandy loam in the level parts of the rock-garden. A top-dressing of sand and leaf-mould is very beneficial, enabling the young shoots to root freely.

A. laricifolia (*Larch-leaved Sandwort*). —The leaves of this are narrow, and arranged in clusters, bearing some slight resemblance to those of the Larch, the flowers white, in clusters of three to six on each stem. This is a native of Switzerland, and should be placed on a rather high ledge.

Arenaria montana (*Mountain Sandwort*).—A handsome plant, having the habit of a *Cerastium*, and large white flowers. It forms spreading tufts, on which the flowers come so thickly in early summer as to obscure the foliage. It is one of the prettiest early summer flowering plants, succeeding the white evergreen Candytufts and like flowers. S. Europe.

A. grandiflora is a large-flowered form of *A. montana*.

A. multicaulis.—From the south of Spain, resembles *A. balearica*, but has more ovate leaves, its flowers higher above the foliage and larger.

quence of the prostrate habit of both shoots and flowers, the plant is seen to much greater advantage when placed on some little bank above the eye. It is a native of the northern parts of Great Britain, and is readily increased by seed.

Of other *Arenarias* in cultivation, the best and most interesting are *A. ciliata*, a rare British plant; *A. triflora*, a neat species in cultivation in some curious collections; and *A. graminifolia*. These, however, and many others are scarcely worth growing, except in botanical collections.

Some of the species above-named will be found in some books under **Alsine**.

Arenaria laricifolia. Syn., *Alsine laricifolia*.

A. purpurascens (*Purplish Sandwort*).—Distinguished from other kinds by its purplish flowers on a densely-tufted mass of smooth, pointed leaves. It is frequent over the Pyrenean Chain; and it should be associated on the rock-garden with the smallest Rockfoils and plants which, though dwarf, are not slow growers.

A. tetraquetra (*Square-stemmed A.*).—This forms compact and singular-looking tufts, in consequence of the leaves, each with a white cartilage along the margin, being in four rows. The sepals are also margined. It is worth a place where the other small Sandworts are grown.

A. verna (*Vernal Sandwort*).—Grows in prostrate tufts, covered in April and May with multitudes of starry white flowers with green centres. In conse-

ARETHUSA BULBOSA.—A beautiful American hardy Orchid, which grows in wet meadows or bog-land, blossoming in May and June. Each plant bears a bright rose-purple flower, showy on its bed of Sphagnum, Cranberry, and Sedge. The little bulbs grow in a mossy mat formed by the roots and decaying herbage of plants and moss. In cultivation it requires the same soil in a shady moist spot, with a northern exposure, the soil a mixture of well-rotted manure and Sphagnum. During winter, protect the bed with some cover, for it is not so hardy in gardens as in its marshy home. Newfoundland to Ontario and southward to N. Carolina.

ALPINE FLOWERS FOR GARDENS

ARMERIA (*Thrift or Sea-Pink*).—Modest perennials, natives of the rocky shore and mountain ground; of much beauty of colour. They are plants of easy culture and increase, and they may be used as carpets and edgings, one or two kinds being native.

Armeria vulgaris (*Thrift*).—This inhabitant of our sea-shores, and also of the tops of the Scotch mountains and the Alps of Europe, is very pretty, with its soft lilac 15 inches to 20 inches high, each bearing a large, roundish, closely-packed head of handsome satiny rosy flowers. It comes from North Africa and S. Europe, and, though hardy on free and well-drained soils, occasionally perishes during a very severe winter, especially on cold soils; it should therefore be placed in a warm position on the rock-garden, and in deep, sandy loam. It is known under various names—*A. formosa, A. latifolia, A. mauritanica, A. pseudo-armeria, Statice lusitanica.*

Thrift on the hills at Anglesey. (Engraved from a photograph by Miss A. Cummings, King's Buildings, Chester.)

or white flowers springing from cushions of grass-like leaves; but it is the deep rosy form of it, rarely seen wild, that deserves a place in rock-gardens. It is like the common Thrift in all respects but the colour of the flowers, which are of a showy rose. It is useful for the spring garden, for covering bare banks or borders in shrubberies, and for edgings. Occasional division (say every two or three years) and replanting are desirable.

A. cephalotes (*Great Thrift*).—From a dense mass of crowded leaves, 4 inches to 6 inches long, spring numerous stems It is, fortunately, easily raised from seed; and, as it is not easily increased by division, it is a good plan to sow a little of it every year. Varies a little when raised from seed; but all the forms I have seen are worthy of cultivation.

ARNEBIA ECHIOIDES (*Prophet-flower*).—Borage-worts, and among the handsomest of flowers, distinct and singular.

A. echioides is 1 foot to 18 inches high, the flowers primrose-yellow, with five black spots on the corolla, which gradually fade to a lighter shade, and finally disappear. It is hardy, succeeds either on the rock-garden or in a well-drained border, and prefers partial shade. It is a native of the Caucasus and Northern Persia, and is best in fine, deep loam. Young plants bloom long, which adds to their charms. Seeds are not freely produced, but it may be increased by cuttings. *A. Griffithi* is a tender annual, and though pretty, not so valuable as *A. echioides*.

ARTEMISIA (*Wormwood*).—Half shrubby and perennial plants of the steppes, arid plains, and mountains; of a bitter flavour and pungent odour, and which give a distinct greyish hue to many arid regions, but are often of secondary interest only for the rock-garden. Among a large number of species known, there are many of slight interest for the rock-garden, and a few are neat in habit and pretty in flower, such as the Silvery Wormwood, *A. frigida, glacialis, nana, sericea,* and *Baumgarteni,* all of easy culture and increase.

ASARUM (*Wild Ginger*).—Curious little plants resembling Cyclamens in their leaves, but of little garden value. *A. canadense* is the Canadian Snake-root, which bears in spring curious brownish-purple flowers, the roots being strongly aromatic, like Ginger. *A. virginicum* is the Heart Snake-root, with leaves thick and leathery, with the upper surface mottled with white. *A. caudatum* is from Oregon, and much like the others in habit, but the divisions of the flower have long tail-like appendages. *A. europæum* is the *Asarabacca,* the flowers being greenish, about ½ inch long, and appearing close to the ground. The plants are only valuable for the effect of the leaves in dry poor spots.

ASPERULA (*Woodruff*).—Dwarf plants of the Bedstraw (*Galium*) order, so far as known of secondary use in the rock-garden.

Asperula odorata (*Woodruff*).—A little wood plant, abundant in some parts of Britain, is worthy of a place in the rock-garden, in localities where it does not occur wild. It is sometimes used as an edging to the beds in cottage gardens, and it mixes prettily with Ivy where that is allowed to clothe the ground. It belongs to a numerous genus of plants, few, however, of which are worth a place among the choicer rock-plants.

A. azurea setosa is a pretty early spring flowering hardy blue annual, flowering in April and May. Sow the previous autumn. *A. cynanchica* is a rosy red perennial, a good rough rock plant.

ASTER (*Starwort*).—A beautiful family of northern plants, chiefly American, but also some, and among the handsomest, European. Although mostly tall and often too vigorous, there are some beautiful mountain kinds, and, to a great extent, the family are found on mountains; but they are rarely suitable for the rock-garden. One of the handsomest plants in the alpine meadows of Europe and other countries is the alpine Starwort, but in cultivation and richer ground it is not so attractive as in the wild state. Nevertheless, in large rock-gardens some of the dwarfer kinds may often be useful, all the more so to those who enjoy their gardens mostly in the autumn.

Among the best of all, however, are the European Starworts, *A. amellus* and *A. acris,* of which last there are dwarf forms, precious for their fine colour and not too tall for the bolder parts of the rock-garden, and for growing among the shrubs near it, as advised elsewhere in this book.

Some of the Indian Starworts are dwarfer and more refined in habit than the American, and in the vast and not yet explored regions, there may be gems for the rock-garden.

The dwarf habit of these Himalayan Daisies makes them valuable for the rock-garden. They are all found in the temperate regions of the Himalayas, a few at high elevations, and are hardy.

Aster stracheyi.—A pretty plant, more or less hairy, and rarely more than 1 inch to 3 inches or 4 inches in height. The flowers are about the size of those of the Michaelmas Daisy, the involucre bracts few, scarcely overlapping, all about one length, and usually narrow and pointed. Native of the Western

Aster stracheyi.

Alpine Himalayas, Kumaon, at 13,000 feet elevation, flowering with us in early summer. It is hardy in the open air, and forms a charming rock-garden plant, thriving best in half-shady spots.

A. alpinus (*Alpine Starwort*).—This might be called the blue Daisy of the Alps, so diminutive is it when met with high up or even in rich green alpine meadows. In a wild state it does not form the sturdy tufts which it does in gardens, and, like the wild Orange Lily, is more beautiful when isolated in the grass. The flower is of a pale blue, with an orange-yellow eye, 2 inches across on plants cultivated in gardens, smaller in a wild state. It forms tufts 8 to 10 inches high, slightly downy, and sometimes velvety. There is a white variety. Easily multiplied by division, thrives well in any sandy soil, and begins to flower in early summer.

Of the very large Aster family there are few dwarf enough for our purpose, one of the best being that known as *versicolor*, which, as it is somewhat prostrate, might be planted with good effect on the lower parts of the rock-garden. *A. altaicus* is also a dwarf species, with mauve-coloured flowers, and *A. Reevesii* is a dwarf kind.

ASTRAGALUS (*Milk-Vetch*).—Perennial and alpine plants of the Pea flower order, the species numerous, but, so far as is now known, not very important for the rock-garden. The Tragacanth plant (*A. Tragacantha*) forms a dwarf grey bush, and is hardy, and may be grown even in towns, but it is not attractive in flower. Some are natives of Britain.

Astragalus hypoglottis (*Purple Milk-Vetch*).—A dwarf, prostrate perennial, and large heads of bluish-purple flowers. In Britain it is found chiefly on the eastern side of the island from Essex and Herts to Aberdeen, and on dry, gravelly, and chalky pastures. It is pretty on level spots, and should always be associated with very dwarf subjects; and though it is not particular as to soil, it will be found to thrive best in open, well-drained, sandy loam, or in chalky soil. A variety has paper-white heads of flowers sitting close upon the dwarf carpet formed by the leaves. It looks showy for such a dwarf white plant, and the flowers look singular from contrast with the short sooty or black hairs. It is so distinct from any other cultivated alpine plant in flower about the same period, that it would be wise to form a little carpet of five or six plants of it in some level spot, as it is not at all difficult to grow.

A. Monspessulanus (*Montpellier Milk-Vetch*).—A vigorous kind, with leaves a span long, the leaflets smooth on the upper surface, and with short whitish hairs thinly but almost quite regularly scattered over their under sides. The flowers are borne on stalks from 6 inches to a foot long, the racemes of bloom being from 2 to 5 inches long, according to the strength of the plant. The closely-set and unopened flowers at the head of the raceme are usually of a deep crimson, but as they open, they become of a pale rosy lilac, with bars of white on the upper petals. The shoots, though vigorous, are prostrate, which causes it to be seen to greater advantage when drooping over rocks, and it grows well in any soil. A native of the South of France, easily raised from seed. There are several varieties.

Astragalus onobrychis (*Saintfoin Milk-Vetch*).—A fine hardy kind, in some varieties spreading, and in others growing about 18 inches high, with pinnate leaves about 4 inches long, the leaflets smooth, and handsome racemes of purplish-crimson flowers. As the individual flowers, when fully open, are a shade more than five-eighths of an inch long, and borne in clusters of from six to sixteen on each raceme, it is an attractive plant, and will thrive well in any good loam. There are several varieties enumerated, three of which, *alpinus, moldavicus,* and *microphyllus,* are prostrate in habit, and would prove valuable. The plant is particularly suited for the rougher parts of the rock-garden, and for positions where a rich effect rather than minute beauty is sought. There are white forms of all the varieties. Europe and N. Asia.

A. pannosus (*Shaggy Milk-Vetch*).—A dwarf kind, with silvery, woolly pinnate leaves, which, growing in compact tufts about a span high, give the plant somewhat the appearance of a silvery fern. Attracted by this appearance, when I saw the plant in cultivation in Switzerland, I brought home some seeds, from which plants have been raised by Mr J. Backhouse and Mr W. Bull. I have not yet seen it in flower, but from the beauty of its leaves alone, it is likely to prove an excellent rock-garden plant. It is easily increased by seeds, and comes from Asia Minor.

ATRAGENE (see **CLEMATIS**).

AUBRIETIA (*Purple Rock-Cress*). — If there were but one family of rock-plants known to us, this which gladdens the rocks of Greece and all near countries with its soft colours in the dawn of spring, would be almost enough to justify the lovers of rock flowers for any extravagance in their behalf. In these plants all difficulties of culture, increase, soil, etc., fly away, and though from the hills above the cities of Greece or the sites ennobled in human story, they are as happy in our British land as the grasses of our fields.

These rock plants will succeed on any soil, and never fail to flower, even should the cutting winds of spring shear all the verdure of the budding Willows. There is hardly a position selected for a rock plant that may not be graced by them. Rocks, ruins, stony places, sloping banks, and walls, suit them perfectly; and no plant is so easily established in such places, nor will any other alpine plant so quickly clothe them with the desired kind of vegetation. Growing in common soil, in the open border, or on any exposed spot, they thrive as well as on the best-made rock-garden, forming round spreading tufts; and on fine days in spring the flowers come out on these in such crowds as to completely hide the leaves, making hillocks of colour. They are quite easy to naturalise in bare rocky places, and often sow themselves on walls. They are easily propagated by seeds, cuttings, or division. Grown together, their affinity is clearly seen, and few things may be more safely united under one species than the *Aubrietia* at present in cultivation.

Among the several varieties, *A. deltoidea grandiflora* and *A. Campbelli* are the best. *Dr Mules* is the richest colour. *A. græca* is simply a variety. *Aubrietias* vary a good deal from seed, but their little differences make them all the more valuable as garden-plants, and they all agree in carpeting the earth with dense cushions of compact rosettes of leaves, profusely clothed with beautiful purplish-blue flowers in spring, and, in the case of young plants, in moist and rich soils, almost throughout the year. There are one or two pretty variegated varieties.

AZALEA (*Swamp Honeysuckle*).—Thinking as I do, that the most satisfying and enduring kind of rock-garden cannot be made without the aid of mountain shrubs, or in which they take the main part, such lovely mountain bushes as the Azaleas cannot be left out of our view, as they are true mountaineers, and of splendid value for their flowers in summer and foliage in the autumn, and even in habit, if naturally grown. Their hardiness, fine colour, and ease of culture, should almost give them the first place with the happy people who have rocks of their own, as so often happens in the north, and in Scotland, Wales, and Ireland. There is scarcely a plant among the Azaleas that is not worth growing, but I am now thinking more of the wild kinds, chiefly American, which deserve to be grown, and grouped each kind by itself, these wild kinds being, I think, more beautiful, and more worthy of a place on the shrubby rock-garden than the hybrids, though all are good. More brilliant than any other shrubs, they are lovely in flower in early summer, in some cases continuing into midsummer, and hardy as the mountain rocks. They are much varied, coming from European, American, Chinese, and Japanese species, both in their wild forms and in the varieties raised. It is not only the often brilliant flowers they give us we have to think of, but the finest leaf hues in autumn, especially when massed in the sun. They are not so difficult to grow as the Rhododendron, owing partly to that being on their own roots they can be grown in a greater variety of soils. From an artistic point of view, their form in winter is better than that of rhododendrons, and they do not run into heavy dark masses like the commoner Rhododendrons.

A great advantage is that they are tender to life below them, and, instead of devouring all other plants, like the Rhododendrons, they are very kind to all sorts of beautiful things, such as Blue Anemones, Trillium, Double Primroses, and a great variety of bulbs and choicer hardy flowers, growing beneath them, the effect of which below the bushes is far better than when by themselves, the inter-relations of colour being so much better than from solid masses of green. It is usual to regard them as only to be grown in peat soils; but it is by no means necessary, and the absence of peat should never be a bar to their growth. Even if they do not on sands or loams grow as rapidly as on good peats, the beauty is none the less, especially on the rock-garden, where we seek beauty of form and colour, shown in no matter how small a scale, rather than the too vigorous vulgarity of shrubbery growth. My Azaleas are grown in soil and situation wholly different from what is usually and rightly supposed to favour Azalea growth, and the growth of my plants is certainly less vigorous than in good peat soil, but I enjoy the beauty of the plants just as much.

Although from a botanical point of view there is no distinct line between Azalea and Rhododendron, and the two genera are merged into one by nearly all botanists, for purposes it may be convenient to treat Azalea as a separate genus. Loudon united it with Rhododendron upwards of forty years ago, and all writers of any weight have followed in his footsteps. Still, as the plants treated of in this article—or, at any rate, most of them—are, almost without exception, mentioned in Catalogues and spoken of by gardeners as Azaleas, it has been thought preferable to keep up the older name.

The introduction of a number of kinds from Japan, China, India, and Borneo, destroyed the old lines of demarcation between the two genera, for the number of stamens in some of the so-called Azaleas is often ten, and in several the leaves are evergreen.

No attempt is made to include here any of the so-called Indian Azaleas, the fact of these succeeding in the open air in some parts of the southwest of England and the Channel Islands not being ground enough to class them in a list of hardy shrubs, though it is likely that most of the beautiful garden plants, so deservedly popular under the name of Ghent Azaleas, are hybrids, derived from *A. calendulacea*, *A. nudiflora*, *A. viscosa*, and *A. pontica*. Of late, however, *A. sinensis* (better known as *A. mollis*), and the Western American, *A. occidentalis*, have been used for crossing, and from the latter a beautiful race of late-flowering forms has sprung. Both double and single varieties, ranging from white through every shade of yellow, orange, and red to crimson, with many uncommon intermediate tints, are to be seen in many gardens, and the beautiful colours assumed by the decaying leaves in autumn make them worth growing, even apart from the flowers.

All the hardy Azaleas thrive best in peat, and like best a moist situation, but it is astonishing how well they will do without peat, provided they have an abundance of leaf-mould, and are well supplied with water during the summer months. They are readily raised from seeds, but if it is desired to increase any particular sort, layering is the best way.

Azalea arborescens (*Tree A.*).—This is a native of the Alleghany Mountains, from Pennsylvania to North Carolina. Its leaves are margined with short hairs, are slightly leathery when mature, bright green and shining above and glaucescent beneath. The corolla is fully 2 inches long, white or tinged with rose, and the long red stamens and style add to the beauty of the plant and give it a fine character. It was introduced in 1818, but was probably lost to cultivation soon afterwards, and not re-introduced until a few years ago. The leaves in dying exhale an odour similar to that of the Sweet Vernal Grass; they are well developed before the flowers appear in June.

A. calendulacea (*Flame A.*).—In this the corolla varies in a wild state from orange-yellow to flame-red; the flowers, not fragrant, appear before or with the leaves in May. It is a native of woods in the mountains of Pennsylvania, Virginia, Kentucky, and varies in height from 3 feet to 10 feet.

A. linearifolia (*Slender A.*). In all probability this is not so hardy as the other species here mentioned, but it has stood for several years without protection in the open air at Kew. It is a small shrub, with slender branches beset with rigid, red-brown hairs; the long, narrow leaves, with wavy margins, crowded at the ends of the twigs. The flowers in clusters at the tips of the branches, with five recurved, red-purple petals.

A. nudiflora (*Pinxter Flower*).—This is the purple Azalea, of the United States, where it occurs in swamps from

STONE PATHWAY IN ROCK GARDEN, FRIAR PARK. *June* 1910.

Massachusetts and New York to Illinois and southward. Flowering in April or May, either before or at the same time as the leaves. In a wild state the more or less fragrant flowers vary from flesh colour to pink and purple. Of this species there are numberless varieties and hybrids, no fewer than forty-three being enumerated in Loddiges' Catalogue in 1836.

Azalea occidentalis (*California A.*).—One of the most beautiful flowers when the glossy leaves are well developed, and after most other Azaleas are past. The species is a native of the western foot-hills of the Sierra Nevada throughout the length of California, and in the coast ranges along streams. This fine distinct kind is a free grower, even where there is no peat.

A. Pontica (*Pontic A.*).—An immense number of varieties and hybrids have been raised from this species both in British and Continental gardens. The wild plant has fragrant flowers of a bright yellow colour, blossoming in May and June. This comes from the same country as the Pontic Rhododendron—the Caucasus and near regions—and is supposed with good reason to be the source of the honey that led to the poisoning of Anophon's soldiers. It is a free and handsome shrub in almost any soil, and in rocky spots in woods or copses quite at home.

A. rhombica.—The near allies of this distinct-looking plant are Chinese or Japanese; it has bright, rose-coloured, bell-shaped flowers, with a very short tube, 1½ inches to 2 inches across, generally in pairs at the tips of the branches. The dull green hairy leaves are in whorls of three, and of the ten stamens the five upper are much the shortest. In autumn the decaying leaves turn a bronzy-purple colour. This is one of the earliest to flower, and the spring frosts frequently disfigure the blossoms. Mountain woods of Japan.

A. sinensis (*Chinese A.*).—A native of alpine shrublands in Japan, but is largely cultivated both in that country and in China. The flowers vary much in colour; ranging in a wild state from a dull, almost greenish-yellow to orange-yellow or orange-red, but many hues have arisen in nur-series from crossing. Loddiges was the first to publish a figure. Upwards of forty years afterwards Regel gave it the name of *A. mollis*, and subsequently the late Dr Gray described it under the name of *A. japonica*. Syn., *A. mollis*.

Azalea vaseyi (*Vasey's A.*).—A pretty shrub from 3 feet to 10 feet high, with leaves 3 inches to 6 inches long, a roseate corolla, the upper lobes spotted towards the base. As a rock shrub it is very precious, and its pink or purple flowers are distinct and beautiful. N. America.

A. viscosa (*Swamp Honeysuckle*).—Is a shrub from 4 feet to 10 feet high, with clammy, fragrant flowers, white or tinged with rose-colour in a wild state. Innumerable varieties of this have originated under cultivation, no less than 107 being given in Loddiges' Catalogue for 1836. Several wild forms have at various times received specific names; of these *glauca* has paler leaves, generally white, glaucous beneath; *nitida* is a dwarf variety, with oblanceolate leaves, green on both surfaces; *hispida* and *scabra* do not require detailed description. N.E. America.

BELLIUM (*Rock Daisy*).—These are nearly allied to the common Daisy. Three kinds are in cultivation: *B. bellidioides*, *crassifolium*, and *minutum*, none of which are so beautiful as the common Daisy, nor so hardy, and therefore scarcely worthy of cultivation, except in large collections. Where grown without protection in winter, they should be planted in sandy warm soil, and in sunny spots, on which I should certainly not be anxious to give them a place, considering the numbers of brilliant plants we have more fitted for the embellishment of the rock-garden.

BERBERIS (*Barberry*).—Of these handsome shrubs having much beauty of foliage and fruit, while the greater number would not be in stature suited for the rock-garden, certain kinds might be useful where the idea of the shrub rock-garden is carried out. The dwarf evergreen, *Thunbergs' barberry*, and *B. stenophylla*, are suitable for giving a good effect among rocks. Nor does the absence of rocks debar us from grouping them near the rock-garden, and enjoying in such positions their beautiful colour in autumn.

Berberis empetrifolia(*Fuegian Berberis*). —A dwarf, shrubby, trailing species, from the Straits of Magellan, well adapted for rock cultivation, provided a good depth of peaty soil be given it for its underground shoots to ramble in. Its flowers are of a bright orange colour, singly along the whole length of the previous year's growth. It has a delicate fragrance.

BERGENIA.—A name used by some Continental botanists for the large-leaved Indian Rockfoils, known in our gardens by the names of *Saxifraga* and *Megasea*.

BETULA (*Birch*).—Though we know the Birch as a forest tree, it may be as well to remember that there are little northern and antarctic Birches, and those from the high mountains, such as the Scrub Birch (*B. glandulosa*), the dwarf Birch (*B. nana*), and the Bog Birch (*B. pumila*), which might be readily used near rock and marsh gardens of the bolder sort.

BLETIA HYACINTHINA.—A tall and graceful hardy Orchid, with slender flower-stems 1 foot or more high, bearing about half a dozen showy flowers of a deep rosy-purple colour. It thrives in sheltered and half-shaded spots in peaty soil, with some leaf-mould added. In some localities it would be advisable to cover the roots with a handful of protective material during severe cold. It is also known as *B. japonica*. A very interesting plant for association with the peat-loving *Cypripediums* in the drier parts of the bog-garden. China.

BORETTA.—One of the recent botanical names for the Irish Heath, which will be found in this book under **ERICA**.

BRACHYCOME SINCLAIRI, according to a writer in the *Garden*, is a gem for the rock-garden, hardy and perennial, bearing little white Daisy-like heads on stems 2 inches or 3 inches high, all the summer months, and having a distinct habit of growth. The plant spreads moderately by short stolons, and the foliage is arranged in tufts or rosettes, and is brownish or bronzy-green, and very downy. Those seeking for beautiful miniature plants should take note of this. I grow it in loam and leaf-mould, mixed with small stones, and in a position where it can have plenty of moisture and sunshine.

The pretty little *B. iberidifolia* (Swan River Daisy), is one of the annual flowers which may be used with good effect to clothe any bare spaces that may occur in the rock-garden from winter losses or other causes.

BRUCKENTHALIA SPICULIFOLIA.—A dwarf-plant, belonging to the Heath family. The flowers are

Bruckenthalia spiculifolia.

pale purple or lilac, on stems rarely more than 9 inches high. It is suited to dry, peaty positions, or in peat or leaf-soil will make itself at home in a half shady spot in the garden.

BRYANTHUS (*Rocky Mountain Heath*).—Alpine bushes of the Heath family, mostly natives of the mountains of North America, and little known in gardens. I brought one handsome species from the sierras of California, but it is lost. They are pretty little shrublets which well deserve introduction, and growing as they do on some of the coldest mountains of the world, I have little doubt that they will prove as easy to cultivate as many other American bushes which thrive in our gardens. Mr Bulley, in the *Gardener's Chronicle*, describes *Bryanthus glanduliformis* as a dwarf, peat-loving plant, not reaching a greater height than 3 inches, and notable for the large size and striking colour of its *Pentstemon*-like flowers. These, which are borne profusely, are $1\frac{1}{2}$ inches long, and of the most vivid magenta-red.

Bryanthus erectus.—A dwarf evergreen bush, from 8 inches to a foot high, bearing pretty pinkish flowers. It is said to be a hybrid. In very fine sandy soil or in that usually prepared for American plants, it grows well, and is worthy of a place in collections of very dwarf alpine shrubs, whether planted in the rock-garden or in peat beds.

B. Breweri.—A neat little plant has been introduced under this name, but is little known in cultivation.

BULBOCODIUM VERNUM (*Spring Meadow Saffron.*)—Grown in our gardens for generations, this very early bulb is one of the earliest of spring bulbs, sending up its large rosy-purple flower buds earlier than the Crocus. The flowers are tubular, nearly 4 inches long, and usually best when in the bud state, the colour being a violet purple, the large buds appearing before the concave leaves, which attain vigorous proportions after the flowers are past. Associated with very early flowering plants like the Snowflake and Snowdrop, it is welcome in the rock-garden, or in warm sunny borders. A native of the Alps of Europe, easily increased by dividing the bulbs, in July or August. *B. Versicolor* is a variety.

BUXUS (*Box*).—The dwarf forms of the common Box are very pretty little evergreens, and the Japanese Box has the merit of being extremely hardy, as it endures the winter in North Germany, where the common Box does not. In dealing with those limestone and other rocks which abound in many parts of the country, I think this and dwarf forms of our native Box might be very well used in giving evergreen effects. Many stony and rocky districts which are now uninhabited will some day be valued as among the most pleasant places to live in, and planting the naturally rocky surface will have to be faced, and I can think of no more beautiful way of adorning it than with such hardy mountain shrubs, among which this is one of the most pleasant of evergreens.

CALAMINTHA GLABELLA is a minute plant, forming neat little tufts about 3 inches high, flowering in summer, tubular, lilac-purple, scented, very numerous and large for its size. May be grown on the rock-garden in sandy loam, and among the very dwarfest plants. Division.

CALANDRINIA UMBELLATA (*Brilliant C.*).—A native of Chili, with reddish, much branched, little stems, half-shrubby, and rarely growing more than 3 or 4 inches high. For brilliancy of colour there is nothing to equal it in cultivation, the flowers being of a dazzling magenta crimson, yet soft and refined. In the evenings and in cloudy weather it shuts up, and nothing is then seen but the tips of the flowers. It does very well in any fine sandy, peaty, or other open earth, is a hardy perennial on dry soils. It is easy to raise from seed, either in the open air in fine soil, or in pots. As it does not like transplantation, except when done very carefully, the best way for those who wish to use it for very neat and bright beds in the summer flower-garden is to sow a few grains in each small pot in autumn, keep them in dry sunny pits or frames during the winter, and then turn the plants out without much disturbance into the beds in the end of April or beginning of May, and it may also be treated as an annual, sown in frames very early in spring, associated with diminutive plants.

CALLA PALUSTRIS (*Bog Arum*).—A small trailing Arum, with pretty white spathes, hardy, and, though often grown in water, likes a moist bog better. In a marsh or muddy place, shaded or otherwise, it thrives, and in a bog carpeted with the dark green leaves of this plant the effect is good, as its white flowers crop up here and there along each running shoot, just raised above the leaves. Those having natural bogs would find it an interesting plant to introduce, while for moist spongy spots near the rock-garden, or by the side of a rill, it is worth a place. N. Europe, and also abundant in cold marshes in N. America, flowering in summer, and increasing rapidly by its running stems.

CALLUNA. (*See* **ERICA**).

CALOPHACA WOLGARICA.—A prostrate half-shrubby plant of the pea flower order, with deep yellow flowers in racemes in summer, and small pinnate greyish leaves. A pretty rock shrub, easily grown and best from seed. Avoid grafted plants, and plant in full sun.

CALTHA (*Marsh Marigold*).—Showy dwarf perennials of essential use in the marsh-garden. The native kind is so frequent in a wild state that there is rarely need to give it a place, except on the margin of water. Its double varieties, however, are worth a place in a moist rich border, or, like the single form, by the water-side. There is a double variety of the smaller creeping *C. radicans*, about half the size of the common plant. In addition to the common species, *C. palustris*, and the rarer variety, *C. radicans*, there are double-flowered forms, *C. monstrosa*, bearing golden rosettes, and *C. minor fl.-pl.*, a small kind. There are also *C. leptosepala*, a Californian kind, and *C. purpurascens*, distinct and handsome, about 1 foot high, with purplish stems, and bright-orange flowers, the outside of the petals flushed with a purplish tinge.

The various forms of the Marsh Marigold are handsome in colour, and in groups or bold masses are effective; and they are easily grown, and increase freely.

CAMPANULA (*Hairbell*).—A large family of northern pasture, mountain and alpine plants, many of these last among the best for the rock-garden, dwarf, graceful in form, lovely in colour, and for the most part easy to grow and increase. The tall perennials are too coarse for the rock-garden, and neither these nor the medium-sized kinds require its aid, growing, as they do, freely in any soil; but the dwarf mountain kinds are essential to its beauty—all the more so, as they rarely demand any special position, but may be grown in chinks or between steps on any aspect. Where there is no good rock-garden they may be grown well and with good effect behind and about stone or flint edgings. Among these plants garden-hybrids are not now uncommon, but it is better on the rock-garden to keep to the wild forms. Some hybrids, however, like *G. F. Wilson*, are pretty. Ordinary garden-soils suit well even the mountain kinds, with a little change in the case of the kinds inhabiting high moraines, and a rather peaty soil for the graceful *C. pulla*. In congenial soils they bear seed freely, and often sow themselves. In a numerous group like this, where beauty of effect is sought, we arrive at it more surely by growing well and placing rightly the more beautiful kinds, than by collecting every kind we can.

The following Hairbells are mostly of dwarf stature, natives of rocky or mountain ground, excluding the more vigorous herbaceous kinds as unfit for the rock-garden and delicate or doubtful species. They will fairly represent in the rock-garden and on walls the beauty of a fine family of northern and high mountain plants—many of which are not in cultivation:—

Campanula Allioni (*Allioni's Hairbell*).—A dwarf kind, the flowers very large for a plant growing seldom more than 3 inches or 4 inches in height, purplish-blue (rarely white), almost erect on a slender stalk. It is an excellent rock-plant, and though requiring plenty of moisture, it should have a well-drained position, and is therefore best grown in a narrow crevice filled with sandy loam with small stones and grit. Flowering summer. Piedmont. Syn., *C. alpestris*.

Campanula alpina (*Alpine Hairbell*).—This is covered with stiff down, which gives it a grey hue, with longish leaves and erect, not spreading, habit, like the *Garganica* group, and with flowers of a fine dark blue, scattered in a pyramidal manner along the stems. It is a native of the Carpathians, hardier than the dwarf Italian Campanulas, and valuable for the margins of borders as well as for the rock-garden. In cultivation it grows from 5 inches to 10 inches high, and may be readily increased by division or seeds.

C. barbata (*Bearded Hairbell*).—One of the blue Hairbells that abound in the meadows of Alpine France, Switzerland, and N. Italy. It is readily known by the long beard at the mouth of its pretty pale sky-blue flowers, nearly 1¼ inches long, nodding from the stems, which usually bear two to five flowers, and rise from rough, shaggy leaves. In high ground in its native country, it grows no more than from 4 inches to 10 inches high, but nearly twice as high in the valleys in Piedmont. There is a white-flowered form, both thriving freely in loam.

C. cæspitosa (*Tufted Hairbell*).—One of the most beautiful plants in the alpine flora, abundant over the high ranges in the central parts of Europe, and thriving in all parts of the British Isles. It grows only a few inches high, and looks the same fresh, purely-tinted, ever-spreading and bravely-flowering little plant in a British garden as it is when seen mantling round the stones and crevices of rocks on the mountains. There is a white variety as pretty as the blue, and both are admirable for the rock-garden or mixed border. It is easily increased by division and also by seed, but as a few tufts may be divided into small pieces, and quickly form a stock large enough for any garden, it is scarcely worth while raising it from seed. As it occurs so freely by the roadsides along the roadways into Italy, it was one of the first alpine plants to be grown in

Britain, and thriving so well in our climate, it is the one so often seen. Syn., *C. pumila.*

Campanula Carpatica (*Carpathian Hairbell*).—This, while bearing cup-shaped flowers as large as those of the Peach-leaved Hairbell, has the dwarf neat habit of the alpine kinds. It is a native of the Carpathian Mountains and other parts of the same region, and fortunately easy of culture, growing from 6 inches to over a foot in height, according to the depth, and richness of the soil. It begins to flower in early summer, and often continues in bloom for a long time, especially if the plants are young, and the seed-vessels be picked off. There is a white variety, *C. c. alba;* a pale blue one, *pallida;* and a white and blue kind,

five lobes. It should have a gritty, stony and moist soil. Alps of Central Europe.

Campanula excisa.—An interesting species, usually found at high altitudes; the flowers pale blue and deeply cut. At the base between each two lobes this incision takes the shape of a round hole, and it is this which suggested the name. The whole plant is not more than 4 inches or 5 inches in height, and likes a position not fully exposed to the sun, but where the air would be cool and moist.

C. fragilis (*Brittle Hairbell*).—In handling this the stems break off as if made of ice. It is a pretty Hairbell, the root-leaves on long stalks heart-shaped in outline, and bluntly lobed, those of the stem more lance-shaped, the rather large blue open flowers somewhat bell-shaped,

Campanula Garganica. (Engraved from a photograph by Mrs Stafford, Waldeck, Ridgeway, Enfield.)

bicolor—names for the most noticeable variations raised from seed.

C. Cenisia (*Mont Cenis Hairbell*).—An alpine growing at very high elevations. I have found it abundantly among the fine *Saxifraga biflora,* at the sides of glaciers on the high Alps, scarcely ever making much show above the ground, but, like the Gooseberry-bush in Australia, very vigorous below, sending a great number of runners under the soil. Here and there they send up a compact rosette of light green leaves. The flowers are solitary blue, somewhat funnel-shaped, but open, and cut nearly to the base into

borne on half prostrate stems, the plant rarely reaching 6 inches in height, smooth and rather fleshy. A native of the South of Italy. Invaluable for the rock-garden in well-drained chinks into which it can root deeply without being too wet in winter; on light soils not requiring this care. *C. fragilis hirsuta* is a form covered with stiff down.

C. Garganica (*Gargàno Hairbell*).—A showy kind, with somewhat of the habit of the Carpathian Hairbell, but smaller; the leaves that spring from the root are kidney-shaped, those from the stem heart-shaped, all toothed and downy. In summer the plant becomes a prostrate mass of bluish-purple starry flowers with white centres, from 3 inches to 6 inches high; it is seen best in interstices on vertical parts of the rock-garden, in warm and well-drained spots. The better and deeper the soil the finer and more prolonged the bloom will be. It is a native of Italy, flowers in summer, and is easily increased by cuttings, divisions, or seeds.

Campanula hederacea (*Ivy Hairbell*).—A fragile, creeping thing, with almost thread-like branches bearing small, delicate leaves, its flowers of a faint bluish-purple, less than half an inch long, and drooping in the bud. It is a native of Britain, creeping over bare spots by the sides of rills and on moist banks, and wherever there is a moist boggy spot near the rock-garden, or by the side of a streamlet, or in an artificial bog, it will be found worthy of a place. It occurs chiefly in Ireland and Western England; less in the East. Division.

C. isophylla (*Ligurian Hairbell*).—A free flowering Italian species, the leaves roundish or heart-shaped, deeply toothed, and nearly all of about the same size, the flowers of a pale but very bright blue, with whitish centre, and protruding styles. It is a charming ornament for the rock-garden, and should be placed in sunny positions in well-drained, rather dry fissures in sandy loam, and then it will repay the cultivator by a brilliant bloom.

C. macrorrhiza (*Ligurian Hairbell*).—"This is one of the most beautiful of the southern plants, and one of the most free-flowering of the Campanulas. The rootstock is thick and woody; it throws out a large number of drooping branches; flowers very numerous, of a fine blue, two to eight in a spreading cluster. I can never forget the impression I received on first seeing it in flower in the walls of the small town La Turbie above Monaco. The little flowers were in myriads, brightening up the dismal streets of this decaying place, and giving it life and colour. It must have a vertical position in full sun, and in a fissure of wall or rock, calcareous if possible. It is increased by cuttings, divisions, or seed."—H. Correvon (in *Garden*).

Campanula mollis.—Though the native home of this Bellflower is on the shores of the Mediterranean, it has nevertheless proved hardy in this country. The flowers are of a dark purplish-blue, borne freely during May and June, the plant from 6 inches to 8 inches high; forming a spreading carpet of glossy leaves even at midwinter. It is a very useful kind of free dwarf habit. S. Europe.

C. muralis (*Wall Hairbell*).—This, a native of Dalmatia, is a pretty and useful plant as a dense carpet, from 6 inches to 8 inches high, with a bell-shaped corolla about ½ inch in length, flowering throughout the summer. The radical leaves are reniform, smooth, dark green, and more than 1 inch in diameter; the cauline leaves smaller, and with coarsely serrated edges. There is also a more robust variety named *C. m. major*. Syn., *C. Portenschlagiana*.

C. pulla (*Violet Hairbell*).—A distinct plant, the stems only bearing one flower, of a deep bluish-violet, the habit very graceful. On the rock-garden it should be placed on a level spot, free from other Hairbells or rampant plants of any kind, and in sandy peat. It spreads underground, and sends up shoots in a scattered manner. A native of the Tyrol and of other mountains in Central and Southern Europe, it is increased by division or by seeds, but in heavy soil is apt to disappear.

C. Raineri (*Rainer's Hairbell*).—One of the most beautiful, quite dwarf in habit, the distinct stems not more than 3 inches long (though it is said to reach twice that height), and quite sturdy, branched, each little branch bearing a large somewhat funnel-shaped erect flower of a fine dark blue. A native of high mountains in the North of Italy, it should be grown in gritty or sandy loam, with a few pieces of broken stone half-sunk in the soil near the plant.

C. rotundifolia (*Common Hairbell*).—There is no fairer flower on the mountains than this, so often adorning roadside and hedge bank. It is well worthy of a place in the rougher part of the rock-garden. There is also a white form. *C. r. Hosti* is a variety distinguished by larger flowers of a deeper blue and by stronger wiry flower-stems, but, according to Mr Correvon, writing in the *Garden*, it is a distinct species and a native of the Eastern Alps. *C. r. soldanellæflora* is another distinct form with semi-double blue flowers split into many narrow divisions.

Campanula turbinata (*Vase Hairbell*). —A neat sturdy showy kind, the leaves rigid, of a greyish-green, toothed and pointed, forming stiff tufts from 2 inches to 3 inches high, and an inch or so above them rise the cup-shaped flowers, of a deep

Campanula turbinata.

purple, and each nearly 2 inches across. It comes from the mountains of Transylvania, is hardy in our islands, not fastidious as to soil, and is one of the best plants for the rock-garden, on which, in deep light soil, the flower stems sometimes reach a height of 6 inches or 8 inches.

C. Waldsteiniana (*Waldstein's Hairbell*).—A pretty little kind, 4 inches to 6 inches high, the flowers in racemes of from five to nine blossoms each, of a pale purplish-blue colour, with lobes spread out almost flat, so as to give the flowers quite a star-like appearance. Forms carpets for the rock-garden. Croatia.

Campanula Zoysi.—This plant grows scarcely more than 3 inches or 4 inches in height and bears pale blue flowers with a rather long tubular corolla. It is not common, perishing in our changeable winters. Alps of Austria.

CARDAMINE (*Ladies' Smock*).— For rock-gardens, there are not many of these attractive, but several deserve cultivation in the marsh garden. The double forms of our Wild Ladies' Smock are pretty in such places, and among other kinds worth growing are *C. latifolia, C. trifolia,* and *C. asarifolia,* all of the simplest culture and easy increase by division.

CASSIOPE (*Arctic Heath*).—Beautiful dwarf alpine and Arctic shrublets; of great interest, but not easy to grow in lowland gardens: they are best in moist sandy peat, and in cool but not shady spots among very dwarf plants. Syn., *Andromeda*.

Cassiope fastigiata (*Himalayan C.*).—A tiny shrub, with the leaves overlapped along the stems, so as to make them square like those of *C. tetragona*, but distinguished from that plant by the leaves having a white, thin, chaffy margin, and a deep and broad keel. The flowers, of a waxy white, produced at the top of each little branchlet, are turned down bell-fashion; the reddish-brown calyx spreads half-way down the waxy flowers. This, one of the most beautiful Himalayan plants, is, happily, not so difficult to grow, though it requires care. It has been successfully grown by Dr Moore, in the Botanic Gardens at Dublin, and should have a sandy, moist peat soil. It thrives best in moist and elevated districts; but, safely planted in deep, moist soil, and guarded against drought during the warm season, it may be grown in cool spots never shaded.

C. hypnoides (*Mossy C.*).—A minute spreading, moss-like shrub, 1 to 4 inches high, with wiry branches, densely clothed in all their parts with minute bright green leaves, and bearing small, waxy, white flowers, borne singly and drooping on slender reddish stems. It is one of the most beautiful of all alpine plants, and one of the most difficult to grow, being very rarely seen in a healthy state even in the choicest collections. Drought is fatal to it. It is a native both of Europe and America, either far north into the coldest regions of these countries, or on the summits of high mountains. It is such a delicate and fragile evergreen shrub, that any impurity in the air is sure to injure it. In elevated and moist parts of these islands, it will succeed in very sandy or gritty moist but well-drained peat, freely exposed to the sun and air, and placed quite apart from more vigorous plants on rockwork. The chief difficulty would seem to be the procuring of healthy plants to begin with; once obtained, it would be desirable to carefully peg down the slender main branches, and to place a few stones round the neck of the plant, so as to prevent evaporation.

Cassiope tetragona (*Square-stemmed C.*).—One of the prettiest of the diminutive shrubs introduced to cultivation, seldom growing more than 8 inches high. When in health, the deep green branches grow so densely that they form compact tufts. The flowers are produced singly, but rather freely; of a waxy white, five-cleft, contracted near the mouth, and drooping. It is not likely to be confounded with any other plant except the much rarer *C. fastigiata*, from which it may be distinguished by the absence of the thin chaffy margin of the leaf. It is a native of Northern Europe and America, quite hardy, requiring a moist peat or very fine sandy peat for its thriving. I have not elsewhere seen it so healthy as in the nurseries near Edinburgh; loves abundance of moisture in summer, and is easily increased by division.

CERASTIUM (*Mouse-Ear Chickweed*).—Tufted rock plants of the pink order, rather numerous, but so far as known in gardens, not among the best rock plants.

Cerastium alpinum (*Shaggy C.*).—A British plant, found on Scotch mountains, and also more sparsely on those of England and Wales. Dwarf, tufted, and prostrate, spreading freely, but seldom rising more than a couple of inches high, with leaves broader than those of the common weedy species, and densely clothed with a dewy-looking down, giving the plant a shaggy appearance, and with rather large white flowers in early summer. It is not, like the common kinds, a plant fitted for forming edgings. Messrs Backhouse say that it flourishes best under ledges that prevent the rain and snow falling on the foliage, but I have found it stand all sorts of weather, and winters in the open border in London. Division, by cuttings, or seeds.

C. Biebersteinii (*Bieberstein's Mouse-Ear C.*).—A silvery species, useful for the same purposes, and cultivated with the same facility, as *C. tomentosum*. It was once expected that it would surpass in utility the common kind, but this it has failed to do. A very good plant for borders or rough rock or root work. A native of the higher mountains of Tauria flowering with us in early summer.

C. grandiflorum (*Large flowered C.*).—This is readily known from either *C. tomentosum* or *C. Biebersteinii* by having narrower and more acute leaves, and being less hoary, and it usually grows somewhat larger than either of the two very silvery kinds, rapidly forming strong tufts, and bearing pure white flowers. A fine plant for the front margin of the mixed border, or for the rougher parts of the rock-garden, but only in association with many fast-growing plants, as it spreads so quickly that it would overrun delicate and tiny plants if placed near them. Like the other cultivated kinds, it is readily propagated by division or by cuttings inserted in the rudest way in the open ground, and is a native of Hungary and neighbouring countries, on dry hills and mountains, flowering with us in early summer.

C. tomentosum (*Common Mouse-Ear Chickweed*).—This was once used in almost every garden for forming silvery edgings to flower-beds, its hardiness, power of bearing clipping, and facility of increase, making it worthy of its work. It is also useful as a border-plant, and for rough rockwork South of Europe, flowering freely with us in early summer.

The preceding include all the kinds that are worth growing, except in botanical collections. The other kinds enumerated in Catalogues are:—*C. incanum, lanuginosum, ovalifolium, ovatum, tenuifolium, Wildenovii*, and *trigynum*.

CHEIRANTHUS (*Wallflower*).— Perennial and biennial plants of pleasant association with our subject, one being the best of wall-gardeners. They are mostly of easy culture and increase.

Cheiranthus cheiri (*Wallflower*).—In a book advocating the culture of alpine plants on walls, we must not forget the old plant that has so long dwelt on walls and ruins, loving a wall better than a garden ; while it grows rank in garden soil, it forms a dwarf enduring bush on an old wall, and grows even on walls that are new, planted in mortar. There is no variety of the Wallflower yet seen that is not worthy of cultivation ; but the choice old double kinds—the double yellow, double purple, double dark orange, are plants worthy of a place beside the finest rock-shrubs. These are the varieties most worthy of a place on dry stony banks near the rock-garden, and also on walls, on which the common kind is likely to find a home for itself. To scatter seeds on any wall we wish to adorn with this plant is enough, using seed of the common dark or yellow Wallflower, or that of the wild plant.

Among other kinds are **C. Marshalli** (*Marshall's Wallflower*).—This, which is said to be a hybrid between *Cheiranthus Ochroleucus* and *Erysimum Peroffskinum*, is a half shrubby plant, 1 to 1½ foot high, with erect angular branches. The flowers appear in spring or early summer, are nearly ¾ of an inch across, of a deep clear orange at first, afterwards becoming somewhat paler. The fine orange-colour of the flowers of this plant makes it a pretty one for the rock-garden, in well-drained soil. It is increased by cuttings, and a young stock should be kept up, as it is not perennial, and is apt to perish in winter.

Cheiranthus mutabilis (*Madeira Wallflower*).—A low bushy plant, distinct, and of much value as a plant for dry walls. The flowers are a soft orange colour, the buds forming a central boss of a dark red. I find it hardy and of easy culture, but it may be delicate in the north. Easily increased by division.

C. alpinus (*Alpine Wallflower*).—This handsome plant forms neat, rich green tufts, 6 to 12 inches high ; in spring covered with sulphur-coloured flowers. The rock-garden is the best home for it ; it does very well on level ground, but is apt to get naked about the base, and may perish on heavy soils in severe winters ; it does best when often divided, and the conditions that best suit it on old walls, or even new walls made against banks, as shown in the first part of this book. Alps and Pyrenees, flowering in spring and early summer. There are several varieties. Syn., *Erysimum Ochroleucum*.

CHIMAPHILA MACULATA (*Spotted Wintergreen*). — A dwarf wood plant of North America, having leathery, shining leaves, the upper surface of which is variegated with white, and bearing whitish flowers— one to five—on rather long stems. The plant attains a height of 3 to 6 inches, and is a very pretty one for a half shady and mossy, but not wet, place in the rock-garden, associating well with such plants as the *Pyrola*, and succeeding best in very sandy decomposed leaf-soil.

C. umbellata, with glossy unspotted leaves, and somewhat larger reddish flowers, is suited for like positions. Both are rare in cultivation, and very seldom seen well grown. They flower in summer, and are increased by careful division.

CHIOGENES HISPIDULA (*Creeping Snowberry*).—A slender creeping evergreen plant, bearing small white flowers, followed by white globular berries. It is like a small cranberry, a native of cold boggy places and wet woods in Newfoundland and Canada to British Columbia and southward on the mountains. It is a plant for the bog bed or a moist corner, with such plants as the *Linnæa*.

CISTUS (*Rock Rose*).—Small shrubs and bushes of distinct beauty; mostly from the sun-burnt hills of Southern Europe, and for that reason none the less welcome to rock-gardeners. Many people complain that the great heat of recent years has affected the culture of alpine plants, especially on dry soils in the south. These Rock Roses enjoy the hot sands, and rocks, and arid places, which are death to the true alpine of the icy fields of the north and of the alpine slopes. The only drawback to their successful culture is our climate, in which certain kinds are tender, and may perish in hard winters; but several are hardy, especially in such positions as we may give them in the rock-garden and on the tops of dry walls or on poor banks. In such soils as the poor sands of Surrey, they are at home. Among other rock plants we have to pick and choose, rejecting many from the rock-garden point of view; but here all are pretty; the larger kinds taking their place among shrubs, and the smaller on the rocks. Some are evergreen shrubs, and have a spicy fragrance of the warm south, grateful to the northerner. I feel sure that in certain districts one might have a pretty rock-garden of the Rock Roses and Sun Roses, and a few other sun-loving shrubs, like Rosemary and the Heaths that love the sun.

Many of the species vary in colour, and not a few appear to hybridise freely. In spite of the fugacious character of the flowers (they do not last more than one day), their bright colours, and the profusion in which a succession is kept up for a considerable time, place them amongst the most welcome of garden shrubs during the summer months.

Cistus albidus (*White Rock Rose*).—The name of this is derived not from the colour of the flowers, for these are a fine rose, but to the whitish tomentum which clothes the leaves and young shoots. It forms a compact bush 2 to 4 feet high; the old branches are covered with a brownish bark. The rose-coloured flowers are nearly 2 inches across, and the style is longer than the tuft of yellow stamens. Southern Europe.

C. Bourgæanus is a native of the Pine woods of Southern Spain and Portugal, where it flowers in the month of April, grows a foot in height, and has somewhat prostrate branches, covered with Rosemary-like dark-green leaves. The white flowers are about an inch across, and it is a charming plant for a sunny spot in the rock-garden.

C. Clusii (*Clusius's Rock Rose*).—In habit this is more erect than the last-named, but the flowers are the same colour and size, as are also the leaves. As a rock plant, or grown for cool house decoration, it is charming. It is met with under the name of *C. rosmarinifolius*.

C. crispus.—This forms a compact bush, 1 to 2 feet high, with tortuous branches, the rose-coloured flowers nearly 1½ inches across. There are some hybrids between this species and *C. albidus* which are nearer the seed-bearing parent than they are to *C. albidus*.

C. florentinus (*Florence Rock Rose*).—A pretty bush, flowering freely and of easy culture. I find it hardy and enduring on soils where other kinds perish. It is evergreen and charming on the tops of high walls and banks; and for the rock-garden one could not desire a prettier or more easily grown plant. It is about 1 foot to 18 inches high, bearing myriads of white flowers.

Cistus formosus (*Beautiful Rock Rose*).—Much-branched, bushy shrub, with leaves greenish when old, but whitish when young, and large bright yellow flowers with a deep purplish-brown blotch near the base of each petal. The plant thrives well in any rich, dry soil, but is apt to succumb in severe English winters. It is, however, such a beautiful plant, that it is well worth the trouble of putting in a pot of cuttings each autumn in a cold frame, planting these out in the open the following spring. If raised from seeds, young shoots and flower-stalks are hairy, as are the leaves on both surfaces; the flowers whitish, smaller than those of *C. glaucous*, and the style is shorter than the stamens. South-Western Europe.

Cistus ladaniferus (*Gum Cistus*).—This is one of the most beautiful of all the *Cistuses*; the leaves, smooth and glossy above, clothed with a dense white wool beneath. The very large flowers are white, in the more handsome forms with a large dark vinous-red blotch towards the base of each petal; in others without

Cistus formosus.

some variation in the colour results. I find it does well on the top of dry walls.

C. glaucus.—A much-branched bush, 1 to 2 feet in height, with reddish-brown bark; the upper surface of the leaves is dull green, glossy, and glabrous, the lower strongly veined and clothed with a hoary down. The flowers are large, white with a yellow blotch at the base of each petal, and the very short style is much exceeded by the stamens. Southern France.

C. hirsutus (*Hairy Rock Rose*), is a shrub from 1 to 3 feet in height; the blotch. It also varies in the size of the leaves, the extreme forms having narrow, almost linear leaves.

C. laurifolius (*Bush Rock Rose*).—This is the hardiest Rock Rose; in some southern shrubberies large plants exist, which have withstood many winters. The flowers are less than those of *C. ladaniferus*, are white with a small citron-yellow blotch at the base of each petal. It requires no protection, and may be raised from seeds, which ripen in abundance, and also by cuttings, which, however, do not strike so freely as in some of the other kinds. This attains a height of about 6 feet; it is a native of South-Western Europe.

Cistus Monspeliensis (*Montpelier Rock Rose*).—A species widely distributed throughout the Mediterranean region; is very variable in size of its leaves and also in stature of plant; in some spots it hardly grows more than 6 inches in height; in others it attains a height of about 6 feet. The flowers are white, about an inch in diameter, each petal bearing a yellow blotch at the base.

C. populifolius (*Poplar-leaved Rock Rose*) is a robust kind, with large rugose, stalked, Poplar-like leaves and medium-sized white flowers, tinged with yellow at the base of the petals. Varieties of *C. salvifolius* are often misnamed *C. populifolius* in Nurseries and gardens. Amongst the numerous garden forms of this species may be mentioned *C. narbonnensis*, with shorter flower-stalks, smaller leaves—altogether a smaller plant than the type—and *C. latifolius*, another with broader leaves. Southern France, in Spain, and Portugal. It is an erect branched shrub, 3 or 4 feet high.

C. salvifolius (*Sage-leaved Rock Rose*).—This is a very variable kind, and of slender habit, with Sage-like leaves and long-stalked, white, yellow-blotched flowers. In a wild state it is found all along the Mediterranean, and a number of slightly varying forms have received distinctive names, but do not appear to have been introduced to gardens.

C. vaginatus is the largest of the red-flowered section; robust, with large-stalked, hairy leaves, and large, deep rose-coloured yellow-centred flowers. The stamens are more numerous in this than in, perhaps, any other *Cistus*, and form a dense, brush-like tuft, overtopped by the long style. It is a native of the Canary Islands. For many years a fine plant flowered freely against the wall of the herbaceous ground at Kew, but the severe winters of several years ago proved too much for it.

C. villosus, a widely-distributed Mediterranean kind, is a very variable plant, an erect bush with firm-textured leaves. The flowers of all the forms are rose-coloured, with long styles. *C. undulatus* is a variety with wavy-margined leaves, *C. incanus* represents what may be regarded as the common typical form. *C. creticus* is another with deeper rose-red flowers than those already mentioned.

CLAYTONIA VIRGINICA (*Spring Beauty*).—A pretty American plant of the Purslane family, sending up in March and April simple stems bearing a loose raceme of rose-coloured flowers marked with deeper veins, which, unlike the flowers of most of the species of this family, remain open for more than one day. Suited for the rock-garden or borders, in loam and leaf-mould. *C. sibirica* and *C. alsinoides*, although only biennials, or perhaps little better than annuals, sow themselves freely in crevices, and so often find a place among alpine plants.

CLEMATIS.—Though the showy hybrids of these climbing shrubs are not the best fitted for the rock-garden, I know nothing more graceful about rocks than the Alpine Clematis (*C. alpina*), and also the common *C. Viticella*, and any of the smaller kinds. The winter-flowering Clematises, which are so pretty along the mild coast districts in Britain in winter and early spring are excellent for scrambling over rocks or banks. These plants, which should always be raised from seed and layers, are more enduring than the hybrid kinds, which are usually grafted, and perish very quickly.

COLCHICUM (*Meadow Saffron*).—Hardy bulbs of the meadows and mountains of Europe and the East. They have not the fine colour of the *Crocus*, but some of the kinds introduced of recent years are very interesting for the rock-garden. Among these more than perhaps any other plants for our purpose, we should seek out the more beautiful among the many-named, and, once found, make effective use of them. The individual flowers do not, as a rule, last long, but, as they are produced in succession, there is a long season of bloom. The flowers are often destroyed through being grown in bare beds, where the splashing of the blooms during heavy rainfalls impairs their beauty. A good way is to plant them in grass, where the soil is well drained and rich. In the rock-garden, too, among dwarf *Sedums* and similar subjects, *Colchicums* thrive, and make a pretty show in autumn, when rock-gardens are often flowerless. They look better in grassy places or in the wild garden than in any formal bed or border. Their naked flowers want the relief and grace of grass and foliage. The plants have a rather wide range, some species extending to the Himalayas; others are found in North Africa; but the majority are natives of Central and Southern Europe. Though there are so many names to be found in Catalogues, the distinct kinds are few, and there is such a striking similarity among these, that they may be conveniently classed in groups. The best known is—

Colchicum autumnale, commonly called the autumn Crocus. The flowers appear before the leaves, rosy-purple, in clusters of about six, 2 or 3 inches above the surface, flowering from September to November. There are several varieties, the chief being the double purple, white and striped; *roseum*, rose-lilac; *striatum*, rose-lilac, striped with white; *pallidum*, pale rose; *album*, pure white; and *atropurpureum*, deep purple. Similar to *C. autumnale* are *C. arenarium, byzantinum, montanum, crociflorum, lætum, lusitanicum, neapolitanum, alpinum, hymetticum;* all, like *autumnale*, are natives of Europe, and, from a garden standpoint, are very similar in effect.

Colchicum Parkinsoni.—A distinct and beautiful plant, readily distinguished from any of the foregoing by the peculiar chequered markings of its violet-purple flowers. It produces its flowers in autumn, and its leaves in spring. Other allied kinds are *Bivonæ, variegatum, Agrippinum, chionense, tessellatum,* all of which have the flowers chequered with dark purple on a white ground.

C. speciosum, from the Caucasus, is large and beautiful, and valuable for the garden in autumn, when its large rosy-purple flowers appear nearly 1 foot above the ground. Like the rest of the Meadow Saffrons, *C. speciosum* is as well suited for the rock-garden as the border, thriving in any soil; but to have it in perfection, choose a situation exposed to the sun, with sandy soil. There are several varieties of it.

C. Bornmulleri.—According to M. S. Arnott, writing in the *Garden,* "this is one of the most handsome of all the *Colchicums,* which is admired by every one who sees it here. It is larger than *speciosum,* and comes pale-coloured when in bud, passing off purple, with a broad white zone in the interior."

C. variegatum (*Chequered Meadow Saffron*).—This is one of the prettiest kinds, and is often grown under the name of, and mixed with, the common meadow Saffron, but is distinguished by its rosy flowers being regularly mottled over with purple spots, and its leaves undulated. Like the common species, it flowers abundantly in autumn, grows well in ordinary soil, and may be associated with the autumn-flowering Crocuses on the rock-garden.

C. Sibthorpii (*Sibthorp's Meadow Saffron*).—Of rather recent introduction to gardens, this is thought by lovers of those plants to be the finest of all. It is an inhabitant of the mountains of Greece, ascending to a height of 5000 feet. Its flowers are distinctly tessellated, the segments of the perianth broad, and the leaves not undulated. It is a good grower in free sandy or gritty loam.

CONANDRON RAMONDIOIDES.
—A small Japanese plant, allied to *Ramondia*, having thick wrinkled leaves, in flat tufts, from which arise erect flower-stems some 6 inches high, bearing lilac-purple and white blossoms. Though said to be hardy, it is better in a sheltered spot in the rock-garden. Plants placed between blocks of stone thrive if there is a good depth of soil in the chink, and the soil is moist. Japan.

CONVALLARIA MAJALIS (*Lily-of-the-Valley*).—So long have we been accustomed to this in our gardens that we can scarcely think of it as an alpine plant. But, as the traveller ascends the flanks of many a great alp, he sees it blooming low among the Hazels and other mountain shrubs; and it grows through Europe and Russian Asia, from the Mediterranean to the Arctic Circle. A few tufts of it taken from the matted and often exhausted beds in which it is usually grown in the kitchen-garden to half shady spots near wood walks, and among low shrubs on the fringes of the rock-garden or hardy fernery, would be quite at home. It might also be planted in tufts among shrubs, and in any of these positions its beauty will be more appreciated than when it is seen grown as prosaically as kitchen Spearmint. There are several good forms, a variety with double flowers, one with single rose flowers, one with double rose flowers, one with the leaves margined with a silvery white, and one richly striped with yellow. Although growing in almost any soil, it flowers best in a free sandy loam, and thrives in poor healthy places better than in rich heavy ground.

CONVOLVULUS (*Bindweed*).—Graceful climbing and creeping plants, some of the more northern kinds of a refined and elegant habit, which makes them welcome on the rock-garden, and having distinct value for draping stones. It is well to keep out vigorous growing kinds which may even be too vigorous for a garden, let alone for the choicer morsel of our earth we call our rock-garden. The kinds of best use for our present purpose are the North African Blue Bindweed, a charming rock-garden plant, and I find it quite hardy even on cool soil. It grows abundantly in walls and rocky banks, and even if the plants perish in hard winters, is so easily raised from seed or cuttings. The silvery *C. Soldanella* of Southern Europe is also worthy of a place on the rocks, and also *Althæoides* and the Sea Bindweed.

Convolvulus althæoides (*Riviera Bindweed*) is one of the commonest plants around the basin of the Mediterranean. It is chiefly found on dry banks and among the Olive terraces, and flowers all through April and May. Although a very variable species, both in the leaves and flowers, the form which grows freely round Mentone seems to be the one in general cultivation. This species and its various forms stand our English climate very well. Being a non-climbing sort, it is at home on the rock-garden, where its large, purplish flowers are pretty. Seed or division of root.

C. cantabricus.—A pink-flowered species from Southern Europe, growing a foot or so high, and producing its blossoms in clusters of two or three. The shaggy or dwarf nature of the peduncles, and the distinctly narrow sepals, are distinguishing features of this kind.

C. Cneorum (*Silvery Bindweed*).—A distinct sub-shrubby kind, having pink blossoms and silvery leaves, and forming a capitate cluster. It is a beautiful plant for a warm position against a rock. In the north it is probably tender, but not so in the southern counties. S. Europe.

Convolvulus lineatus (*Dwarf Silvery Bindweed*).—This is quite a pigmy, the whole plant often showing nothing but a tuft of small silky, rather narrow, and pointed leaves above the ground. Among these appear in summer delicate flesh-coloured flowers more than an inch across, and in full perfection at less than 3 inches high, though in warmer soils and districts than those on which I have seen the plant, it sometimes grows an inch or two higher. Few plants are better for embellishing some arid part of the rock-garden near, and somewhat under the eye as its beauty is not of a showy order. Mediterranean region. Better increased by dividing the root.

C. mauritanicus (*Blue Rock Bindweed*).—A beautiful plant, without the rampant growth of many of its race, but withal throwing up graceful shoots, which bear numbers of clear, light-blue flowers. It is quite distinct from any other plant, and, happily, is hardy in sunny chinks. It is seen to the best advantage in a somewhat raised position, so that its free-flowering shoots may fall freely down, though it may also be used with good effect on the level ground in the flower-garden, or as a vase plant. Mountains of North Africa; readily increased by cuttings and by seeds.

C. scammonia (*Scammony*).—A twining kind of slender growth, and bearing in summer creamy-white flowers. Although doing well in any position, it seems to want plenty of sun, and thrives best in a light deep sandy soil, as the large roots go a long way down. Syria.

C. soldanella (*Shore Bindweed*).—This is recognised by its leathery, roundish leaves, and by its stems being short, heavy, and without the twining habit so common in the family. The flowers are large, of a light pink colour; thrives and flowers freely in ordinary soil far away from the seaside, and therefore the plant is worthy of a place among the trailers of the rock-garden. A native of maritime sands, in many parts of the world; not uncommon on our own coasts, and flowering in summer. Where difficult to establish, plenty of coarse river sand might be mixed in the soil.

Convolvulus tenuissimus.—A pretty climbing species from the Mediterranean region, much in the way of *C. althæoides*, but in the present kind the foliage is much more divided. A marked feature is the way the leaf segments radiate around a common centre, the central leaf being of considerable length and of long linear lance-shaped outline.

The plant known in gardens as *Calystegia pubescens fl.-pl.* is really a Bindweed, and a pretty kind, with double flowers of white and pale rose. In warm or light stony soil this plant grows apace, and in summer for a long time the twining stems are thickly studded with the flowers.

COPTIS TRIFOLIATA (*Gold Thread*).—A little evergreen bog-plant, 3 inches or 4 inches high, with three-leafleted or trifoliate shining leaves. It derives its common name from its long bright yellow roots. It is occasionally grown in botanic gardens. A native of the northern parts of America, Asia, and Europe, flowering in summer; white, and easily grown in moist peat or very moist sandy soil. Division.

CORIS MONSPELIENSIS (*Montpelier C.*).—A rather pretty dwarf, branching plant, about 6 inches high, usually biennial in our gardens. Thrives on dry and sunny spots of the rock-garden, in sandy soil, and among dwarf plants. Seed. South of France.

CORNUS (*Dogwood*).—Hardy and valuable shrubs with, so far as yet grown, few kinds dwarf and compact enough for our purpose.

Cornus canadensis (*Canadian Cornel*).—A very pretty but neglected miniature shrub, of which each little shoot is tipped with white bracts, pointed with a tint of rose. I know nothing prettier than this *Cornus* when well established, and it is not at all difficult to grow, but rarely comes in for a proper situation.

It is lost among coarse herbaceous plants, and totally obscured by ordinary shrubs, and should therefore be planted in the bog-garden, or near the edge of a bed of dwarf Heaths or American plants. Wherever placed, rather damp sandy soil will be found to suit it best. N. America, in damp woods.

Cornus suecica is a native of Northern and Arctic Europe, Asia, and America. In Britain it occurs on high moorlands from Yorkshire northwards, and is a charming little plant, flowering in summer, with conspicuous, rather large white bracts, followed by red fruit. It grows but a few inches high, and has unbranched stems from slender creeping rootstocks. It should be grown in light soil or in peat under the shade of bushes.

CORONILLA (*Crown Vetch*).— Pretty shrubs, herbaceous and alpine plants of the Pea-flower family, one or two shrubs interesting and hardy in the warmer districts, but the smaller kinds hardy and free everywhere, and in any soil.

Coronilla iberica (*Caucasian Crown Vetch*).—A plant with glaucous foliage and decumbent habit, not rising 4 inches from the ground, and producing freely umbels of yellow blossoms. Somewhat similar in appearance, but much larger than our own familiar *Lotus corniculatus*. It flourishes admirably with its woody roots well bedded in the rock-garden, and will cover completely 2 or 3 square feet of rock surface, when so placed. The Caucasus.

C. minima (*Dwarf Crown Vetch*).—A small evergreen herb, prostrate, glaucous green, with many rich yellow flowers, six to twelve in each crown, in April and May. It is a plant of easy culture, and well worthy of a warm spot on the rock-garden, where its tiny shoots may lap over the stones. Deep light soil in sunny fissures will suit it best, and in such places its diffuse little stems will be best seen. Division and seeds. S. Europe.

C. varia (*Rosy C.*).—A handsome and graceful plant, with many rose-coloured flowers, frequent on many of the railway banks in France and Northern Italy. It forms low dense tufts, sheeted with rosy pink, and the most graceful use that could be made of it would be to plant it on some tall bare rock, and allow its vigorous shoots and bright little coronets to flow over and form a curtain. It is also admirable for chalky banks, or for running about among low trailing shrubs. When in good soil, the shoots grow 5 feet long, and therefore it should not be placed near the smaller alpine plants, but rather among the shrubs on banks near. Seeds.

Coronilla montana is from 15 to 18 inches high, and bearing many yellow flowers, is somewhat too large for association with small alpine plants, but, being a showy species, is excellent for the rougher parts of the rock-garden or among its shrubs.

CORTUSA MATTHIOLI (*Alpine Sanicle*).—Somewhat like the tender *Primula mollis*, with large seven- or nine-lobed leaves, the leaf-stalks and the leaves covered with colourless short hairs. A wiry thread of vascular matter runs through the stem leaves, and may be drawn through the blades as well as footstalk of the leaves, without breaking. The flowers, borne on stems about 15 inches high, are pendulous, and of a peculiarly rich and deep purplish crimson, with a white ring at the base of the cup, six to twelve being borne on a stem. It does well in the angle formed by two rocks, where its leaves cannot be torn by the wind. Flowers in early summer, and comes from the Alps. Increased by careful division of the root, or by seed sown soon after being gathered.

CORYDALIS (*Fumitory*). — All these plants are attractive in some way or other, and several kinds are valuable, and as such deserving a place according to their kind. The following are among the more important :—

Corydalis bracteata, a distinct kind, with sulphur-yellow flowers produced in a nearly horizontal manner on the stems, that attain nearly 1 foot high. A distinct feature is the long spur, this frequently exceeding the length of the foot-stalk. More erect than some other kinds, the flowers cluster together at intervals, but by no means in a crowded manner. The leaf growth is not abundant, and the segments of the leaves being cut, render the leafage only more thin-looking. The plant is of quite easy culture, and may be best used around the base of the rock-garden. It is a native of Siberia, and quite hardy.

C. cava is one of the dwarfest of this race, flowering early in the year. The purplish blossoms, however, are not very attractive. A prettier kind is the variety *albiflora*, which is in every respect similar, save the colour of the flowers.

C. Ledebouriana is distinct and pretty, the glaucous leaves being divided into several rather small segments, the main leaves keeping quite close to the soil. The blossoms are of a pinkish hue, and have a dark spot on the upper portion of the sepals. The plant rarely exceeds 6 inches or 8 inches in height, and is best suited for sunny positions in the rock-garden.

C. lutea (*Yellow Fumitory*).—This plant is not so much grown as it deserves, for not only are its graceful masses of delicate pale-green leaves dotted over with yellow flowers, but it grows to perfection on walls. I have seen it in the most unlikely spots on walls in hot as well as in cold countries, and know nothing to equal it for ruins, walls, stony places, and poor bare banks, the tufts often looking as full of flower and vigorous when emerging from some old chink where a drop of rain never falls upon them, as when planted in good soil. It also makes a handsome border-plant, and is well suited for the rougher kind of rock and root work. A naturalised plant in England, and widely spread over Continental Europe. Readily increased by seeds; in any stony position it spreads about with weed-like rapidity.

C. nobilis (*Noble Fumitory*).—A handsome plant, the flower-stems stout and leafy to the top, and bearing a massive head of flowers, composed of many individual blooms in various stages. The open flowers are of a rich yellow, with a small protuberance in the centre of each, of a reddish-chocolate colour; and this, with the yellow and the green rosette when the bloom is young, makes the plant very ornamental. It is easy of culture in borders, but is rather slow of increase, and, where it does not thrive as a border plant, should be planted in light, rich soil on the lower flanks of the rock-garden. It is a native of Siberia. Increased by division, and flowers in early summer.

Corydalis solida (*Bulbous Fumitory*).—A dwarf tuberous-rooted kind, from 4 to 6 or 7 inches in height, with dull purplish flowers. It has a solid bulbous root, is quite hardy, of easy culture in almost any soil, pretty, and is good for rougher portion of the rock-garden, or for naturalising in open spots in woods. It is naturalised in several parts of England, but is not a true native, its home being the warmer parts of Europe; easily increased by division, flowers in April. Syn., *Fumaria solida*.

C. Semenovii.—A rather pretty kind from Turkestan. The flowers, which are rich yellow, cluster together in the upper part of the stem and assume a somewhat pendent position. The spur in this kind is very short. It flowers usually in early spring.

C. thalictrifolia.—A new kind from China that promises to make a very charming addition to rock-garden plants. Barely 1 foot high, tufted, and spreading in habit of growth, it is distinct in various ways from the other species of the genus. The thin, wiry stems each carry two pairs of oppositely-placed leaves on pedicels an inch long, and a terminal leaflet, all being distinctly and rather deeply notched and rounded at the top. The blossoms are yellow, each about an inch long, horizontal or slightly ascending, and produced somewhat after the manner of *C. lutea*. The leaf character is a most distinct feature of this kind. The plant flowers profusely from May to October, and in autumn the foliage assumes a reddish tone.

COTONEASTER)*Rockspray*).—One

of the most interesting and brilliant of the shrubs which adorn the rocks, and every year seems to add to their variety and beauty. They are so hardy and so pretty in habit, in flower, and fruit, that we cannot associate any better shrubs than these with our larger rock-gardens. Some kinds are very small and earth-clinging in growth. They are mostly natives of India, and of the mountains of China, as well as Northern Europe, and one is a native of our own country. In gardens, generally, these plants are often neglected. Their best use is for banks near the rock-garden, and all the dwarf and bushy kinds are worth a place.

Cotoneaster buxifolia (*Box-leaved Rockspray*).—A free-growing bush that at times attains the height of 6 feet, the branches clothed with deep-green box-like leaves; the crimson berries, nestling in profusion among the leaves, are pretty in autumn.

C. horizontalis (*Plumed C.*).—In this the branches are frond-like and almost horizontal, while the small leaves are regularly disposed along the thick sturdy branches. The berries are bright vermilion, and the flowers large and pretty. I find this one of the best of shrubs for rocky banks. China.

C. microphylla (*Wall Rockspray*).—An evergreen clothed with tiny deep-green leaves, in the spring crowded with whitish blossoms, the berries crimson, and remaining on the plants for a long time. There are some well-marked varieties of *C. microphylla*, one of which—*thymifolia*—is smaller in all its parts, while *congesta* is even more of a procumbent habit. *C. microphylla* is useful for stony banks, and its variety, *congesta*, is more at home when draping a large stone than in any other way. Himalayas.

C. rotundifolia is like the preceding, but with thicker branches and rounder leaves, while the berries are of a brighter tint.

COTYLEDON UMBILICUS (*Wall Navelwort*).—A native of Britain and Ireland and many parts of Western Europe, in some districts common on walls. Of little importance for cultivation, except perhaps now and then as a hardy fernery or bog plant.

CROCUS.—Some ordinary kinds of Crocus are very easily grown, and are so free in the common soil of many gardens, that there is no occasion to make rock-garden plants of them. But some wild species are so refined and beautiful in colour, and in many cases so rare, that the rock-garden would be improved by them, and there we could easily give them the kind of soil that suits them best, usually open warm soil, and also get them out of harm's way a little. The autumn kinds, too, are among the most lovely of wild flowers, and in little groups on our rock-gardens they would be most at home, until we got them plentiful. The very late-flowering kinds of delicate colour are best in a sheltered part of the rock-garden. In the case of the pretty autumn Crocuses, their beauty is best seen when the flowers rise from a groundwork of some creeping rock plant. The midwinter blooming species, charming in their own country, will rarely bloom well in our winters. Only the kinds known to be pretty and free under rock-garden conditions are named here.

Crocus biflorus (*Cloth-of-Silver Crocus*).—A very dwarf early and free kind which varies much. In var. *estriatus*, from Florence, the flowers are a uniform pale lavender, orange towards the base. In var. *Weldeni*, from Trieste and Dalmatia, the outer segments are externally flecked with bright purple. In *C. nubigenus*, a small variety from Asia Minor, the outer segments are suffused with brown; *C. pestalozzæ* is an albino of this variety. In *C. Adami*, from the Caucasus, the segments are pale purple, either self-coloured or feathered with dark purple.

Crocus chrysanthus.—A vernal Crocus, flowering from January to March, according to elevation, which varies from a little above the sea-level to a height of 3000 or 4000 feet. The flowers are usually of bright orange, but occasionally bronzed and feathered externally. A white variety is also found in Bithynia and on Mount Olympus above Broussa; this species also varies with pale sulphur-coloured flowers, occasionally suffused with blue towards the ends of the segments, dying out towards the orange throat. There are several varieties of this Crocus.

C. Imperati (*Naples Crocus*).—This is very early flowering, and one of the very best kinds, even in this large family. Excepting *C. vernus* and its varieties, it is one of the most variable species we have in the colour markings and size of its flowers. It is splendid for lawns, useful on the rock-garden as being an early and certain flower, while it will remain in condition without lifting, as long or longer than any other species. *Majus* is a large form of it. In addition to being one of the most free-flowering, it is one of the easiest to manage, and flourishes where many of the others would fail. It is admirable to grow among shrubs near the rock-garden, or in the grass around, flowering in the earliest days of spring.

C. iridiflorus.—Bears in September and October bright-purple flowers before the leaves. Remarkable for purple stigmata and the marked difference between the size of the inner and the outer segments of the perianth.

C. aureus (*Yellow C.*).—One of the commonest and most vigorous of all our garden Crocuses, a native of Eastern Europe, and, it need hardly be added, at home everywhere in Britain. "It is observable that all the wild specimens of this species seem to have grown with the bulbs 5 inches or more underground. Depth is very necessary to their preservation, for mice, which I have found usually to meddle with no other species, will scratch very deep in quest of them. All the varieties of this species seem to prefer a very light soil upon a clay subsoil." (Herbert, in "Trans. Hort. Soc.").

C. nudiflorus (*Purple Autumn C.*).—A beautiful bright purple Crocus, flowering in autumn after the leaves of the year are withered, thriving freely in any light soil, and naturalised in meadows about Nottingham and Derby. Flower with the tube from 3 inches to 10 inches, and the segments 1½ inch to 2 inches long; stigmas reddish-orange, cut into an elegant fringe. It is very beautiful in colour, and groups charmingly on the rock-garden.

Crocus Orphanidis (*Orphanides' C.*).—Lovely soft lilac-blue flowers, having yellow throats, 2½ inches in diameter, and opening in autumn. The bulbs are large, nearly 2 inches long, "closely covered with a bright chestnut-brown tissue." The leaves appear with the flowers, exceeding them in length, and getting much longer afterwards. A native of Greece, and, till plentiful, should be exclusively planted on warm slopes of the rock-garden.

C. pulchellus.—An autumnal species, invaluable for the garden. The pale lavender flowers, with bright yellow throat, are freely produced from the middle of September to early in December. Seed.

C. reticulatus (*Cloth-of-Gold C.*).—This is the little rich golden Crocus with the exterior of its flowers of a brownish black. It is the earliest of the commonly cultivated spring Crocuses, and a native of South-Eastern Europe. There are several varieties, and among them a lilac and a white, but these I have never seen in cultivation. Suitable for association with the earliest and dwarfest flowers of the dawn of spring, thriving in ordinary soil. It is generally known as *C. susianus*.

C. sativus (*Saffron C.*).—This species was formerly cultivated in England for the production of saffron, which is made from the fringed and rich orange style. Its native country is not known with certainty, but it is probably from S. Europe. It blooms in autumn from the end of September to the beginning of November, according to position and soil. The flowers have a delicate odour. The bulbs of the Saffron Crocus should be planted from 4 to 6 inches under the surface, and it loves a sandy loam and a warm position.

Where the natural soil is too cold for this plant, it will be best to give it a home on sunny parts of the rock-garden.

Crocus Sieberi (*Sieber's C.*).—A small species, from the mountains of Greece. We have Crocuses that flower in Spring, and Crocuses that flower in autumn ; but this hardy mountaineer flowers in winter and earliest spring, anticipating all the others. Very dwarf, with pale violet flowers ; is not at all difficult to cultivate, and should be placed on some little sunny ledge or other spot where it may be safe from being overrun

C. speciosus (*Showy Autumn C.*).—The finest of the autumn-flowering Crocuses, and coming into beautiful bloom when the wet gusts begin to play with the fallen leaves, at the end of September or beginning of October ; the flowers, bluish-violet, striped internally with deep purple lines, smooth at the throat, the divisions most deeply veined near their base ; the stigmas of a fine orange colour, cut so as to appear as if fringed ; the leaves appearing about the same time as the flowers, but not attaining their full development till the following spring. It seeds freely in this country, and may be readily increased in that way, and by division. Crimea and neighbouring regions.

C. vernus (*Spring Crocus*).—One of the earliest cultivated species. Alps, Pyrenees, Tyrol, Italy, and Dalmatia. Naturalised in several parts of England. Remarkable for its range of colour, from pure white to deep purple, endless varieties being generally intermixed in its native habitats, and corresponding with the horticultural varieties of our gardens. Flowers early in March at low elevations, and as late as June and July in the higher Alps. The parent of nearly all the purple, white, and striped Crocuses grown in Holland.

C. versicolor (*Striped C.*).—This is a distinct spring-flowering kind, which has spread into a good many varieties, and is abundantly grown in Holland. The ground colour of the flower is white, but richly striped with purple, the throat sometimes white, sometimes yellow, the inside being smooth, by which it can be readily distinguished from *Crocus vernus*, which has the inside of the throat hairy.

Dean Herbert says this "likes to have its corm deep in the ground. If its seed is sown in a three-inch pot plunged in a sand-bed, and left there, by the time the seedlings are two or three years old, the bulbs will be found crowded and flattened against the bottom of the pot ; and, if the hole in the pot is large enough to allow their escape, some of them will be found growing in the sand under the pot." It, however, thrives in any ordinary garden soil.

Crocus zonatus.—Bright vinous-lilac flowers golden at the base, abundant about the middle of September ; highly ornamental and free-flowering, and easy of culture. The flowers come before the leaves, which do not appear till spring.

CYANANTHUS LOBATUS (*Lobed C.*).—A distinct Himalayan rock plant, about 4 inches high, flowering in late summer ; purplish blue with a whitish centre. It thrives in the rock-garden in sunny chinks in a mixture of sandy peat and leaf-mould, with plenty of moisture during the growing season. Increased freely by cuttings. The seed requires a dry, favourable season to ripen it ; in wet weather the large erect, persistent calyx becomes filled with water, which remains and rots the included seed vessel.

Cyananthus incanus.—This quite differs from *C. lobatus*, flowering more freely ; like that species, it should be planted in a dry, sunny, well-drained position, as, if the situation be too damp, the fleshy root-stock is liable to rot. It is a good plan to place something over the plant during the resting season. The flowers are not so large as those of *lobatus*, but they are more charming in colour, which is enhanced in effect by the white tuft of hairs in the throat of the corolla.

CYCLAMEN (*Sowbread*).—Except the Persian kind, these are as hardy as Primroses ; but they love the shelter and shade of low bushes or hill copses, where they may nestle and bloom in security. In the places they naturally inhabit there is usually the friendly shelter of Grasses or branchlets about them, so that their large leaves are not torn to pieces by wind or hail. The Ivy-leaved Cyclamen is in full leaf through winter and early spring, and for the sake of the beauty of the leaves alone, it is best to place it so that it may be safe from injury. Good drainage is necessary to their open-air culture. They grow naturally among broken rocks and stones mixed with vegetable soil, grit, etc., where they are not surrounded by stagnant water. The late Mr Atkins, of Painswick, who paid much attention to their culture, and succeeded in a remarkable degree, thought that the tuber should be buried, and not exposed like the Persian Cyclamen in pots. His chief reason was that in some species the roots issue from the upper surface of the tuber only. They enjoy plenty of moisture at the root at all seasons, and thrive best in a friable, open soil, with plenty of leaf-mould in it. They are admirably suited for the rock-garden, and enjoy warm nooks, partial shade, and shelter from dry, cutting winds. They may be grown on any aspect if the conditions above mentioned be secured, but an eastern or south-eastern one is best.

Perfect drainage at the roots is indispensable for the successful culture of all Cyclamens, growing as they often do in their native habitats amongst stones, rock, and *débris* of the mountains, mixed with an accumulation of vegetable soil—the tubers being thereby often covered to a considerable depth, and not exposed to the action of the atmosphere, as is too often the case under culture if placed on the surface of the soil. This practice is in most instances injurious, drying up the incipient young leaf and flower buds when the tubers are apparently at rest: for I find in most species that, though leafless, the fibres and young buds for the ensuing year are still making slow but healthy progress under favourable circumstances. Collectors from abroad should be specially careful in this particular. We seldom find tubers of some of the species that have been much dried or exposed to the air vegetate freely or sometimes at all. I have now by me some roots imported nearly six years since (I believe from the Greek Isles), that were thus exposed, and though the tubers have remained sound and sent out tolerably healthy fibres, they have not until this season made healthy leaves. In *C. hederæfolium* and its varieties the greater portion of their fibres issue from the upper surface and sides of the tuber, indicating the necessity of their being beneath the soil. The habit in *C. coum, C. vernum*, and their allies, of the leaf and flower stalks, when in a vigorous state, running beneath the soil, often to a considerable distance from the tuber, before rising to the surface, points in the same direction.

Cyclamens generally like a rich soil, composed of friable loam, well-decayed vegetable matter, and cow manure, reduced to the state of mould, and rendered sweet by exposure to the atmosphere before use. They are all admirably adapted for the rock-garden; they enjoy warm nooks, partial shade from mid-day sun, and shelter from the effects of drying, cutting winds. An eastern or south-eastern aspect is best, screened from cutting winds, but a northern one will do well, and they love an open yet sheltered spot.

Cyclamens are best propagated by seed sown as soon as it is ripe, in well-

GENTLY RAISED ROCK GARDEN, THE HOLT, HARROW WEALD. *May* 1910.

drained pots of light soil. I generally cover the surface of the soil after sowing with a little moss, to ensure uniform dampness, and place them in a sheltered spot out of doors. As soon as the plants begin to appear, which may be in a month or six weeks, the moss should be gradually removed. As soon as the first leaf is half developed, they should be transplanted about an inch apart in seed pans of rich light earth, and encouraged to grow as long as possible, being sheltered in a cold frame, with abundance of air at all times. When the leaves have perished the following summer, the tubers may be planted out or potted, according to their strength.

From the earliest times there appears to have been great difficulty felt by our best botanists in clearly defining the species of Cyclamen, from the great variation in shape and colouring of the leaves both above and below. Too much dependence on these characters has been the cause of much confusion and an undue multiplication of species. Some of the varieties of this genus become so fixed, and reproduce themselves so truly from seed, as to be regarded as species by some cultivators. The following are some of the more important synonyms—*æstivum* (*europæum*); *anemonoides* (*europæum*); *autumnale* (*hederæfolium*); *Clusii* (*europæum*); *hyemale* (*coum*); *littorale* (*europæum*); *neapolitanum* (*hederæfolium*); *repandum* (*vernum*); *vernum* of Sweet (*coum*, var. *zonale*). *Anemonoides*, *Clusii*, and *littorale*, are southern varieties of *C. europæum*, quite distinct from the northern type.

Cyclamen coum (*Round-leaved C.*).—Tuber round, depressed, smooth, fibres issuing from one point on under side only. Leaves of a plain dark green, cordate, slightly indented; these, with the flowers, generally spring from a short stem rising from the centre of the tuber. Corolla short, constricted at the mouth; reddish purple, darker at the mouth, where there is a white circle; inside striped red. Flowers from December to March, and is a native of the Greek Archipelago. This, with the others of the same section —viz. *vernum* of Sweet, *ibericum*, *Atkinsii*, and the numerous hybrids from it— though hardy, and frequently in bloom in the open ground before the Snowdrop, yet, to preserve the flowers from the effects of unfavourable weather, will be the better for slight protection, or a pit or frame devoted to them, in which to plant them out.

C. vernum of Sweet is considered by many as only a variety of *coum*, and for it I would suggest the name of *C. coum*, var. *zonale* (from its marked foliage). I was for a long time unwilling to give it up as a distinct species, but now doubt there being sufficient permanent specific distinction to warrant its being retained as such, especially after seeing the many forms and hues the leaves of other species of this genus assume. Though this, as well as *C. coum*, retains its peculiarities as to markings very correctly from seed, so do some undoubted varieties of other species of Cyclamen.

Cyclamen Ibericum (*Iberian C.*).—This also belongs to the *coum* section. There is some obscurity respecting the authority for this species and its native country; but there are specimens of it in the Kew and Oxford herbariums, marked "ex Iberiâ." Leaves very various. Flowers: corolla rather longer than in *coum*; mouth constricted, not toothed; colour various, from deep red-purple to rose, lilac, and white, with intensely dark mouth; produced more abundantly than by *coum*.

C. europæum (*European C.*).—Tuber of medium size and very irregular form, sometimes roundish or depressed and knotted, at other times elongated. The rind is thin, smooth, yellowish, sometimes "scabby." The underground stem or rhizome is often of considerable length and size, sometimes even more than a foot in length. The leaves and flowers originate from stalks or branches, which emerge from all parts of the tuber. The root fibrils spring from the lower surface of the tuber as freely as from the upper, and there are usually two or three stems springing from different parts, and growing in different directions, from which the leaves and flowers arise. The leaves appear before and with the flowers, and remain during the greater part of the year. Flowers from June to November, or, with slight protection, until the end of the year. The petals rather short, stiff, and of a reddish-purple colour. I have often seen them luxuriate in the *débris* of old walls, and on the mountain-side, with a very sparing quantity of vegetable earth to grow in.

Cyclamen hederæfolium (*Ivy-leaved C.*).—A native of Switzerland, South Europe, Italy, Greece and its isles, and the north coast of Africa. Tuber not unfrequently a foot in diameter when full-grown; its shape somewhat spheroidal, depressed on the upper surface, rounded beneath. It is covered with a brownish rough rind, which cracks irregularly, so as to form little scales. The root fibres emerge from the whole of the upper surface of the tuber, but principally from the rim; few or none issue from the lower surface. The leaves and flowers generally spring direct from the tuber without the intervention of any stem (a small stem, however, is sometimes produced, especially if the tuber be planted deep); at first they spread horizontally, but ultimately become erect. The leaves are variously marked, and the greater portion of them appear after the flowers, continuing in great beauty the whole winter and early spring, when they are one of the greatest ornaments of our borders and rockeries, if well grown. I have had them as much as 6 inches long, 5½ inches in diameter, and a hundred to a hundred and fifty leaves springing from one tuber. The flowers begin to appear at the end of August, continuing until October. Mouth or base of the corolla ten-toothed, pentagonal, purplish red, frequently with a stripe of lighter colour, or white, down each segment of the corolla. There is a pure white variety, and also a white one with pink base or mouth of corolla, which reproduce themselves tolerably true from seeds. Strong tubers will produce from two hundred to three hundred flowers each. The varieties from Corfu and other Greek isles are very distinct. They generally flower later, and continue longer in bloom. Their leaves rise with or before the majority of the flowers, both being stronger and larger than the ordinary type, with more decided difference of outline and markings on the upper surface of the leaves, the under surface being frequently of a beautiful purple. Some of them are delightfully fragrant. They are quite hardy, but are worthy of a little protection to preserve the late blooms, which often continue to spring up till the end of the year.

This species is so hardy as to make it essential for the rock-garden. It will grow in almost any soil and situation, though best (and it well deserves it) in a well-drained place on the rock-garden. It does not like frequent removal. It has been naturalised successfully on the mossy floor of a thin wood, on a very sandy, poor soil.

C. græcum is a very near ally, if more than a variety; it requires the same treatment. The foliage is more after the southern var. of *C. europæum* type than most of the *hederæfolium* section; the shape of corolla and toothing of the mouth the same. *C. africanum* much larger in all its parts than *C. hederæfolium*, otherwise very nearly allied, is hardy in warm sheltered situations.

Cyclamen vernum (*Spring C.*).—Tuber round, depressed, somewhat rough or russety on outer surface; fibres issue from one point on the under side only; under cultivation it has little or no stem, but leaves and flowers proceed direct from the upper centre of the tuber, bending under the surface of the soil horizontally before rising to the surface. Corolla long, segments somewhat twisted, mouth round, not toothed; colour from a delicate peach to deep red purple, very seldom white; fragrant. Flowers from April to end of May. Native of South Italy, the Mediterranean and Greek isles, and about Capouladoux, near Montpellier. Leaves rise before the flowers in the spring; they are generally marked more or less with white on upper surface, and often of a purplish cast beneath; fleshy; semi-transparent whilst young. For many years I believed this species to vary in the outline and colouring of the foliage less than any other, but I have now received imported tubers from Greece, with much variety in both particulars, some of the leaves quite plain and dark green, others dashed all over with spots of white, others with an irregular circle of white varying much in outline.

This, though hardy, is seldom met with cultivated successfully in the open air. It is impatient of wet standing about the tubers, and likes a light soil, in a nook rather shady and well sheltered from winds, its tender fleshy leaves being soon injured. The tubers should also be planted deep, say not less than 2 inches to $2\frac{1}{2}$ inches beneath the surface. I have grown them for many years in a border and on rocks without any other protection than a few larch-fir boughs lightly placed over them, to break the force of the wind and afford a slight shelter from the scorching sun. Some authorities give *C. repandum* as a distinct species, but I consider them identical, the only difference being in the shape and markings of the leaves, which are very variable. It is generally cultivated in England under the name of *repandum*, but most of the best continental botanists adopt the name of *vernum* for it, and it is, no doubt, the original *C. vernum* of L'Obel.

CYPRIPEDIUM (*Lady's Slipper*).—Beautiful Orchids, the northern species of which are prettier in colour than the tropical ones, and of the highest interest for the rock-garden. In it the variety of surface and aspect offer means of growing these charming hardy Orchids better than borders. As most of them come from the coldest countries, it is not our climate that is against them, but the soil, when not of the leafy, moist, and nearly always open soil of the moist woods in their native countries.

The best plants of *C. acaule, C.*

Calceolus, and *C. spectabile* I have ever had were grown in the flanks of a piece of rootwork under a canopy of Beech. In preparing a station for them, the soil should be taken out to a depth of 20 inches, and if the upper spit consists of fairly good fibrous loam, it may be laid aside for mixing with the compost. Place a good layer of rough stones or broken brick in the bottom, and fill in with about equal parts of rough fibrous peat, leafmould, and loam, the leaf-mould to be only partly decayed. A little limestone grit, gravel, or similar material may be added with advantage, as some species delight in it, while it will do no harm to any. The roots should be planted from 4 inches to 6 inches deep as soon as received, and a soaking of water given to settle the soil. They may then have a light mulch of rough material, and usually no more water will be required until the leaves are pushing up. The time for lifting and potting varies a little in different species, but, as a rule, the best time is just as the growth has died off. One of the finest species is :—

Cypripedium macranthum, a large and handsome species, but it is rare. It thrives in sound fibrous loam of good texture broken into lumps, with some finely broken charcoal and crocks suiting it well, not disturbing the roots oftener than is necessary. The downy flowerspikes are about 1 foot high, and each bears a single large flower of a rosy pink, streaked with red and white.

C. parviflorum is an old and useful American species that thrives well in a very moist, shady position, or it may be grown in pots in a frame. The sepals and petals are narrow, twisted, shining brown, lined with deep purple; the lip large, drooping, lemon-yellow, spotted with red. It is one of the best.

C. japonicum (*Japanese Lady's Slipper*). A graceful plant about 1 foot high, its hairy stems, which are as thick as one's little finger, bearing two plicate fan-shaped leaves of bright green, rather jagged round the margins. The flowers are solitary, the sepals being of an apple-green tint; the petals, too, are of the same colour, but are dotted with purplish-crimson at the base; the lip large, and folded in front; the colour of the lip is a soft creamy yellow, with bold purple dots and lines. Thrives in half-shady spots, with plenty of leaf-soil.

Cypripedium spectabile (*Noble Lady's Slipper*).—A noble hardy Orchid; a native of meadows, peat bogs, and woods, in the Northern, and on mountains in the Southern, United States. When grown in the open air, I know of no hardy plant to surpass this in delicate purity of colour. The best plants I have ever seen were at Glasnevin, on the cool side of one of the ranges of plant-houses there, planted close against the wall in deep rich soil—a mixture of free moist loam and peat. Wherever there is any kind of a rock- or marsh-garden, there should be no difficulty in succeeding with this fine plant. It should be placed on the lower flanks, and in different positions and aspects, mostly sheltered ones; and if it does not in all cases attain the stature of the Glasnevin plants, it will command admiration as the finest of hardy Orchids.

C. calceolus (*English Lady's Slipper*).—The handsomest of our native Orchids, and therefore an object of much interest. When grown under tolerably favourable conditions, the stem rises to a height of from 16 to 20 inches, with large pointed leaves, and bearing large flowers; the lip yellow, variegated with purple; the long sepals and petals of a brownish-purple. Although reputed to be extinct in Britain, it is known to exist yet in a wild state with us, but in very few places, and let us hope the last remaining plants may long remain undisturbed; it is abundantly distributed over Continental Europe, and should not be difficult to obtain. I have never seen this fine plant nearly so well grown as by the late Mr James Backhouse, of York. He planted it on an eastern shaded aspect of his rock-garden, in deep, fibrous loam, in narrow, well-drained fissures, between limestone rocks. The condition in which this and other Orchises are obtained, has a great influence on their well-being. The roots are often dried up, and nearly or quite dead when obtained; and in this condition they would have but a poor chance of surviving, even if planted in the wilds most favourable to their natural development. Given good sound roots, there will not be the least difficulty in establishing plants in deep loam, in any well-drained, half shady spot, with some shelter afforded by low bushes and plants to prevent the leafy growth of the plant from being destroyed or injured by wind. It is propagated by division of the root, but should not be disturbed for that purpose till the plants are well established, and have begun to spread about.

Cypripedium acaule (*Moccasin Flower*).—A handsome, fragrant, hardy dwarf Orchid, with a large purplish-rose flower, blooming in summer nearly 2 inches long, with a deep fissure in front. It is common in North America, usually growing in sandy or rocky woods under evergreens, and the best position for it in cultivation is in some sheltered and half shaded spots on the lower flanks of the rock-garden, or among shrubs planted near it in sandy loam, with plenty of leaf-mould. It also succeeds in sheltered and shaded spots. It is found with pale, and, more rarely, with white flowers.

C. guttatum (*Spotted Lady's Slipper*).—A beautiful Siberian plant, growing from 6 to 9 inches high, flowering in June; solitary, rather small snow-white flowers, blotched with deep rosy-purple. The flower-stem rises from a single pair of broadly-ovate downy leaves. It requires a shady position in leaf-mould, moss, and sand, and should be kept rather dry in winter. In heavy soil the roots soon perish, and it does not care for lime, but if planted shallow and kept moist, it will usually thrive in the leafy soil.

C. hirsutum (*Yellow Lady's Slipper*) is a tall-growing, handsome Orchid. The flowers are large and handsome, the sepals and narrower petals pale yellow, streaked and spotted with brown; the lip pale yellow. A far northern kind, Nova Scotia and Canada, in copses and woods, also ascending high on the mountains of the southern country. Syn., *C. pubescens.*

Cypripedium arietinum is a beautiful little Orchid, difficult to grow, and liking much moisture. The upper sepal and petals are greenish-white, lined with red-brown, the lip white in the throat, suffused with rose in front and streaked with red. Canada and the colder parts of the United States in cool damp woods.

CYSTOPTERIS (*Bladder Fern*).—The cultivated kinds of this native group are small elegant Ferns of delicate fragile texture, growing on rocks and walls, chiefly in mountainous districts. The best known are : *C. fragilis,* which has finely-cut fronds about 6 inches high. It is of easy culture, succeeding in an ordinary border, though seen to best advantage on shady parts of the rock-garden in a well-drained soil. There are two or three varieties, *Dickieana* being the best. *C. alpina* is much smaller, and, when once established, not difficult to cultivate or increase, but more affected by excessive moisture than *C. fragilis.* A sheltered situation in a well-drained part of the rock-garden suits it. *C. montana* is another elegant plant requiring the same treatment as *C. fragilis.*

CYTISUS (*Broom*).—These graceful and brilliant shrubs, though mostly too large for our purpose in the select rock-garden, wherever we deal with the natural rocks are valuable shrubs, being so free and easily raised by merely shaking the seed on the ground. Even the most arid railway-bank may be adorned by shaking a few seeds over them ; and of course the natural rock would be the very place for them. The purple Broom is naturally a trailing shrub with purplish flowers, but is generally seen grafted mop fashion on Laburnum stems. It is really an alpine shrub, and its place is among rocks and boulders, where its wiry branches can fall over and make dense cushion-like tufts. *C. Ardoini* is a pretty alpine shrub a few inches high, and suitable for the rock-garden ; its tufted growth is covered in summer with yellow flowers.

Cytisus albus is the graceful white Portuguese Broom, an aid where our rocks are bold. The Montpellier Broom is only hardy in mild sea-shore districts, and various other kinds are not hardy in our country. *C. scoparius,* the common Broom of Britain, is one of the most beautiful shrubs, and well worth naturalising where it is not common wild. *C. andréanus* is a fine broom variety of it. The Spanish Broom is a very fine plant like the above, but it is put under the name of *Spartium Junceum.* Some of these are so free and vigorous that one can sow the seed out of hand on poor and stony places and in a very short time see the plants arise (even without covering the seed) on such surfaces as railway banks, sandy slopes, and thin copsey places, rough hedge banks and road-sides. The common Broom comes freely in this way, and also the Spanish Broom, though not a native plant, is superb on railway-banks, coming later than our own Broom. I have raised many in this way by merely shaking the seed in passing, and in the spring of this year (1902) sowed over half a hundredweight of seed of the common Broom in young woods, on rail-banks, and the most likely places for it in or near my own place. The seed is saved in quantity by all the great seed houses, so there should be no difficulty in obtaining it. I recommend the pastime to gentlemen who have had enough of more fashionable forms of amusement. It has even claims from the musical side, as one may hear the nightingale when sowing of an evening in May.

DALIBARDA REPENS (*Violet-leaved Dalibarda*).—A low tufted plant, about 2 inches in height, with white blossoms shaded with the most delicate rose-colour. It loves a deep, peaty soil; and, though hardy, by no means of rapid growth. Nova Scotia and N.E. America.

DAPHNE (*Garland Flower*).—Alpine and mountain shrubs, some dwarf as well as beautiful, fragrant, and of can scarcely hope to witness in our gardens under ordinary treatment. They have but few roots, and are best transplanted when young. The best soil is a mixture of free loam and decayed leaf-mould, with some old road sand added. None of the *Daphnes* require a rich soil, and some of them even prefer old road sand to

Daphne Blagayana.

the highest value for the rock-garden. Where the bushy rock-garden is made, the larger kinds will be useful; the smaller may go with the choicer and more diminutive alpine plants. They are chiefly natives of Europe, and in cultivation do best when shaded in summer from the mid-day sun, and in winter screened from cold winds. If nurtured by the fallen leaves of trees, they will grow with a vigour that we any other; this is especially the case with the *Mezereon*.

Daphne alpina (*Mountain Mezereon*).—A dwarf summer-leafing and distinct rock shrub, reaching 2 feet high, the flowers yellowish-white, silky outside, fragrant, in clusters of five from the sides of the branches. It is a low, branching shrub, flowering from April to June, and bearing red berries in September. Central and S. Europe.

D. Blagayana (*The King's Garland*

Flower).—A dwarf alpine shrub, 3 inches to 8 inches high, of straggling growth, the leaves forming rosette-like tufts at the tips of the branches, encircling dense clusters of fragrant, creamy-white flowers. It blooms in spring for several weeks, and thrives in the rock-garden in well-drained spots surrounded by stones for its wiry roots to ramble among. It is hardy, and in open spots thrives in any good soil; increased by layers pegged down in spring and separated from the plants as soon as roots are emitted.

Daphne Cneorum (*Garland Flower*).—A little spreading shrub, growing from 6 inches to 10 inches high, and bearing rosy-lilac flowers, the unopened buds crimson, and so sweet that, where much grown, the air often seems charged with their fragrance. It is a native of most of the great mountain chains of Europe, and is one of the best of all plants for the rock-garden, thriving in peaty and very sandy soils, but in stiff soils often fails. Wherever the soil is favourable, it should be much used, and is usually increased by layers.

D. Collina (*Box-leaved Garland Flower*).—The leaves of this much resemble in shape and size those of the Balearic Box, the upper surface of a dark glossy green. The flowers are in close groups, and of a light lilac or pinkish colour, the tubes rather broad and densely coated externally with silky white hairs. It forms a low, dense, evergreen shrub, the branches of which always take an upright direction, and form a level head, covered with flowers from February to May. S. Europe. *D. Neapolitana* is a variety of it.

D. Fioniana (*Fion's Garland Flower*).—A compact shrub, not uncommon in gardens, the heads of bloom are in clusters, five fragrant flowers in each, of a pale lilac colour, the tubes densely covered externally with short silvery hairs. This shrub flowers from March to May, and is hardy about London.

D. Genkwa (*Lilac Garland Flower*), is a summer-leafing shrub of from 2 feet to 3 feet in height, with downy branches and fragrant violet-coloured flowers thickly set on the leafless branches in early spring, giving the plant the appearance of a small Persian Lilac. There appear to be several varieties of *D. Genkwa*, some with much larger flowers than others, and some of a darker shade of purple. It is not quite hardy in cold districts. Syn., *D. Fortunei.*

Daphne Houtteiana (*Van Houttes Mezereon*).—This singular kind forms a robust spreading bush, 3 feet or 4 feet high, with all the leaves collected on the young branches, while the old ones are naked. It is a distinct bush, hardy, flowering in the spring before the leaves appear, and is said to be a hybrid, which originated in one of the Belgian Nurseries, between the common *D. Mezereum* and Spurge Laurel. Its leaves are from 3 inches to $3\frac{1}{2}$ inches long, and 1 inch broad, stained with purple on the upper side when fully developed, and when quite young and in the bud state, of a dark purple colour. The flowers are small, dark purple, quite smooth, and are borne along the shoots of the previous year, before the young leaves appear.

D. Mezereum (*Mezereon*).—A wild plant in English woods, is a charming and fragrant bush, and the earliest to flower, often in February. Where the shrubby rock-garden is carried out, nothing is more lovely for its adorning than a group of this. Though quite hardy, it is slow, and not so pretty on some cold soils; but on such soils as we use on the rock-garden it will thrive. It is best to begin with little plants; and it is easily raised from seed.

D. odora (*Sweet Daphne*).—A fragrant and beautiful kind, in mild and southern districts hardy on the rock-garden, usually best on western aspects. It is a greenhouse plant of exceptional merit when well grown. We know no fragrance more pleasant than that emitted by the pinkish flowers of this Daphne. There are varieties called *alba*, *rubra*, *Mazeli*, *punctata*. *Mazeli* is, according to Max Leichtlin, hardier than the older kind. Syn., *D. indica*. China.

D. rupestris (*Rock Garland Flower*) is a neat little shrub, with erect shoots forming dense, compact tufts, 2 inches high and 1 foot or more across, often covered with flowers of a soft-shaded pink, in clustered heads. It is essentially a rock plant, growing wild in fissures of limestone in peaty loam, but is of slow growth, and it takes some years to form a good tuft. It seems to thrive in very stony and peaty earth, with abundance of white sand, and should be planted in a well-drained but not a dry position.

Daphne striata (*Striated Garland Flower*).—A sweet-scented hardy trailing species. It forms dense, twiggy, spreading masses, 1 foot to 3 feet across, which, in June and July, are covered with rosy-purpled, scented flowers in clusters. The trailing and spreading habit of this plant recommends it for covering bare spots. France.

DARLINGTONIA CALIFORNICA
(*Californian Pitcher-plant*).—A most singular plant, resembling the North American pitcher-plants, but distinct; the leaves, which rise to a height of 2 feet or more, are hollow, and form a curiously shaped hood, from which hang two ribbon-like appendages, the hood often a crimson-red, and the flowers are almost as curious. Found to grow in our climate, if care be taken with it, and it would be difficult to name a more interesting plant for a bog garden. It is less trouble out of doors than under glass; indeed, it only requires a moderately wet bog in a light spongy soil of fibrous peat and chopped Sphagnum Moss. Place by the side of a stream, in any moist place, the plants fully exposed to direct sunlight, but sheltered from the cold winds of early spring when they are throwing up their young leaves. They require frequent watering in dry seasons, unless they are in a naturally wet spot. When they become large they develop side shoots, which, taken off and potted, soon make good plants. The plant is also raised from seed, but this requires several years. I found it on the Californian Sierras about little springs on the hills thickly tufted among the common Rush.

DENTARIA (*Toothwort*).—Pretty and interesting perennials of the Stock family, of which there are some half a dozen species in cultivation, all worth growing in half-shaded positions in peat beds among rock shrubs. They grow best in a light sandy soil, well enriched by decayed leaf-mould, or in soil of a peaty nature. Their flowers are welcome in the early days of spring, and remain in beauty for some time. They are of easy propagation by the small tuber-like roots. Some, such as *D. bulbifera*, bear bulblets on the stem, and from these the plant may be increased. The species are— *D. bulbifera*, 1 foot to 2 feet high. Flowers in spring; purple, sometimes nearly white, rather large, produced in a raceme at the top of the stem. *D. digitata*, a handsome dwarf kind, about 12 inches high, flowering in April; rich purple, in flat racemes at the top of the stem. A native of Europe. *D. diphylla* is a pretty plant, growing from 6 inches to 12 inches high, and bearing but two leaves. The flowers are purple, sometimes white, and occasionally yellowish. Woods in North America. *D. enneaphylla* is about 1 foot high; flowers creamy white, produced in clusters in April and June. A pretty plant for a shady border. Mountain woods in Central Europe. *D. maxima* is the largest of all the species, growing 2 feet high. Flowers pale purple in many flower-heads. N. America. *D. pinnata* is a stout species, at once distinguished by its pinnate leaves; 14 inches to 20 inches high; flowering from April to June; large, pale purple, lilac, or white, in a terminal cluster. Switzerland, in mountain and sub-alpine woods. *D. polyphylla*, similar to *D. enneaphylla*, is about 1 foot high, with flowers in clusters, cream coloured. A handsome plant; from woods in Hungary.

DESFONTAINEA SPINOSA.—A brilliant flowering shrub in favoured gardens along the sea-coast, this beautiful ever-green shrub from Chili may be flowered out-of-doors. It is of moderate growth, having foliage like a Holly; flowers are in the form of a tube, scarlet tipped with yellow. It usually flowers about the end of summer, and in some parts of Devonshire it blooms freely, thriving in a light, loamy, or peaty soil. It may here and there thrive among rock-garden shrubs, and it is not a high temperature that seems to help it, but nearness to the sea, as one may see it thriving even in the north of Ireland within a few miles of the sea. A few miles inland, and it fails.

DIANTHUS (*Pink*). — Usually dwarf evergreen herbs, alpine rock, shore, or heath dwellers, many beautiful, and among them two which have given us the many garden Pinks and Carnations we now have. The Pinks, especially the alpine kinds, are moisture-loving plants, and during spring and summer water must be given in such a way as to interfere as little as possible with the tufted crowns, as moisture about the neck or stagnant soil is often fatal. This can best be done by half-buried stones around the plants. The wireworm is the deadliest enemy of this family, and when an affected tuft is found, lift it, wash off all the soil, and replant in a fresh mixture.

The higher and rarer Pinks, such as the alpine and glacier Pinks, deserve the best places in the rock-garden, and in cool stony ground. More lowland kinds, like our common Pink, are much more free than the others, and may be used in bold ways for edges and groups, and the same may be said of certain hybrid kinds, which are often good in colour. Some mountain kinds, like the Cheddar Pink and also kinds like it in habit, are easily established on old walls and bare stony ground. Many Pinks are easily increased by division, but of the rarer kinds the seed should be saved, and sown where we desire the plants to grow on the rock-garden. In this very large family there are many annual kinds, such as the Chinese Pinks, and probably some brilliant species not yet introduced from the large area of distribution of the genus in Europe, Asia, and N. Africa. A cool but open soil of sandy loam and a little leaf-mould suits the alpine kinds best. The alpine kinds are apt in our warmer gardens to get a little drawn and leggy, and a good way is to top-dress the tufts with a fine leaf-mould with river sand or grit among it, gently working it among the shoots. The following is a selection from a large number of kinds of the best for the rock-garden. There as in other cases where the aim is not to have a botanical collection only, we can best enjoy the beauty of the plants by cultivating well and grouping effectively the more distinct kinds. The various races of garden flowers derived from the wild *Dianthus* : Pinks, Carnations, Picotees, Cloves, variously coloured double forms of the Pink, so much grown in our gardens and as cut flowers for market; the many forms of the Chinese Pink, so much grown among annual flowers, and the mule Pinks, effective border flowers, do not rightly belong to the alpine garden, and are not included here.

Dianthus alpinus (*Alpine Pink*). — A distinct and lovely plant, with dense green and obtuse leaves, each stem bearing a solitary flower, deep rose spotted with crimson, and often so freely, as to hide the leaves. In poor, moist, and sandy loam it thrives and forms a dwarf carpet, the flower-stems little more than 1 inch in height. Wireworms, rather than unsuitable soil, often cause its death. It should be in an exposed spot, and guarded against drought. It comes true from seed, and is not difficult to increase in that way, or by division. Alps of Austria, flowering in summer.

Alpine Pink (*Dianthus alpinus*).

Dianthus Atkinsoni.—This is one of the richest coloured of all the family, its flowers crimson and very striking in the early summer. Owing to its flowering so freely, shoots for cuttings are very few, and it is well to reserve some plants for stock, not allowing them to flower.

D. arenarius (*Sand Pink*).—A neat, compact rock plant, about 6 inches high, with very dense foliage, and white, fimbriated or fringed flowers. It blooms in May and June, and should have a dry sunny position. North Europe.

Dianthus cæsius (*Cheddar Pink*).—One of the best of the dwarf Pinks with which rocky places are studded over so great an area of northern regions. The short leaves are very glaucous, and the fragrant rosy flowers borne on stems 6 inches in height in summer. In winter it may perish in the ordinary border, but thrives and flowers abundantly on old walls, as at Oxford. It is a native plant, and grows on the rocks at Cheddar, in Somersetshire. To establish it, the best way is to sow the seeds on the wall in a little cushion of Moss or a little earth in a chink. It may also be grown in calcareous or gritty earth, and placed in a chink between stones. Increased by seeds.

D. callizonus is one of the most distinct of the alpine Pinks, a native of Transylvania, and has the habit of *D. Plumarius*, with the flowers of *D. Alpinus*, but larger. It strikes readily from cuttings, and may be raised from seed, which, however, it ripens sparingly. The flowers are bright rose-purple.

D. caryophyllus (*Carnation*).—The parent of all the races of Carnations, Picotees, and Clove Pinks, so variously coloured, so fragrant and profuse in flower, as to make them among the most valuable of our hardy border flowers. The plant occurs in a wild state on old castles or city walls in various parts of England, and more abundantly in similar places in the West of France, the flowers of the wild form being usually red or white. The wild plant is worth a place on the rock-garden or on walls.

D. Caucasicus.—Flowers bright rose, on stems 12 inches high; foliage glaucous, very compact.

D. cruentus.—This European Pink has sparse foliage, but its crowded heads of deep crimson fragrant flowers are attractive. It is one of the easiest to grow in the border or rock-garden. Seeds freely, and by this means the plant may be grown to any extent in gritty loam. Height 15 inches.

D. deltoides (*Maiden Pink*).—This native of Britain forms close spreading tufts of smooth, pointless leaves, and bright pink-spotted or white flowers on stems from 6 inches to 12 inches long

Although the flower is little more than ½ inch across, there is a bright look about it which makes it welcome. It will grow almost anywhere, not appearing to suffer from wireworm, as most other Pinks do, and often flowers several times during the summer. Seed or by division.

Dianthus dentosus (*Toothed Pink*).—A distinct and pretty Pink; dwarf, with violet-lilac flowers, more than an inch across, the margins toothed at the edge, the base of each petal having a regular dark-violet spot, giving the effect of a dark eye, nearly ½ an inch across. It comes readily from seed, and should be raised periodi-

Flowers in July and August; native of Bosnia.

Dianthus glacialis, a brilliant alpine Pink. It does best in crevices of the rock-garden, as high up as possible, in peaty or leafy soil, well mixed with granite chips. It forms compact tufts of narrow leaves which, during the summer, are thickly studded with rosy-tinted flowers. In the variety

D. glacialis gelidus the habit is much the same, the flowers being rich rosy-purple spotted white in the throat.

D. Knappi.—Distinct by reason of the sulphur-yellow flowers in clustered heads

The Cheddar Pink (*D. cæsius*) in the Royal Botanic Gardens, Edinburgh.

cally, as the once-flowered plants often die. A native of southern Russia, flowering in May and June, and thriving in sandy soil.

D. Fischeri (*Fischer's Pink*).—A beautiful, and as yet rare, species from Russia, 3 inches to 4 inches high, blooms in summer; of a light rose colour, with the petals not much cut, and solitary. Deserves a good position in the rock-garden, in moist, sandy, or gritty loam.

D. Freynii.—A dwarf alpine species, with linear glaucous leaves and purplish flowers about ¾ of an inch in diameter.

after the manner of *D. Cruentus*. The species attains 12 inches or 15 inches high, grows and flowers freely, and gives seeds in plenty also. By this latter means the plant may be grown in quantity.

D. monspessulanus.—Flowers somewhat resemble those of *D. superbus*, but not quite so deeply cut. A useful rock plant.

D. neglectus (*Glacier Pink*).—Forms, close to the ground, tufts like short, wiry Grass, of glaucous leaves, from ½ inch to 1 inch long, the flowers on stems from 1 inch to 3 inches high. The petals are level and firm looking, with the outer margins slightly notched, and the flower about an inch across of brilliant rose. In a wild state, plants of it may be seen in bloom at 1½ inches high, and even less; but when cultivated in deep, sandy loams, it is larger, is surpassed by no alpine plant in vividness of colouring, and is easily grown. Alps and the Pyrenees. Division and seed.

Dianthus superbus (*Fringed Pink*).—A fragrant Pink, its petals cut into lines or strips for more than half their length, which gives the plant a singular effect. It inhabits many parts of Europe from the shores of Norway to the Pyrenees, and is a true perennial, though it perishes so often in our gardens, when very young, that many regard it as a biennial. It is more apt to perish in winter on rich and

Dianthus neglectus.

Dianthus petraeus (*Rock Pink*).—With short sharp-pointed leaves, forming hard tufts an inch or two high, and fine rose-coloured flowers in summer. It seemed to escape the attacks of wireworm when nearly every other species was destroyed. A dry and sunny position is most congenial to this species. Hungary.

D. plumarius (*Pink*).—This plant, the parent from which the varieties of Pinks have sprung, has single purple flowers, rather deeply cut at the margin, and is naturalised on old walls in various parts of England, though not a true native. It is rather handsome when grown into healthy tufts, but on the level ground it is not long-lived.

D. proliferus.—Flowers of a beautiful reddish-purple, of easy culture, and very useful.

D. pungens.—Flowers rosy-pink, plant forming nice tufts; leaves glaucous.

D. rupicola.—Flowers deep red, late, and very useful.

D. Seguieri.—Flowers large, deeply cut, rosy-purple, with a deeper ring at base of each petal. Flowers late in summer.

D. subacaulis.—Of tufted growth, with glaucous leaves; flowers small, pink, solitary.

moist soil than on that which is somewhat light and well-drained, and it should be planted in fibrous loam, well mixed with sand or grit. Unlike some of the other kinds, it comes quite true from seed, generally grows more than a foot high, flowering in summer or early autumn.

D. tymphresteus.—A free and continuous blooming species from Northern Greece, growing from 15 inches to 18 inches high, with deep rosy flowers; makes a good perennial and showy border plant.

D. vaginatus—This belongs to the clustered-flowered section of this genus, the flowers carmine, on stems only 6 inches high. It is a rare species, continuing in bloom for nearly two months.

DIAPENSIA LAPPONICA (*Lapland D.*)

—A sturdy and dwarf little evergreen alpine shrub, rarely seen even in botanic gardens, and considered impossible to cultivate, but which may be grown well on fully exposed spots in deep sandy and stony peat, kept moist during the dry season. It grows in very dense rounded tufts, with narrow closely packed leaves, and

ALPINE FLOWERS FOR GARDENS

solitary white flowers about ½ an inch across, with yellow stamens the whole plant being often under 2 inches high. A native of N. Europe and N. America, on high mountains or in arctic latitudes, flowering in summer, and most easily increased from seed.

DICENTRA (*Bleeding Heart*).—Graceful perennials of the Fumitory Order, including about half a dozen cultivated species, of which the following are the finest:—

Dicentra chrysantha.—A handsome plant, forming a spreading tuft of glaucous foliage, from which arises a stiff leafy stem, 3 feet to 4 feet high, bearing long panicles of bright yellow blossoms, each about 1 inch long. It flowers in August and September; the seedlings do not bloom till the second year. California.

D. eximia combines the grace of a Fern with the flowering qualities of a good hardy perennial. It grows from 1 foot to 1½ feet high, and bears its numerous reddish-purple blossoms in long drooping racemes. Thrives in a rich sandy soil, but it will grow anywhere. N. America. Division.

D. formosa is similar to the preceding, having also Fern-like foliage, but is dwarfer in growth, the racemes are shorter and more crowded, and the colour of the flowers is lighter. California.

D. spectabilis is a beautiful plant, too well known to need description, nearly every garden in the country being embellished with its singularly beautiful flowers, which open in early summer, gracefully suspended in strings of a dozen or more on slender stalks, and resemble rosy hearts. It succeeds best in warm, rich soils, in sheltered positions, as it is liable to be cut down by late spring frosts. Besides a position in the mixed border, it is of such remarkable beauty and grace that it may be used with the best effect near the lower flanks of the rock-garden, or on low parts where the stone or "rock" is suggested rather than exposed.

There is a "white" variety, by no means so pretty as the common one. Propagated by division in autumn.

Dicentra cucullaria (*Dutchman's Breeches*) and **D. thalictrifolia** are less important, belonging more to the curious garden.

DIPHYLLEIA CYMOSA.—An interesting perennial of the Barberry family, about 1 foot high, having large umbrella-like leaves in pairs. Flowers white, in loose clusters in summer, and succeeded by bluish-black berries. N. America, on the borders of rivulets and on mountains, thriving in peat borders and fringes of beds of American plants, in the most moist spots. Hitherto only seen as single weak specimens, this plant, if more plentiful, might be made good use of in a rock-garden. Division.

DODECATHEON (*American Cowslip*).—Graceful and distinct perennials, quite hardy and charming for the rock-garden, where they usually grow well in soils of an open nature. They are plants of wide distribution in North, Western, and Eastern America, and also the Pacific coast, and they vary without end, according to the region. They are very often found towards the arctic circle, and also on the high mountains and even the islands of the Behring Straits. The American botanists consider these plants to be varieties of the same, but this, from the garden point of view, is of little moment, as there is considerable distinction among them when cultivated. There are a number of cross-bred forms, which are pretty.

The American Cowslips are perennial and hardy, requiring a cool situation and light, leafy, or open sandy soil. In cool spots on the rock-garden, where *Primulas* and *Sol-*

danellas thrive, *Dodecatheons* will be found to grow well, and form lovely and attractive objects. All the species and varieties grow freely, and soon form large tufts, which require dividing every third or fourth year. The best time is the latter end of January or beginning of February, when the roots are becoming active, taking care not to divide them into too small pieces, as in that case there is danger of losing the plants. They may also be easily raised from seed, but this can only be obtained in very favoured situations.

Dodecatheon Meadia (*American Cowslip*).— Bright, graceful, and perfectly hardy, is second to none of our old border-flowers, supported in umbels on straight slender stems from 10 to 16 inches high, each flower drooping elegantly, the purplish petals springing up vertically from the pointed centre of the flower, much as those of the greenhouse Cyclamen do. It inhabits rich woods in North America, from Maryland and Pennsylvania, in the North, to North Carolina and Tennessee, in the South, and far westward, loves a rich light loam, and is one of the most suitable plants for the rock-garden. In deep light loams, the plant flourishes without any preparation, but where a place is prepared for it, as is often necessary, it is well to add plenty of leaf-mould. In a somewhat shaded position, it attains its greatest size, and beauty, though it thrives in exposed borders, and is best increased by division when the plants die down in autumn; when seed is sown, it should be soon after being gathered.

D. Integrifolium (*Small American C.*). —A lovely and gaily-coloured flower, deep rosy crimson, the base of each petal white, springing from a yellow and dark orange cup, and appearing in May on stems from 4 to 6 inches high. The leaves are much smaller than those of *D. Meadia*, oval, and quite entire. A native of the Rocky Mountains, a gem for the rock-garden, planted in sandy loam with leaf-mould, and increased by careful division of the root and by seed, which it ripens freely in this country. It is easily grown in pots, plunged, in the open air, in some sheltered and half-shady spot during summer, and kept in shallow cold frames during winter.

Dodecatheon Jeffreyanum (*Great American C.*).—A noble kind, which I have grown as high as 2 feet in favourable soil, and have known to grow much larger even in London gardens than the old American Cowslip. It has much larger and thicker leaves, of a darker green, and with very strong and conspicuous reddish midribs, the flower being like that of the old kind, except that it is somewhat larger and darker in colour. It is a hardy and first-class plant, flourishing freely in light deep loam, and thriving in a warm and sheltered spot, where its great leaves may not be broken by high winds.

DONDIA EPIPACTIS (*Dwarf Masterwort*).—A most unusual form of the umbel-bearing plants, and amongst our earliest flowers. It grows only some 3 or 4 inches high, and though the blossoms individually are small, they are surrounded with a bright yellow involucre, retaining its fine colour for nearly two months of the spring. It is a strong-rooted plant, likes a good stiff loam, and is perfectly hardy. Carinthia and Carniola. Syn., *Hacquetia*.

DRABA (*Whitlow Grass*).—Minute alpine plants, most of them having bright yellow or white flowers, and leaves often in neat rosettes. They are too dwarf to take care of themselves among plants much bigger than Mosses, and therefore should be grouped with the dwarfest plants.

In addition to the golden colour of the flowers of one section, the plants are characterised by a dwarf compact habit, and by much neatness in the arrangement of the bristly ciliated hairs, which not unfrequently become bifurcate; thus the attractive appearance in the matter of colour is enhanced on a closer inspection by the beauty of form. In another section we find white to be the predominant colour, and though in many cases the flowers are small, still, in the mass, filling up a nook or crevice, and contrasted with the dark-green leaves, they become very effective. They should be placed in the sunniest aspects; the more effectually the plants are matured by the autumn sun the more freely will they return these favours by an abundant bloom in early spring. The third section, which includes plants of a purple and violet colour, is chiefly, if not altogether, confined to the high mountain lands of South America. Of these we have but one in cultivation, *Draba violacea*, and of so recent introduction that it may be considered rash to pass any opinion on it beyond the fact that it is a remarkably beautiful plant, of doubtful hardiness.

Draba aizoides (*Seagreen Whitlow-Grass*).—This may be taken as typical of the Golden *Draba*; it is indigenous to Britain, but only found in one locality in South Wales. In growth it does not exceed 3 inches in height, and when planted on the slope of a sunny border, in sandy soil, which it loves, it forms a dense yellow carpet in the early part of March. It does not ripen seed freely, but increases readily by division.

D. aizoon (*Evergreen Whitlow-Grass*).—A native of the mountains of Carinthia, and a vigorous grower; the leaves of a dark green, and arranged so as to form a complete rosette, not unlike the *Sempervivums*. From the centre of this rosette it sends up a stem 5 or 6 inches long, bearing numbers of bright-yellow flowers, and ripens its seeds freely. *Draba bœotica* I am disposed to consider a narrow-leaved form of the above. In the cultivation of both it must be borne in mind that, unlike *D. aizoides*, the old stems will never throw out roots, consequently they cannot be classed as spreading plants. They increase freely from seed, some of which it would be interesting to sow on old walls.

Draba alpina (*Alpine Whitlow-Grass*).—An arctic plant, with dark green, smooth leaves, growing about 2 inches high, and bearing bright golden flowers. It is rather a delicate plant, and best adapted for pot culture, or well-drained chinks in the rock-garden. The true species is somewhat scarce in cultivation. It, like *D. tridentata*, is liable to suffer from slugs, and both should be carefully guarded against their attacks, especially during the winter months. Allied to this is *Draba aurea*, a Danish plant, with flowers produced in a dense corymb, on a leafy stem some 8 or 9 inches high; the habit is not neat, otherwise it is a well-defined species.

D. ciliata (*Eye-lashed Whitlow-Grass*).—This is a good white *Draba*, not unlike a diminutive specimen of *Arabis albida*. The leaves are sparsely but distinctly ciliated, in loose rosettes. Flowers in early spring; pure white, about eight on a stem, the whole plant when in bloom not being more than 2 inches high. Mountains of Croatia and Carniola.

D. cinerea (*Grey Whitlow-Grass*).—This native of Siberia, frequently called *D. borealis*, is the most effective of the white-flowering *Drabas*. Of dwarf habit, bearing many clear white flowers in the earliest spring, well relieved by the dark green leaves, and of a free-growing habit, it should be in every collection. Seeds abundantly, and by that means, as well as by division, it may readily be increased.

D. cuspidata (*Pointed Whitlow-Grass*) —A native of the highest mountains in Spain, with the points of each of the ciliated leaves, of which the dense little rosettes are formed, somewhat incurved, and for close examination it is the gem of the yellow *Drabas*, forming a thick woody stem. It is only to be increased by seed.

D. lapponica, a native, as the name indicates, of the arctic regions, though bearing the aspect of *D. rupestris*, is dwarfer in habit, and devoid of the ciliated hairs on the leaves; it forms dense tufts, and flowers freely in early spring, producing an almost equally abundant bloom in the autumn; it also seeds freely.

D. rupestris, frigida, and *Chamæjasme,* are three very dwarf plants, closely allied, in fact so much so that they may be considered as varieties. The flowers in each case are small, but are produced abundantly. Considering the neat habit of the plants, every collection should possess at least one of them.

D. nivalis, a native of the Swiss Alps, is the most diminutive of the genus. The leaves are of a whitish-green, owing to the presence of minute stellate hairs. The plant, when in flower, is not over 2 inches high, of nice compact habit, but rather a shy grower, and is rarely met with.

Draba glacialis (*Glacier Draba*).—A very dwarf kind, forming dense little cushions 1 to 2 inches high, which in April are covered with bright golden-yellow flowers. Leaves linear, smooth, ciliated, forming small rosettes closely packed in pincushion-like masses. The plant very much resembles a small specimen of *D. aizoides,* and is considered by Koch to be a variety of that, growing at a higher elevation; but it differs from it by having a few-flowered stem, pedicels shorter than the pod, and a short style. It is found on the granitic Alps of Switzerland, and is suited for exposed spots in the rock-garden, in moist and very gritty soil, and associated with the dwarfest alpine plants.

DRACOCEPHALUM (*Dragon's-head*).

—Plants of the Sage family, among which are a few choice perennials, suitable for the rock-garden, succeeding in light garden soil, and increased by division or seed.

Dracocephalum Austriacum (*Austrian D.*).—A showy species, with blue flowers more than an inch and a half long, in whorled spikes, the plant of rather a woody texture, spreading into masses about a foot high, the floral leaves velvety, and with long fine spines. A native of nearly all the great mountain chains of Europe, thriving in light soil, and increased by seed or division. Quite free to grow in most garden soils, but, like many other mountain plants, only attaining ripeness of texture on well-drained, warm, and sandy soils.

Dracocephalum grandiflorum (*Betony-leaved D.*).—A plant rarely seen in our gardens; it is distinct, not diffuse or procumbent, in habit more like a dwarf Betony; the flowers, handsome, blue, in whorled oblong spikes, 2 to 3 inches long; the plant little more than half a foot high, though it varies from 2 inches to a foot high. Native of Siberia, and thriving in sandy and thoroughly-drained loam, it should be guarded against slugs, which may quickly destroy young and small plants. Flowers in early summer, and increased by division.

D. **Ruyschianum** (*Ruysch's D.*).—Flowers in rather close spikes at the summit of the stem; the floral leaves also entire. A pretty perennial, flowering rather late in the summer, and thriving on slightly elevated spots, for which it is well fitted by its spreading, somewhat prostrate, habit, forming tufts about a foot high. Division or seed.

Other kinds (omitting the taller, more herbaceous kinds) are: *Botrioides,* with purple flowers, *Ruyschianum, japonicum, argumense,* and *Ruprechtii;* but though likely to thrive, seldom effective in southern gardens.

DROSERA (*Sundew*).

—Interesting little bog-plants, of which all the hardy species but one are natives of Britain and characterised by leaves, their surfaces covered with dense glandular hairs. When the native kinds are grown artificially, the condition of their natural home should be adopted as far as possible. In a bog on a very small scale it is not easy to secure the humid atmosphere they have at home, but they will grow wherever *Sphagnum* grows. The native kinds are *intermedia, longifolia, obovata,* and *rotundifolia.* The North American Thread-leaved Sundew (*D. filiformis*) is a beautiful plant, with very long slender leaves covered with glandular hairs, the flowers purple-rose colour, half an inch wide, opening only in the sunshine. Quite hardy, but difficult to cultivate.

DRYAS OCTOPETALA (*Mountain Avens*).—Few have travelled in alpine districts without seeing how abundantly the mountains are clothed with the creeping stems and large creamy-white flowers of this plant. An evergreen, good in habit as well as handsome in bloom, it ought to be grown in every collection of rock-plants. Widely distributed through the mountain region of Europe, Asia, and North America, and very abundant in Scotland. Easy of culture in moist peat soil, in which it grows so freely about Edinburgh, that it is used for edgings to beds in some nurseries. Seed, or by cuttings and division. The var. *minor* is dwarfer and dense in habit. *D. tenella* is a rare species from Labrador. *D. Drummondi*, very like it, but with yellow flowers, is also in cultivation.

ECHINOCACTUS SIMPSONI.—A beautiful little Cactaceous plant, native of Colorado, high on the mountains, and hardy enough for our climate. It grows in a globular mass, 3 or 4 inches across, covered with white spines. Flowers early in March in this country, the blossoms large, pale purple, and very beautiful. The natural conditions should be imitated as far as may be. It enjoys a dry climate, and is, more or less, protected from the effects of frost by a covering of snow. In this country it has withstood 32° of frost without injury, and, therefore, if in a dry spot, it may escape and flourish.

EMPETRUM NIGRUM (*Crowberry*). —A small evergreen Heath-like bush, of the easiest culture. May be planted with the dwarfer and least select rock shrubs. It is a native plant, and the badge of the Scotch clan M'Lean.

EPIGÆA REPENS (*May Flower*).— A little trailing evergreen bush, found in sandy or rocky soil, especially in the shade of pines, common in many parts of North America, with delicate rose-coloured flowers in small clusters, exhaling a fine odour, and appearing in early spring. It is a plant very seldom met with in good health in this country, and, in planting it, it would be well to bear in mind that its natural habitat is under trees, and plant a few in the shade of pines or shrubs. In New England it is known as the May Flower. It is so common in the cold sandy woods of Eastern America in poor sandy soils, that it is not easy to see why it should not thrive with us.

EPILOBIUM (*Willow Herb*).— Some of these perennials are occasionally grown among alpine plants, but are usually too large for the rock-garden, with the exception of *E. obcordatum*. This, which is by far the dwarfest of the alpine Willow-herbs, forms handsome little tufts, 3 or 4 inches high, and bearing late in summer large rosy-crimson blossoms. Coming from the summits of the Sierra Nevada, it is hardy, and one of the most attractive of rock plants, thriving in sandy loam.

EPIMEDIUM (*Barrenwort*).—Interesting and graceful perennials with finely formed leaves, evergreen in favourable conditions, and precious for the rock-gardener; all the more so, for those who think with me that the hard-and-fast idea of a rock-garden should give way to the more natural one of the association of the mountain shrub, and the best perennials with the smaller alpine plants. The *Barrenworts* thrive nowhere so well as among the peat-loving shrubs. We should use in such a case the partial shade of the shrubs as well as the soil suiting them. They also form beautiful carpets below the shrubs, covering the ground as it always should be.

Epimediums are typical of many garden plants, which, though of great beauty when naturally grown, are rarely artistically used—I mean by the word rightly used, both as regards culture and placing. It should never be forgotten that good culture and effect may often go together. Such plants as these are often dotted singly, and among other and coarser things, and they suffer and eventually may disappear under some coarse shrubs or plants. But if we plant them so that they form an effective group, we are not so likely to forget them, and it is then better worth our while to give them the shade and position they want. I have seen these plants grown in the open in botanic and other gardens without a bit of shade; but place them in partial shade of what we call American shrubs, in peaty soil, and within good broad groups, and the effect will be one of the best we can see in the garden. It would be a case of cultivation and effect and simplicity of culture going well together, because, if we know that one place is given up to a certain group, we are not likely to make mistakes with it, as in the general muddle of the mixed border.

E. pinnatum is a hardy perennial from Asia Minor, 8 inches to 2½ feet high, with handsome leaves, and bearing long clusters of yellow flowers. The old leaves remain until the new ones appear in the ensuing spring. It is not well to remove them, as they shelter the buds of the new leaves during the winter, and the plants flower better when they are allowed to remain. Cool peaty soil and a slightly shaded position are most suitable. They thrive in half-shady spots in peat, or in moist sandy soil. None are so valuable for general culture as the first-mentioned. The other species are *E. alpinum*, Europe ; *concinnum*, Japan ; *elatum*, Himal. ; *macranthum*, Japan ; *Musschianum*, Japan ; *Perralderianum*, Algeria ; *pteroceras*, Caucas. ; *pubescens*, China ; *pubigerum*, Caucas. ; and *rubrum*, Japan.

EPIPACTIS PALUSTRIS (*Marsh E.*).—A pretty hardy Orchid, 1 to 1½ feet high, flowering late in summer, with handsome purplish flowers. A native of moist grassy places in all parts of temperate and Southern Europe. A good plant for the bog-garden, or for moist spots near a rivulet, in moist peat. In wet districts, it thrives very well in ordinary soil.

ERANTHIS HYEMALIS (*Winter Aconite*).—A small plant, with yellow

Eranthis Hyemalis (*Winter Aconite*).

flowers, surrounded by a whorl of shining-green divided leaves, with a short, blackish, underground stem resembling a tuber; the flowers, an inch or more across, being thrown up on stems from 3 to 8 inches high. It is naturalised in woods and copses in various parts of the country, but has probably escaped from cultivation, and is not considered a native, its true home being shady and humid places on southern continental mountains. It is pretty well known, being frequently sold by our bulb merchants, and is too common a plant for the choice rock-garden.

Eranthis cilicica is another kind of like use, but which may for a time deserve a better place on the rock-garden than the easily-grown winter Aconite, as free as a weed in any open and chalky soils.

ERICA (*Heath*).—Wiry and usually rather dwarf hill and moor shrubs of much native charm. Some of the prettiest inhabit our own country, and these break into varieties of distinct value for the garden. If there were no other plants than these, we could make pretty rock or moor gardens of them, even in hot and poor soils, and these and a few other plants, such as Brooms, Sun Roses and Rock Roses, might adorn many a hot slope of poor ground, the smaller kinds the rock-garden, the larger coming into the shrubby parts near. Even some of the tender ones of Southern Europe are very happy in mild districts in our climate. Several of the taller and less hardy Heaths are here omitted—the best kinds for the rock-garden given.

Erica Australis (*Southern Heath*).—A pretty bush Heath of the sandy hills and wastes of Spain and Portugal, 2 feet to 3 feet high, flowering in spring in Britain. The flowers are rosy purple and fragrant.

Erica carnea (*Alpine Forest Heath*).—A jewel among mountain Heaths, and hardy as the Rock Lichen. On many ranges of Central Europe at rest in the snow in winter, in our mild winters, it flowers early, and in all districts is in bloom in the dawn of spring—deep rosy flowers, the leaves and all good in colour. Syn., *E. herbacea*.

E. cinerea (*Scotch Heath*).—A dwarf Heath, common in many parts of Britain, very easily grown, and with pretty varieties of white and various colours. Its flowers of reddish purple begin to expand early in June. Among its varieties are *alba, bicolor, coccinea, pallida, purpurea,* and *rosea*.

E. ciliaris (*Dorset Heath*).—A lovely dwarf Heath, and as pretty as any Heath of Europe. A native of Western France and Spain in heaths and sandy woods, it also comes into Southern England. The flowers are of a purple-crimson, and fade away into a pretty brown, thriving also in loamy as well as in peaty soils, and flowering from June to October.

E. hybrida (*Hybrid Heath*).—A cross between *E. carnea* and *E. mediterranea*. It is a charming bush, and flowers freely in winter and far into the spring, thriving in loamy soil almost as well as in peat.

E. hibernica (*Irish Heath*).—Mr Boswell Syme, whose knowledge of British plants was most profound, considered this Irish plant distinct from the Mediterranean Heath, "the flowering not taking place in the Irish plant till three or four months after the Mediterranean Heath;" a fine shrub in Mayo and Galway, growing from 2 to 5 feet high.

E. lusitanica (*Portuguese Heath*).—This is for Britain the most precious of the taller Heaths, 2 to 4 feet high, and, hardier than the Tree Heath, it may be grown over a larger area. Even in a cool district I have had it in a loamy soil ten years, and almost every year it bears lovely wreaths of flowers in mid-winter, white flowers with a little touch of pink, in fine long Foxbrush-like shoots. In about one year in five, it is cut down by frost, but usually recovers, and is a shrub of rare beauty for sea coast and mild districts. Syn., *E. codonodes*.

Erica mediterranea (*Mediterranean Heath*).—A bushy kind, 3 to 5 feet high, best in peat, and flowering prettily in spring. Although a native of Southern Europe, it is hardier in our country than the Tree Heaths of Southern Europe. Of this species there are several varieties.

E. stricta (*Corsican Heath*).—A wiry-looking shrub, compact in habit, about 4 feet high, and a handsome plant. A native of the mountains of Corsica, flowering in summer.

E. tetralix (*Marsh or Bell Heather*).—This beautiful Heath is frequent throughout the northern, as well as western, regions, thriving in boggy places, but also in ordinary soil in gardens. This Heath has several varieties, differing in colour mainly. *E. Mackaiana* is thought to be a variety of the Bell Heather. *E. Watsoni* is a hybrid between the bell heather and Dorset Heath.

E. vagans (*Cornish Heath*) is a vigorous bush Heath, thriving in almost any soil, 3 to 4 feet high. A native of Southern Britain and Ireland, and better fitted for bold groups in the pleasure ground or covert than the garden. There are several varieties, but they do not differ much from the wild plant.

E. vulgaris (*Heather: Ling*).—As precious as any Heath is the common Heather and its many varieties, none of them prettier than the common form, but worth having, excluding only the very dwarf and monstrous ones, which are useless, except in the rock-garden, and not of much good there. Heathers are excellent to clothe a bare slope of shaly soil, not taking any notice of the hottest summer in such situations. Among the best varieties are *alba, Alporti, coccinea, decumbens, Hammondi, pumila, rigida, Searlei,* and *tomentosa.* Syn., *Calluna.*

E. dabœcii (*Dabæcs Heath*).—A beautiful shrub, 18 inches to 30 inches high, bearing crimson-purple blooms in drooping racemes. There is a white variety even more beautiful, and one with purple and white flowers, called *bicolor.* I have had the white form in flower throughout the summer and autumn on a slope fully exposed to the sun, and in very hot years, too. Syn., *Menziesia polifolia,* also *Dabœcia* and *Boretta.* West of Ireland.

ERIGERON (*Fleabane*).—Michaelmas Daisy-like plants of dwarf growth, somewhat alike in general appearance, and having pink or purple flowers with yellow centres, and a few of the dwarfest suited for the rock-garden. Of these, *E. alpinum grandiflorum* is the finest. It is similar to the alpine Aster, having large heads of purplish flowers in late summer, and remaining in beauty a long time. *E. Roylei*, a Himalayan plant, is another good alpine, of very dwarf tufted growth, having large blossoms of a bluish-purple, with yellow eye. *E. mucronatus,* known also as *Vittadenia triloba,* is a pretty Daisy-like flower, compact, and for several weeks in summer is a dense rounded mass of bloom about 9 inches high. The flowers are pink when first expanded, and afterwards change to white. All are easily increased by division in autumn or spring.

ERINUS ALPINUS (*Wall E.*).—A pretty and distinct little plant, with many violet-purple flowers in short racemes, over very dwarf tufts of downy, toothed leaves. A native of the Alps of Switzerland, the Tyrol, and the Pyrenees, perishing in winter on the level ground in most gardens, but permanent on old walls or ruins. I have seen brick garden walls with every chink between the bricks filled with this plant, so as to look at a distance as if covered with moss in winter, and in summer becoming covered with masses of lovely colour. It is easily established on old walls, by scattering the seeds in mossy or earthy chinks, and is of course well suited for the rock-garden, growing thereon in any position, often flowering bravely on earthless mossy rocks and stones, naturalised on the Roman remains at Chesters, Northumberland

On my own walls there is a pretty variety of colour, purple, white, and a pretty rose. Do not try to cover the seed in.

ERIOGONUM.—North American plants which, as seen on the Rocky Mountains and alpine regions in California, are of much beauty, but which I have never seen good in cultivation, except, perhaps, *E. umbellatum*, which, from a dense spreading tuft of leaves, throws up numerous flower-stems, 6 inches to 8 inches high, with yellow blooms in umbels 4 inches or more across, forming a pretty tuft. It is worthy of a place on any rock-garden or border, in light, sandy soil, in which it has never failed to bloom profusely. Other species are *E. compositum, flavum, racemosum, ursinum*.

ERITRICHIUM NANUM (*Fairy Forget-me-not*).—An alpine gem, closely allied to the Forget-me-nots, which it far excels in the intensity of the azure-blue of its blossoms. Though reputed to be difficult to cultivate, a fair amount of success may be ensured by planting it in broken limestone or sandstone, mixed with a small quantity of rich fibry loam and peat, in a spot in the rock-garden where it will be fully exposed, and where the roots will be near masses of half-buried rock, to the sides of which they delight to cling. The chief enemy of this little plant, and indeed of all alpine plants with silky or cottony foliage, is moisture in winter, which soon causes it to damp off. In its native mountains it is covered with dry snow during that period. Some, therefore, recommend an overhanging ledge, but if such protection be not removed during summer, it causes too much shade and dryness. A better plan is to place two pieces of glass in a ridge over the plant, thus keeping it dry, and allowing a free access of air, but these should be removed early in spring. Alps of Europe, at high elevations.

ERODIUM.—Dwarf, greyish rock plants of the Geranium order, but less vigorous, and suited for warm and sunny spots on the rock-garden, also for dry walls where such are made.

Erodium carvifolium (*Caraway-leaved Heronsbill*).—A good perennial species, 6 to 10 inches high, producing red flowers larger than those of *E. romanum*, the whole plant being more vigorous, and more decidedly perennial than that species. A native of Spain.

E. macradenium (*Spotted Heronsbill*).— Allied to the rock Heronsbill, but distinguished from it by the two upper petals being marked with a large blackish spot, the lower petals being larger and of a delicate flesh-colour, veined with purplish rose, two to six flowers being borne on stalks from 2 to 6 inches high. The flowers are pretty, and the entire plant has an agreeably aromatic fragrance. It is easily grown in chinks and dry spots, in warm rather than rich soil, and is increased from seeds, and also by division. Pyrenees.

E. manescavi (*Noble Heronsbill*).—A showy kind, with long, much divided leaves, from which spring many stout flower-stems, each bearing an umbel of from five to fifteen handsome purplish flowers, each more than an inch across. It is distinct, and deserves a place in every collection, flourishing on the level ground, and being a vigorous grower, it should be associated with the strongest rock plants only. A native of the Pyrenees, flowering in summer, and, when the plants are young and in fresh soil, for a long time in succession. Easily raised from seed, and in cultivation grows from 10 inches to 2½ feet high.

E. petræum (*Rock Heronsbill*).—A small kind, with much divided, somewhat velvety leaves, and rather large, lively rose, or white-and-veined, but not spotted flowers, from 3 to 6 inches high, and thriving in warm and dry chinks or nooks on the sunny sides of rock. It is a plant to try on old walls; on the level ground the leaves grow fat at the expense of the flowers, and the softness of tissue resulting, causes them to perish in winter. There is a smooth variety, *E. lucidum*, and one with more curled and downy leaves, *E. crispum*; all are natives of dry rocky places in the Pyrenees and Southern Europe, and are increased by seed or division.

Erodium Reichardi (*Reichard's Heronsbill*).—A tufted stemless plant, a native of Majorca. The heart-shaped little leaves rest upon the ground, and the flower-stems attain a height of 2 or 3 inches, each bearing a solitary white flower, faintly veined with pink. It flowers freely, and usually from spring or early summer till autumn; is quite easy of culture in moist sandy soil, on bare exposed spots or in chinks.

E. Romanum (*Roman Heronsbill*).—A pretty species, with gracefully cut leaves like those of the British *Erodium cicutarium*, to which it is allied; but it differs in having larger flowers, in being stemless and a perennial; the flowers purplish, in the end of March or beginning of April. It is easily grown, and comes up thickly from self-sown seeds, at least in light and chalky soils; would thrive on old walls. S. France and Italy.

ERPETION RENIFORME (*New Holland Violet*).—This mantles the ground with a mass of small leaves, has slender, creeping stems, and blue and white flowers of exquisite beauty, rising not more than a couple of inches from the ground. A Violet it is indeed, but a Violet of the southern hemisphere, and without the vigour and depth of colour of our northern sweet Violet. It is good for planting out over the surface of a bed of very light earth, in which some handsome plants would be put out during the summer in a scattered manner, and the little Violet allowed to creep over the surface. Being small and delicate as well as pretty, it should not be used under or around coarse subjects. It must of course be treated like a half-hardy plant—taken up in autumn, and put out in May or June. In every place where alpine plants are grown in pots, it should find a home; and in mild parts of these islands, say the south and west coast, it would probably maintain its ground without perishing during winter. Syn., *Viola*.

ERYNGIUM (*Sea Holly*).—Though some of the plants of this are beautiful, and some inhabit alpine lands, they are almost, without exception, too large for the rock-garden, though they may be grown with good effect among shrubs near it. The same remarks, however, apply to many fine perennials.

ERYSIMUM.—This is a little genus of alpine plants, very much resembling alpine wall-flowers, but of much less value, though one or two are pretty for the alpine garden.

Erysimum pumilum (*Liliputian Wall-flower*).—Resembling in the size and colour of its flowers the alpine Wallflower, but without the rich green foliage of that, but with flowers large for the size of the plant, often only an inch high, above a few narrow leaves barely rising above the ground. I have seen it in bloom with flowers nearly as large as those on the alpine Wallflower, and yet flowers and all could be almost covered by a thimble. In richer soil and less exposed spots it is larger. A native of high and bare places in the Alps, it should be grown in an exposed spot in very sandy loam, surrounded by a few small stones to guard it from drought and accident, and associated with the smallest alpine plants.

E. Rhæticum (*Rhætian Wallflower*).—A pretty mountain flower which, though rare in cultivation, is a common alpine in Rhætia and the neighbouring districts, where in early summer its broad dense-tufted masses are aglow with clear yellow blossoms. *E. canescens*, a South European species, with scentless yellow flowers, is also a good alpine plant, and so is *E. rupestre* easy to grow, and thriving in gritty soil and well-drained spot.

ERYTHRÆA (*Centaury*).—A small genus of rather pretty dwarf biennials, belonging to the Gentian family. The native species, *E. littoralis*, common in some shore districts, is worth cultivating. It is ↳ to 6 inches high, and bears an abundance of rich pink flowers, which last a considerable time in beauty, and will withstand full exposure to the sun, though partial shade is beneficial. The very beautiful *E. diffusa* is a similar species. It is a rapid grower, with a profusion of pink blossoms in summer.
Erythræa Muhlenbergi is another beautiful plant. It is neat and about 8 inches high, putting out many slender branches. It bears many flowers, and the blossoms are 3½ inches across. They are of a deep pink, with a greenish-white star in the centre. Seeds should be sown in autumn, and grown under liberal treatment till the spring; the plants will then flower much earlier and produce finer flowers than spring-sown plants. They are excellent for the rock-garden and the margins of a loamy border, but the soil must be moist. On account of their duration or other peculiarities, they are of more botanical than garden interest.

ERYTHRONIUM (*Dog's - Tooth Violet*).—Graceful and distinct bulbous plants, dwarf, hardy and well suited for our purpose. The European kind is a charming flower with handsome spotted leaves and drooping flowers, of which there are various coloured varieties. No need to speak here of its cultivation, as it is one of the easiest plants to naturalise in grass. The most interesting of the family are the American kinds recently come to us; these have a graceful habit and beauty. Like so many other plants, they are best in warm light soils.
Erythronium dens-canis (*Dog's-tooth Violet*).—One of the hardiest of the mountain plants, its handsome oval leaves pointed above, marked with patches of reddish brown, the flowers singly on stems 4 to 6 inches high, drooping, and cut into six rosy purple or lilac divisions. There is a variety with white, one with rose-coloured, and one with flesh-coloured flowers. It is one of the best plants for the spring or rock-garden, and will grow in any ordinary soil. The bulbs are white and oblong; hence its common name; and it is increased by dividing them every two or three years, replanting rather deeply. Alps.
E. Americanum.—The commonest kind in the Eastern United States of North America, narrow in foliage, with bright yellow pendent flowers. It is a good and free growing plant, but in our country fails to flower on some cold and heavy soils. To ensure its doing so, plant in warm open sandy soil. The main interest of these plants, however, is centred in the fine kinds from the North Pacific coast, including the Rocky Mountains and a vast region of tree-clad mountains, a thousand or more miles across, from which all of these plant treasures are not yet gathered. In some soils in our countries they do not thrive, requiring soils of a leafy and open nature, which accounts for their slow and uncertain growth in heavy soils, like some of those around London. The following by one who knows them well in their native homes, is invaluable as a guide:—

Erythroniums are woodland plants, and need some shade to develop the leaves and stems. Partial shade by trees will answer. I give my beds a lath shade. I have for several years been experimenting with soils for them. While often found in heavy soils, they make better growth in a soil of rocky *debris* mixed with leaf mould. Much of the charm of Dog's-Tooth Violets is in their large leaves and tall slender stems. Rocky *debris* has not been available, and I have tried several substitutes, but have discarded all for a soil of one-half to one-third half-rotten spent tan-bark with sandy loam. Our tan-bark here is the bark of the Tanbark Oak (*Quercus densiflora*), and is ground at the tannery. This gives a soil rich in mould and always loose and porous. It suits the needs of *Erythroniums* exactly, and answers well for many other bulbs. They should always be planted early, as, with few exceptions, the bulbs are not good keepers after the fall, and the sooner they are in the ground after the first of October the better. I plant them so that the top of the bulb is about 2 inches from the surface. The drainage should be good. With these essentials, shade, drainage, and a loose soil, success is very probable. Although quite hardy, a heavy coat of leaves, such as Nature protects them with in their woodland home, would be a wise precaution in cold climates. They do not seem to have any peculiar disease, and growing and flowering as early as they do, artificial watering is not necessary. In the region including the Rocky Mountains and the country westward to the Pacific, fifteen forms are now known, classed as species and varieties. A more charming group of bulbous plants does not exist. Their leaves show a variety of mottling, and in the flowers delicate shades of white, straw-colour, and deep yellow, deep rose, pink, light and deep purple are represented. To describe all of these forms, so that even a botanist could readily identify them by the descriptions, would be difficult, but in the garden each has some charm of leaf, of tint, or of form. In their native homes they grow throughout a wide range as to climate and altitude, and in cultivation they maintain their seasons, so that the display which is opened by *E. Hartwegi* with the Snowdrops and earliest Narcissi, is closed by *E. montanum* and *E. purpurascens* when the others have flowered and become dry. *E. Hartwegi* can be propagated freely by offsets; all of the others come from seed. A bulb may have an offset occasionally, and sometimes a clump of four or five will form in some years, but the contrary is the rule. It is all important in handling the bulbs of *Erythroniums* that they should not be allowed to dry out. Many of the failures are owing to lack of care in this respect. If properly handled, they can be kept in good condition out of the ground until midwinter or even February, although early planting is always advisable. The bulbs should be kept in a cool place in barely moist earth or peat or Sphagnum, and in shipping carry best in Sphagnum barely moistened. In dry or hot air, they soon become hollow, and their vitality is impaired.

Erythronium grandiflorum.—The species is not to be confused with *E. giganteum*, which has straw-coloured flowers and richly mottled leaves. Nearly all of the bulbs grown heretofore as *E. grandiflorum* are really *E. giganteum*. The true *E. grandiflorum* has light green leaves, entirely destitute of mottling, the filaments slender and the style deeply three-cleft. There are four strong-coloured forms, each of which has a wide distribution. Mr Watson, in his revision, only mentions two of these, and is incorrect as to localities. They are

(1) The type of the species, one to five-flowered, stout, flowers a bright clear yellow. This is the species which was exhibited recently in London as *E. Nuttallianum*. Eastern Oregon.

(2) Var. **Nuttallianum.**—This only differs from the type in having red anthers.

(3) On the high peaks of Washington, there is a form with white flowers with yellow centres. It is one to five-flowered, and from very low to 18 inches, according to soil and situation. Watson's var. *parviflorum* is accredited to the same localities, but, acccording to him, is bright yellow.

(4) Var. **album**, a form having pure white flowers with a yellow centre and a greenish cast, one to five-flowered. This handsome form grows in the Pine forests in a low rolling region of Eastern Washington. In cultivation I find some difficulty with *E. grandiflorum*, from its tendency to flower too quickly. The plants will often come through the ground with the flower half expanded. In the cooler climate of Northern Europe, which is more similar to that of its native home, it will do much better. Rocky Mountains, Colorado, and British America.

Erythronium Hartwegi is not only the earliest but also the most easily grown of all, and unique in its habit. Its leaves are mottled in dark green and dark mahogany-brown. The two to six flowers are each borne on a separate slender scape, and form a sessile umbel. The general effect of a well-grown plant is of a loose bouquet with the two richly mottled leaves as a holder. The segments recurve to the stalk, and are light yellow with an orange centre. Well-grown flowers measure 2 inches to 2½ inches across. Its bulbs are short and solid, producing small offsets, and, unlike most sorts, they retain their vitality until late in the season, and are in good condition in February, when bulbs planted earlier are in flower. Sierra Nevada of California.

E. montanum.—This is an alpine species from the high peaks of the Cascades, in Oregon and Washington. The leaves are without mottling, and alone among *Erythroniums* are abruptly contracted at base with a slender unmargined petiole. The flowers are pure white with an orange centre, resembling in shape those of *E. giganteum*. Its bulbs are peculiar in having the old rootstock persistent, and showing the annual scars of many years. Often it forms a spiral around the bulbs.

E. purpurascens has unmottled leaves, which in the earlier stages of growth are strongly tinged with purple and become dark green. The segments of the perianth are not reflexed, as in all the others, but spreading white to creamy, with orange centre, and turning purplish. The flowers are small and crowded in a raceme, style not divided. A very distinct species, growing in the higher regions of the Sierra Nevada, in California. As a garden plant it is not to be compared to the others here described. Bulbs obtained from high altitudes flower very late with *E. montanum*; from lower altitudes they flower a little earlier than *E. giganteum*.

E. revolutum is a widely scattered species, extending along the coast from Sonoma County, California, to the central part of British America, usually in deep forests. It is a plant of low altitudes, the leaves always mottled; filaments broad and awl-shaped, the style large and prominent and three-cleft; the scape stout and usually one-flowered, but sometimes three to five-flowered. *E. revolutum* can always be identified by the broad filaments and prominent appendages. I have seen six well-marked variations.

(1) *The Species.*—This has broad leaves mottled with white or seldom with light brown, scapes stout, 6 inches to 15 inches high. The petals are narrow; at first white to delicate pink, they soon become purple. This form was the first *Erythronium* collected, being found by Menzies in British Columbia over a hundred years ago, and described as *E. revolutum*. It was lost sight of until a year ago I found a form in the Redwood forest of Mendocino County, California, which is identical with the original. These two points are 1000 miles apart, but I have since found several intermediate habitats, and it stretches along the coast the entire distance in a long narrow band.

(2) Var. **Bolanderi.**—This seems to be a local low-growing form very similar to the last, but the flowers are white, only tardily becoming purplish. Eel River Valley, Mendocino County, California.

(3) Var. **Johnsoni** (*E. Johnsoni*, Bolander).—This exquisitely beautiful kind has broad leaves mottled with white, and looking as if varnished. The flower is of a delicate reddish tint with orange centre. Well illustrated in a *Garden* plate, 20th February, 1897. North-Western Oregon.

(4) **Erythronium revolutum** (*Creamy Form*).—This, according to Mr Watson, is the type of the species, but as variety No. 1 is proved to be the original, it becomes a variety. The leaf is more darkly mottled than in either of the others with brown or dark brown. The petals are broad and of much substance, and become reflexed more tardily than most *Erythroniums*, although at length closely reflexed. In colour it is from light to dark cream, with a greenish cast, and a yellow centre. It is one of the best in cultivation. Coast ranges, Oregon and British Columbia.

(5) **E. revolutum var. albiflorum.**—This beautiful variety is like the preceding, except that the ground colour is pure white, with a slight greenish cast. It was

described in Europe as *E. grandiflorum* var. *albiflorum* in *Gardeners' Chronicle*, 1888, t. 77. It had also been described as *E. giganteum* var. *albiflorum*. It is one of the most beautiful of *Erythroniums*.

Erythronium giganteum has long been known and grown as *E. grandiflorum*. While its flowers are no larger than in the other kinds, it excels all in height and number of blossoms. I have often seen it with eight or ten flowers, and once with sixteen. The leaves are mottled with white and brown, or deep brown; the flowers light yellow, with a deeper centre, and often banded with brown. The filaments are very slender, and the style three-cleft. It can be distinguished from *E. grandiflorum* by its mottled leaves, from *E. revolutum* by the slender filaments and small appendages. Its range is a broad belt in the coast ranges from San Francisco Bay north to Southern Oregon.

E. citrinum resembles *E. giganteum*, but has an undivided style. The leaves are mottled, the flower light yellow, with an orange base. Southern Oregon.

E. Hendersoni is another species also closely resembling *E. giganteum*, but easily distinguished by its undivided style and purple flowers with an almost black centre. Southern Oregon.

E. Howelli.—This alone of the western *Erythroniums* has no appendages at the base of the petals. By this character, with its undivided style, it can always be identified. The flowers are pale yellow with an orange base. Southern Oregon.

CARL PURDY, in *Garden*.

FRAGARIA (*Strawberry*).—The wild strawberry is very pretty on banks, and occasionally most useful on old mossy garden walls, where it establishes itself. One kind, *F. monophylla*, is a beautiful rock-garden plant, with large white flowers. The Indian strawberry, *F. indica*, is a pretty little trailer, bearing many red berries and flowering late. All are of the easiest culture in any not too wet soil, and of facile increase by division.

FRANKENIA LÆVIS (*Sea Heath*).—A very small Evergreen, with crowded leaves like a Heath. Common in marshes by the sea in many parts of Europe and on the east coast of England. Best for the rock-garden, but mainly of botanical interest.

FRITILLARIA (*Snakeshead*).—These distinct and graceful bulbous flowers are so hardy and free in many soils, that there is no need of rock-garden luxuries for them. But in this large group of plants there are rare and beautiful kinds which the variety of surface and of aspect in a well-formed rock-garden may be very welcome to, and some American and European plants of this race are very striking and deserving of our best care. Their singular grace is charming on a carpet of rock plants, which can be easily established on any aspect of the rock-garden. The lovely yellow kinds, although long in cultivation—I have seen them admirably drawn in Dutch pictures two hundred years old—are slow to establish in gardens, and I found *aurea* tender in Sussex. This, no doubt, arises from the fact that in their own countries they lie under the snow until the winter is quite gone.

Mr Carl Purdy, writing to the *Garden* from California, says that some American kinds, including those of most striking beauty, are woodland plants, and, therefore, in planting them, we ought never to omit plenty of leaf mould. The shrubby rock-garden I so heartily advocate will give us for these plants the little shelter and half shade which is desirable.

The following are a few of the more select for the rock-garden, omitting our handsome native Snakeshead, which grows so freely in grass in any moist field. In so large a family, there are no doubt other alpine and choice kinds worth seeking by rock gardeners, and not a few yet to be introduced to gardens.

Fritillaria alpina is a pretty species, of dwarf growth, and bearing drooping flowers, chocolate on the outside and yellow within, while its margin of brighter yellow gives the flower a pretty effect. It blooms quite early in spring, and is of easy culture.

F. armena, from Asia Minor, is a dwarf form, with soft yellow bell-shaped blossoms on frail stems less than 6 inches high. This kind is best suited for sunny spots in the rock-garden or for planting freely in pots or pans for very early flowering. A soil of peat and loam suits this admirably. Next in order is

F. aurea (*Golden Snakeshead*).—A large and beautiful flower, though the plant is quite dwarf, and perhaps the gem of the family. I have often found it stricken with frost in my garden, owing, no doubt, to its coming from a country where the snow protects it long, and, therefore, I think it is safer to put it on the cool side of the rock-garden where it might flower later. A dwarf carpet of Sandwort or Rockfoil above looks well, and may be otherwise a gain.

F. Burnati, a handsome hardy plant about 9 inches high, with solitary drooping blossoms, 2 inches long, which are of a plum-colour, chequered with yellowish green. Alps. Flowers with the Snowdrop, and is as easy to grow.

F. Moggridgei is a beautiful kind, with handsome drooping blossoms of golden-yellow, prettily chequered with chestnut-brown on the inner surface. It is a dwarf kind, requiring treatment like *F. aurea* above noted. Maritime Alps.

F. pudica, a lovely kind with blossoms of a clear yellow, about three-quarters of an inch across, of much substance, and lasting long. Not the least attractive part of the plant is the fragrance of its golden bells. It is quite hardy, and, grown in mixture of loam and leaf soil with plenty of sand and a little manure, gives a charming effect. California.

F. Whittalli, a recent introduction, is beautiful and quite distinct, the blossoms of a red-brown on a yellow ground, tessellated on both surfaces.

GALANTHUS (*Snowdrop*).—Of late a host of forms of the Snowdrops have come into gardens, many of them with Latin names, and some as beautiful in their way as the old Snowdrop. There is reason to believe that these are not species, but varieties from very different localities, but this cannot affect their garden value. They are, however, so easily grown in any open soil, that there is no occasion to devote the rock-garden space to them, fair as they are, springing here and there in groups through moss-like rock plants. Usually, however, the Snowdrops are best naturalised in grass.

GALAX APHYLLA (*White Wand Plant*).—A distinct Evergreen perennial from North America, forming a thick matted tuft of scaly creeping rootstocks, thickly set with fibrous red roots, from which it sends up a number of roundish, shining leaves (about 2 inches wide) on slender stalks. The flowers appear in June, and form a wand-like spike of small and minutely-bracted white flowers, on the summit of a slender stem, 1 to 2 feet high. Useful for the rock-garden, in loam and leaf-mould.

GAULTHERIA PROCUMBENS (*Creeping Wintergreen*).—This plant barely rises above the ground, on which it forms dense tufts of shining leaves, with small drooping white flowers in June, succeeded by a multitude of bright red berries about the size of peas. The neat little shrub is of itself pretty, and the berries give it a charm through the winter months. A native of North America, in sandy places and cool damp woods, often in the shade of Evergreens, from Canada to Virginia. Loudon says it is difficult to keep alive, except in a peat soil kept moist; but I have never seen it prettier or so full of berries as on stiff loam. The plant was thoroughly exposed, and the only advantage it had corresponding to those usually mentioned as necessary was that the soil was moist. It thrives also in moist peat. There are few other plants of these important for the rock-garden, except *G. nummulariæfolia*, a dwarf creeping Evergreen. The large Partridge Berry of the Rocky Mountains (*G. shallon*) is too strong a grower for any but the roughest of stony banks in woods or elsewhere.

GENISTA (*Rock Broom*).—These shrubs are dwarf and very often tufted in growth, bearing yellow flowers of some beauty. They are easily grown and raised, and, being good in habit, should be worth a place in hot sandy soils where the true alpine flowers are despaired of. They would go well with the Rock Roses, Heaths, and Rosemary, which might be happy in such soils. From the following selection, we omit those that are too large for the rock-garden, or that have been found to be tender in the neighbourhood of London.

Genista anglica (*Heather Whin*) is a dwarf spiny shrub, not often growing to a height of 2 feet. It is widely distributed throughout Western Europe, and in Britain occurs on moist moors from Ross southwards. The short leafy racemes of yellow flowers appear in May and June.

G. anxantica, found wild in the neighbourhood of Naples, is very nearly allied to our native Dyer's Greenweed (*G. tinctoria*). It is very dwarf in habit, and its many racemes of golden-yellow flowers come in late summer. A beautiful rock-garden plant.

Genista aspalathoides, a native of South-western Europe, makes a densely branched, compact, spiny bush from 1 foot to 2 feet in height. It flowers in July and August (the yellow blossoms are somewhat smaller than those of *G. anglica*), and is a good shrub for the rock-garden.

G. ephedroides, a native of Sardinia etc., is a much-branched shrub, 2 feet in height, bearing yellow flowers from June to August.

G. germanica, a species widely distributed throughout Europe, makes a bright rock-garden shrub, not more than a couple of feet in height. It flowers very freely during the summer and autumn months and the stems are inclined to arch when 1 foot or more high.

G. hispanica, a native of South-Western Europe, is a compact undershrub, evergreen from the colour of its shoots. It scarcely attains more than 1 foot or 18 inches in height, and the crowded racemes of yellow flowers are borne at the tips of the spiny twigs from May onwards.

G. pilosa, a widely distributed European species, is a dense, prostrate bush and a rock-garden plant. In Britain it is rare and local, being confined to gravelly heaths in the south and south-west of England. It grows freely, flowering in May and June. Like the rest of the British species of the genus, it has bright yellow blossoms.

G. præcox is a garden name for *Cytisus præcox*, a beautiful hybrid between the white Spanish Broom (*Cytisus albus*) and *C. purgans*, a golden-flowered species.

G. radiata, a native of Central and Southern Europe, is 3 feet or 4 feet in height, evergreen from the colour of it much-branched spiny twigs. The heads of bright yellow flowers appear throughout the summer months. It is quite hardy at any rate in the south of England.

G. ramosissima.—A native of Southern Spain, and one of the best garden plants in the genus, grows about 3 feet high, and the slender twigs are laden in July with bright yellow flowers. This also passes under the name of *G. cinerea*.

G. sagittalis (*Winged Rock Broom*).—A singular plant, its branchlets winged (by the stem expanding into two or three green membranes), and bearing rich yellow flowers in summer; the shoots are usually prostrate, and the plant is rarely more than 6 inches high. It is met with in the grass in the mountain pastures of many parts of Europe. In cultivation, it is hardy and vigorous in the coldest soil, forming profusely flowering tufts when fully exposed. Seed.

Genista tinctoria (*Dyer's Greenweed*).—A dwarf native shrub, with numerous slender branches, forming compact tufts from a foot to a foot and a half high, pretty yellow flowers in early summer. It is grown in many of our Nurseries, and merits a place among rock-shrubs. There is a double variety. Not unfrequent in many parts of England, but rare in Scotland and Ireland.

G. tinctoria var. elatior is a tall-growing form from the Caucasus, which under cultivation frequently grows from 4 feet to 5 feet high, and bears huge paniculate inflorescences.

GENTIANA (*Gentian*).—Alpine and mountain pasture plants of classic beauty and variety, some herbaceous, some evergreen herbs, some annual plants. Beautiful as the Gentians are on the mountains of Europe—and it is not easy to describe their beauty at its best, as, say, of a plateau of acres of the vernal Gentian on the Austrian Alps, or of the Bavarian Gentian along the side of an alpine streamlet—I think I was even more struck with the beauty of the American, fringed, and other Gentians which do not seem easy of cultivation in Britain. There is no serious difficulty as to the culture of the best European kinds, save, perhaps, *bavarica*, but the American kinds are more liable to perish in some of our soils. Gentians are not all worthy of cultivation on the rock-garden. I never could see any beauty from that point of view, in the tall Gentian of the Alps (*G. lutea*), and some of the annual kinds are of no value for the rock-garden, but there are not a few kinds among the fairest of known rock plants.

If any plants justify the formation of a good rock-garden, it is these; and we should seek to get their best effect from an artistic point of view by, if possible, grouping them in a natural way. There will be no difficulty in this as regards some kinds, particularly *Gentianella*, which is very effective on some soils, and in its various forms might be grouped well when sufficiently increased. The Willow Gentian also lends itself to good effect among the bushes in the rock-garden, and is readily increased. One or two good kinds, well grown and grouped, will be better than a dozen dotty examples of ill-grown kinds, however rare or curious.

It is curious in growing the vernal Gentian how little way is made, with perhaps the most brilliant of alpine plants that flower on the higher mountains in late summer. There we see acres of it in every sort of position; in banks by streams, in open grassy places, in little green vales; sometimes in wide peaty flats, almost blue with its fine colour. In gardens it is too much coddled, wanting nothing really but moist, peaty, or fine loamy soil, not shallow, and the plants never cocked up on the ridiculous "rockwork" of the garden, but kept on low ledges or borders, and never placed near herbaceous or any other vigorous plants.

Gentiana acaulis (*Gentianella*).—Among the most beautiful of the Gentians, easily cultivated, except on dry soils. In some places edgings are made of it, and where the plant does well, it should be used in every garden to some extent in this way. It is at home on the rock-garden, where there is moist loam into which it can root. It is sometimes sold in Covent Garden in pots, when in flower in spring, and is readily propagated by division, and also by seeds; but these are so small, and so slow in germinating, that its propagation in this way is never worth the trouble. It is abundant in many parts of the Alps and Pyrenees. I have grown this plant very well in "battered" walls, and it flowered freely thereon. My friend, M. Francisque Morel of Lyons, tells me that the form of this fine plant, which is cultivated in British Gardens, is unknown on the Savoy mountains and those near. He thinks it is an Italian form, but there are other handsome Gentians among its allies on those mountains and others near, which are well worth the attention of rock gardeners. As the old plant we have is so easily grown in Britain, there is no reason why these should not be equally so. I think they would all do grown on walls in the way described in the first part of this book—that is to say, on "battered" walls against earth banks, with the stones so set that they will catch all the rainfall.

According to M. Correvon, there are four or even five well-marked forms of *G. acaulis*, viz.:

Gentiana angustifolia.—A stoloniferous plant, emitting underground runners. Flowers large, handsome, of a fine deep sky-blue colour, and spotted on the throat with sprightly green. This is the handsomest species. It flowers in May and June, and is found on calcareous parts of the Alps at an altitude of 3000 feet to 4000 feet.

G. a. Clusii.—The flowers of this are of a fine dark blue colour, and have no green spots on the throat. The plant blooms in May and June, and is found on calcareous rocks of the Alps and the Jura range at an altitude of 3000 feet to 5000 feet.

G. a. Kochiana.—Flowers of a violet-blue colour, marked on the throat with five spots of a blackish-green colour in May and June. Common in pastures on the granitic Alps.

G. a. alpina.—Leaves small, of a sprightly green colour, glistening, curving inwards and imbricated, forming rosettes which incurve at about the middle part of their length. Blooms in May and June. Found on the granitic Alps at an altitude of 6000 feet to 9000 feet; also on the Pyrenees and the Sierra Nevada. The two last-named species require a compost of one-third crushed granite, one-third heath soil, and one-third vegetable loam, and should be planted half exposed to the sun.

Gentiana a. dinarica (*Beck.*)—This is a form of *G. acaulis* with broad, thick leaves and erect, slender, almost cylindrical flowers of a dark blue colour. Alps of Southern and Eastern Austria.

G. Andrewsii (*Blind Gentian*).—The kinds of Gentian which attract so much attention for their beauty on European mountains open their flowers wide when the sun shines. This does not do so, having closed tubes each about an inch long, in clusters, and of a deep dark blue. Then, instead of spreading low and mantling the ground with rosettes of leaves, the shoots grow erect, and a foot or more high. It is handsome, thrives in a sandy peat, but has been hitherto so little grown, that experiences of its likes and dislikes are not yet obtainable. The flowers are closely set in clusters near the tops of the shoots. A native of moist rich soil in North America, flowering in autumn, and increased by division and by seed.

G. asclepiadea (*Swallow-wort*).—A true herbaceous plant, *i.e.*, dying down every year, thus keeping out of danger in winter time, and easily cultivated in almost any soil. It grows erect, with shoots almost willow-like, and from 15 inches to 2 feet high, according to the nature of the soil; bearing numerous large purplish-blue flowers, arranged in handsome spikes. Little need be said of its culture, as it is not fastidious, but in a deep sandy loam or peat it will grow twice as large as in a stiff clay. In a wild state it inhabits pine woods. In consequence of its tall habit, this species is best adapted for the bushy parts of the rock-garden, or in the borders near at hand. It is a native of European mountain woods. Division.

G. bavarica (*Bavarian Gentian*).—In size this resembles the vernal Gentian, but has smaller Box-like leaves of a yellowish green, all its tiny stems being thickly clothed with foliage, forming close, dense little tufts, from which spring flowers of the most lovely blue, which seems occasionally flushed with a slight tinge of purplish-crimson. The plant is a native of the high Alps of Europe. *G. verna* occurs abundantly in the same localities; but, while it is found on ground not overflowed by water, *G. Bavarica* is in bloom in very boggy spots, where some diminutive rill has left its course and spread out over the Grass, not covering it, but saturating it so that, when walked upon, the water bubbles up around. The best thing to do with it is to plant it near the margin of a rill, taking care to let no Carices, Cough Grass, Cotton Grass, or other strong-growing subjects get near the spot, or they would soon cover and destroy the plant. It may also be grown in pots, plunged in sand during the summer; sandy loam to be the soil used, the plants to have repeated and abundant waterings from early spring till the heavy autumnal rains set in, or be placed standing half-plunged in water, with free exposure to light.

Gentiana ciliata.—A rare and beautiful species, with flexuose, almost simple, stems, about 1 foot high, bearing large, solitary, azure-blue, deeply fringed flowers, each from 1 inch to 1½ inches long. It is a native of the Alpine regions of Central and Southern Europe, and the Caucasus in dry pastures, and requires to be planted in a mixture of rich fibrous loam and broken limestone, in sunny fissures of rock; or it may be grown in well-drained pots, using the same compost. In all cases it should be kept rather dry in winter. Young plants flower freely when only 2 inches or 3 inches high.

G. crinata (*Fringed Gentian*).—A singularly beautiful plant, frequenting wet ground and river sides, about 1 foot in height, with the loveliest fringed deep indigo-blue flowers I ever saw. It is a biennial plant, very beautiful for the bog garden, if we could get it established in our country from seed. It grows in moist woods and pastures, and also near rivers and streams, and has a wide range in N. America and Canada.

G. cruciata (*Cross-wort*).—This species has somewhat erect, spreading leaves, arranged at right angles or cross-like on simple ascending stems, which are from 6 inches to 1 foot in height, the flowers blue, and in whorls. It is a native of dry pastures in Central and Southern Europe. In growing this plant, fibrous loam should be plentifully mixed with small pieces of broken limestone.

Gentiana decumbens.—Stem erect, 12 inches to 16 inches high. Flowers numerous, of a fine blue colour, and borne in terminal spikes, from June to August. Native of Siberia, at an altitude of 2000 feet to 3000 feet. Syn., *G. adscendens.* There is a good white variety of this.

Gentiana decumbens alba.

G. gelida.—Forms dense tufts, or carpets, a foot high, with bent, ascending stems, and blunt leaves, closely set, the flowers very nearly 2 inches long, in large heads of a brilliant blue colour. A native of alpine districts in the Caucasus and Armenia, thriving in rich, moist loam. Division or seed.

G. Kurroo.—One of the most beautiful

of the Himalayan Gentians, and one of the easiest to cultivate. In the south of Scotland it does well, but then alpine Indian plants find there a congenial home. Near London, on a north aspect, it has flowered well. The compost in which it grows is a rich peaty mixture, and it receives copious waterings during the summer months. It forms a tuft, or rosette, of smooth leaves about 3 inches long, from the base of which rises the flower-stalk, and from the upper joints short stalks bearing single flowers, each an inch broad, and of the brightest azure-blue, in July and August. Himalayas.

Gentiana macrophylla.—A taller kind, with lower leaves from 10 inches to 12 inches long. Flowers blue, small, numerous, borne in closely set heads. It comes very near *G. cruciata*, from which it is distinguished by the size and shape of its leaves, and, lastly, by the lobes of the corolla standing erect instead of spreading out.

Gentiana, G. macrophylla. (Engraved from a photograph sent by Miss Willmott.)

Gentiana pneumonanthe (*Marsh Gentian*).—A British perennial, scarcely less beautiful than any alpine Gentian, with tabular flowers, an inch and a half or more long, of a beautiful blue within, with five greenish belts without, the lobes of the mouth short and spreading; on stems 6 inches to a foot high. A native of boggy heaths and moist pastures, and in cultivation requiring moist peat. It is not recorded from Scotland or Ireland, though not rare in some parts of England. Few plants are more worthy of a place on the rock-garden, and where the plant occurs wild, it might well be guarded against extermination.

G. Punctata.—A free, rather bold, dark yellow kind, growing plentifully in Alpine meadows, the flowers very distinct in colour and form too.

G. pyrenaica (*Pyrenean Gentian*).—Somewhat like the vernal Gentian in size, but with narrow, sharp-pointed leaves, and dark violet almost stalkless flowers, the flat portion of the flower being formed of five oval lobes, with a triangular appendage between each, nearly as long as the lobes. It requires much the same treatment as *G. verna*, flowering in early summer, and is well worthy of a place in the choice rock-garden, though not of such a vivid hue as *G. verna*.

G. septemfida (*Crested Gentian*).—A lovely plant, bearing on stem 6 inches to 12 inches high flowers in clusters, widening towards the mouth, of a beautiful blue and white inside, greenish-brown outside, having between each of the larger segments of the flowers one smaller and finely cut. A native of the Caucasus, and one of the best for cultivation on the rock-garden, thriving well in moist sandy peat. Division.

G. verna (*Vernal Gentian*).—This covers the ground with rosettes of small leathery leaves, often spreading into tufts from 3 inches to 5 inches in diameter, and producing in spring, flowers that even the botanist calls "beautiful bright blue," though botanical books are usually above taking any notice of colour at all. Sometimes the blooms barely rise above the leaves, and at other times are borne on stems 2 inches or 3 inches high. A few things are essential to success in its cultivation, and far from difficult to secure. They are good, deep, gritty loam on a level spot, perfect drainage, abundance of water during the dry months, and full exposure to the sun. Grit or broken limestone may be advantageously mingled with the soil, but if there be plenty of sand, they are not essential; a few pieces half buried on the surface of the ground will help to prevent evaporation and guard the plant till it has taken root and begun to spread about. It is so dwarf that, if weeds be allowed to grow around, they soon injure it. In moist districts, where there is a good, deep, sandy loam, it may be grown on the front edge of a border carefully surrounded by half-plunged stones. It may also be grown in pots or boxes of loam, with plenty of rough sand, well drained and plunged in beds of sand, well exposed to the sun, and well watered from the first dry days of March onwards till the moist autumn days return. In all cases, good, well-rooted specimens should be secured to begin with, as failure often occurs from half-dead plants that would have little chance of surviving, even if favoured with the air of their native wilds. In a wild state this plant is abundant over mountain pastures on the Alps of Southern and Central Europe, and those of like latitudes in Asia.

GERANIUM (*Cranesbill*).—Showy hardy perennials, for the most part too rampant for the rock-garden, and in no need of its soils or other refinements. Therefore we should keep in this case to the dwarfer and more alpine kinds, such as the following:

Geranium argenteum (*Silvery Cranesbill*).—A lovely alpine plant, with leaves of a silvery white, and large pale rose-coloured flowers, on stems seldom more than 2 inches high, and nearly prostrate. It comes from the Alps of Dauphiny and the Pyrenees, is hardy, flowering in early summer, and is a gem for association with the choicest plants. It loves a firm, sandy, and well-drained soil, and should, as a rule, be placed near and somewhat below the eye, as, though the plant is of a high, it is not of a conspicuous, order of beauty. Increased freely by seed.

Geranium cinereum (*Grey Cranesbill*).—A beautiful dwarf plant, with five- or seven-parted leaves, clothed with a slightly glaucous pubescence, and bearing very large and handsome pale pinkish flowers, veined with red. A native of the Pyrenees, 2 to 5 or 6 inches high, growing freely, and easily propagated by seeds. On the rock-garden it is at home, and fitted for association with the choicest kinds. It seeds abundantly, and may be easily raised from seed.

G. sanguineum (*Blood Cranesbill*).—A native plant, forming spreading close tufts from 1 to 2 feet high; the flowers are large, nearly or quite 1½ inches across, of a deep crimson purple. Its close habit instantly distinguishes this plant from any other Geranium, and the flowers being more beautiful than those of any other, it deserves to have a place in every rock-garden, among the larger and more easily grown plants. It grows on any soil, is readily propagated by division or seeds, and occurs in a wild state in some parts of Britain, though not a common plant.

There are two forms or varieties of the Blood Geranium. One, the common or "true" species, with ascending stems matting into vigorous but compact tufts; the other more hairy, less vigorous in its growth, and usually prostrate in habit. This last form usually occurs on sandy sea-shores. A form of this variety, with pale pink flowers veined with red, was found at Walney Island, in Lancashire, and has been distinguished as a species under the name of *G. lancastriense*, but it differs only in colour from the sea-shore variety. Both these forms, being smaller and less vigorous than the common one, are worth having for the rock-garden. There is also a white variety, a good plant.

G. striatum (*Striped Cranesbill*).—An old and charming plant, still to be seen in many cottage gardens. "This beautiful Cranes-bill," says Parkinson, writing nearly three hundred years ago, "hath many broad yellowish green leaves arising from the root, divided into five or six parts, but not unto the middle as the first kinds are: each of these leaves hath a blackish spot at the bottom corners of the divisions: from among these leaves spring up sundry stalks a foot high and better, joynted and knobbed here and there, bearing at the tops two or three small white flowers, consisting of five leaves apeece, so thickly and variably striped with fine small red veins that no green leafe that is of that bigness can show so many veins in it, nor so thick running as every leaf of this flower doth." It is a native of Southern Europe, growing very freely in warm sandy soils, and is easily increased by seed or division.

GEUM (*Avens*).—Perennial herbaceous plants with red or yellow, rarely whitish, flowers, some of which are too vigorous in growth for the rock-garden.

Geum montanum (*Mountain Avens*), which is found on turfy declivities and pastures on the Alps, Pyrenees, Apennines, Carpathian Mountains, the Sudetic Range, and Mount Scardo, in Macedonia. The plant has a thick root-stock and large leaves of a cheerful bright green colour; the flowers are of large size, on stalks from 4 inches to 10 inches high, and are succeeded by a cluster of feathery awns of a reddish-brown colour. It thrives well on any kind of rock-garden, and also on walls.

G. reptans.—A handsome kind, found in clefts of rocks and in rocky *debris* on the Upper Alps at an altitude of 2000 mètres to 2500 mètres, also on the Pyrenees, the Carpathian Mountains, and the high mountain ranges of Macedonia. It is the rock form of *G. montanum*, and requires to be grown in the full sun. The flowers are very large (sometimes nearly 2 inches across), and of a pale yellow colour. The leaves are more deeply incised than those of *G. montanum*, and are of a greyish-green colour, velvety and not glistening. Moreover, the plant sends out long thread-like runners, bearing at their extremities small buds or shoots, which take root often at a distance of more than 10 inches from the plant.

GLOBULARIA NANA (*Dwarf G.*).—A dense trailing shrub, forming a firm mass of thyme-like verdure, about half an inch high, and dotted over with compact heads of bluish-white flowers, with stamens of a deeper blue or mauve. The flower heads are not half an inch across, and barely rise above the foliage. It should be planted in sandy or gritty soil, and so that it may crawl some little way over the face of the surrounding stones, and in a very open sunny spot in such a position, it will not be so liable to be overrun by coarse plants. A native of the Pyrenees, and increased by division. There are several other Globularias in cultivation: *G. nudicaulis, trichosantha*, and *cordifolia*, but these are scarcely worthy of a place except in large collections.

GOODYERA PUBESCENS (*Rattlesnake Plantain*) is a beautiful little Orchid, with leaves close to the ground, delicately veined with silver. It thrives in any shady spot, such as may be found in any good rock-garden, planted in moist peaty and leafy soil, with here and there a bit of soft sandstone for its roots to cling to and run among. It is quite hardy. Native of Eastern United States. *G. repens* and *Menziesi* are less desirable.

GYPSOPHILA.—Perennials and annuals of the Stitchwort family. The larger kinds are elegant, bearing tiny white blossoms in myriads on slender spreading panicles. These are mostly too vigorous for our purpose, but *G. prostrata* is a pretty species for the rock-garden. It grows in spreading masses, and has white or pink small flowers, borne on slender stems in loose graceful panicles from midsummer to September. It is a very useful plant, and blooms for a long season. *G. cerastioides* grows about 2 inches high, and has a spreading habit, bearing small clusters of blossoms, which are half an inch across, white with violet streaks. It is from Northern India, and unlike any of the group now in our gardens, being dwarfer, and having larger flowers. It is a rapid grower, and soon spreads into a broad tuft if in good soil, and in an open position on the rock-garden. Increased by seeds or cuttings in spring.

HABENARIA (*Rein Orchis*).—Terrestrial Orchids from N. America, some of which are pretty and interesting, and all grow from 1 foot to 2 feet high. To succeed in out-door culture, a spot should be prepared with about equal parts of leaf-mould, or peat, and sand, with partial shade; the soil should be well mulched with leaves, grass, or other material to protect the roots from the heat of the sun, and to keep it moist. *H. blephariglottis* flowers in July, in spikes, white and beautifully fringed. *H. ciliaris* is the handsomest, the flowers bright orange-yellow, with a conspicuous fringe upon them. *H. fimbriata* flowers in a long spike, lilac-purple, beautifully fringed. *H. psycodes*, flowers purple, in spikes 4 inches to 10 inches long, very handsome and fragrant. They are charming plants for the bog-garden, or for a quiet nook with moist, peaty soil.

HABERLEA RHODOPENSIS.—This is a pretty little rock-plant, resembling a *Gloxinia* in miniature. It forms dense tufts of numerous small rosettes of leaves, which somewhat resemble those of the Pyrenean *Ramondia* (*R. pyrenaica*), each rosette bearing in spring from one to five slender flower-stalks, with two to four blossoms each, nearly 1 inch long, of purplish-lilac colour, with a yellowish white. Messrs Frœbel, of Zurich, who grow it well, write to us : "We have treated this plant in the same manner as the Pyrenean *Ramondia, i.e.* we have planted it on the north side of the rock-garden; therefore, the sun never directly reaches it. We grow it in fibrous peat, and fix the plants, if possible, into fissures, so that the rosettes which it forms hang in an oblique position, just as they do in their native country. It succeeds well in this way; but if no rock be at hand it may be grown equally well on the north side of a Rhododendron bed. We have it thus situated quite close to a stone edging, a way in which we also grow the *Ramondia*, and the *Haberlea* flowers profusely every year in May and June. The plant is very hardy, having withstood several very hard winters, without any protection, quite unharmed." It is a native of the Balkan Mountains, where it is found growing among moss and leaves on damp, shady, steep declivities at high elevations.

HABRANTHUS PRATENSIS.—A brilliant bulbous plant of the *Amaryllis* family, hardy, at least in the southern and eastern parts of the country. It has stout and erect flower-stems, about 1 foot in height, and flowers of brightest scarlet, feathered here and there at the base with yellow. The variety *fulgens* is the finest form of the plant. It grows freely in loam, improved in texture by the addition of a little leaf-mould and sand. Its propagation is too easy, for in many soils it breaks up into offsets, instead of growing to a flowering size. A choice plant for the rock-garden. Chili.

HEDYSARUM OBSCURUM (*Creeping-rooted H.*).—A handsome, creeping, vetch-like plant, with large purplish-violet flowers in long spikes, from 6 to 12 inches high, and sometimes more in rich soil. Readily increased by division or seeds, grows freely in ordinary garden-soil on level ground, and is a valuable rock-plant. A native of the Alps of Dauphiny and the Tyrol.

HELIANTHEMUM (*Sunrose*).— Mostly dwarf and wiry shrubs, inhabiting rocky, sandy, and heathy places; of much beauty of colour, for the most part hardy, and easy to grow, and, therefore, very useful for the rock-garden, or for dry walls or banks. If we had only the varieties of our native Sunrose, they would be a precious aid; but there are also other species of much beauty, and well deserving the care of the rock gardener. It is not a group in which we have to pick and choose, as every known kind is worth growing.

Helianthemum canum (*The Hoary Sunrose*).—A native of limestone rocks in Britain, but somewhat rare, is much dwarfer than the common kind, and produces small pale yellow flowers. The whole plant does not grow more than 3 inches high, and forms a pretty rock-shrublet.

H. guttatum (*Annual Sunrose*).—The pretty annual spotted Sunrose, found in the Channel Islands, on the Holyhead Mountain, in Anglesea, and widely on the Continent, deserves a place in the curious collection, and indeed has beauty enough to recommend it. It is quite easily grown, but is best raised in pots in spring, and then planted out in May. Once established, it sows itself annually.

H. ocymoides (*Basil-like Sunrose*).—A native of dry rocky hills in Spain and Portugal, with bright yellow purple-eyed flowers nearly an inch and a half across, and hoary leaves an inch to an inch and a half long; and very useful on the warmer and drier parts, among the stronger alpine shrubs. Increased by seed or cuttings. Syn., *Cistus algarvensis*.

ALPINE FLOWERS FOR GARDENS

Helianthemum Pilosella (*Downy Sunrose*).—A dwarf kind, with a woody prostrate stem; about 6 inches high, flowering in summer; small, yellow, in clusters. Pyrenees. The rock-garden and margins of dry borders, in ordinary soil. Seed and cuttings.

H. polifolium is also a native of our country, though rare. It seems to me that there are many plants of this genus not yet in cultivation, worth introducing, especially for sandy and poor hot soils.

H. rosmarinifolium (*Rosemary-leaved H.*).—A neat, erect little bush, about 1 foot high, flowering in summer; white, on short stalks, bearing each from one to three flowers. North America. Pretty in the rock-garden, in sandy loam. Cuttings and seed.

H. Tuberaria (*Truffle Sunrose*).—A distinct and beautiful rock-plant, bearing flowers like those of a single yellow rose, 2 inches across, and with dark centres, drooping when in bud, and on stems about 9 inches high. It is quite distinct from all the other cultivated Sunroses in not having woody stems, but sending up large hairy leaves, somewhat like plantain-leaves, from the root, and scarcely looking like a Sunrose. It flowers in summer, and continuously, if in good health and in good soil. It is said to grow abundantly where truffles abound, and is well worthy of a position in a well-drained spot, or dry fissure on the sunny side of the rock-garden. S. Europe.

H. vulgare (*Common Sunrose*).—A well-known British under-shrub, growing in dry pastures and heaths, with bright yellow flowers, on stems from a few inches to nearly a foot long. In a cultivated state this plant varies a good deal in colour, and numerous plants passing under different names in our gardens are really forms of this species, and some well worthy of cultivation. While thriving in almost any soil, they attain ripest health, and flower most profusely, on chalky and warm ones, and on soils of this description they may be used with good effect on the margins of shrubberies, especially the copper-coloured and red varieties. They are only suited for the rougher parts of the rock-garden. The best way to obtain varieties of different colours is by seed, which is offered in most of the Catalogues; but some of the named varieties are very bright, and should be secured, such as *amabilis, sunbeam, venustum,* and *Ball of Fire.*

HELICHRYSUM ARENARIUM (*Yellow Everlasting*).—This is the beautiful little plant which affords the "everlasting flowers" so much used for immortelles. The grey leaves are closely covered with long down, and the flower-stems, ascending from 4 to 10 inches, are clothed all the way up with narrow hoary leaves, having their edges turned backwards, and support a number of flowers, of a bright, glistening yellow. To preserve the flowers, they should be gathered when fresh and newly-blown, as, if allowed to mature, they are apt to fall away. A native of sandy and sunny places in Central and Southern Europe, and in this country on warm, sandy, and drained soils. Division.

HELLEBORUS (*Christmas Rose*).—Though these plants are not usually included among alpine and rock-plants, they are true mountaineers: being often slow to grow in our gardens I think that the advantages of aspect and improved soil and good drainage, which a well-made rock-garden gives, might be well for these noble plants. In any case, where we work with mountain shrubs, these will come in well, and there can be nothing more attractive in winter, in warm or chalky soils, than the winter kinds, or in spring, when the eastern kind blooms so early.

Helleborus Niger (*Christmas Rose*).—Although this familiar old plant may be thought too vigorous for association with the often minute gems to which this book is chiefly devoted, yet its fine evergreen foliage and handsome large flowers entitle it to a place. Although hardy enough to grow almost anywhere, yet, as it flowers at the dreariest season, when low ground is often saturated with cold rain, it always repays for being planted in slightly elevated spots, and where it may enjoy as often as possible the faint wintry sun, by giving clearer and larger flowers, and finer foliage. And as in the warmer and more sunny countries it misses the shade of the big rocks, it is often well to group any of its fine forms on the cool side of the rock-garden.

The following are some of the best-known varieties of this fine plant:—

H. n. altifolius is the most vigorous of the group. It is early in bloom, often commencing to expand its flowers in autumn. The flower stems are mottled with red, and the backs of the petals faintly rosy.

St Brigid's Christmas rose is a very beautiful flower, the blossoms pure white, and cupped in form.

The Riverston variety is a very free-blooming one. Its flower-stems are apple-green, but the leaf-stalks are red-spotted, the leaves themselves being of a pale green.

The Bath variety is the form perhaps most generally in use for providing blooms at Christmastide. It is larger and taller than the type.

H. n. Madame Fourçade is in habit of growth not unlike *H. n. altifolius*, but flowers a full month later, and the blooms are whiter and more cup-shaped.

These fine plants deserve a better fate than they often meet with in gardens. The full exposure of the ordinary plain, and perhaps cold and wet soil, does not always suit them. In the lowly mountain valleys they come from, there are "many mansions," so to say, and, although the heat is greater than ours in summer, the shades of the rocks often give them relief, and there is the open, gritty soil, and other advantages. In certain parts of our country, where the natural soil is warm and good, they do well, but in others they fail, and require a well-made soil that has plenty of sand or grit and some leaf-mould. We may also have to think of the aspect. I have known them succeed in the shade of walls when they failed in the open.

It is all the better to group them so that as they flower in the middle of winter, the flowers may be easily protected with a few bell-glasses or hand-lights.

Besides the true Christmas Rose, there are other species of *Helleborus* well worthy of cultivation; and among the best is *H. atrorubens*, with flowers of a dark purple. The colour, though somewhat dull, by turning up the usually pendent flower is seen to greater advantage, being then contrasted with the yellow stamens. It has the quality of throwing its flowers well above ground to a height of 9 to 12 inches, and is a free grower, but rather scarce, requiring, as all the *Hellebores* do, a considerable time to establish itself after being disturbed. *H. olympicus*, with large rose-coloured flowers, and good habit, is very similar. *H. argutifolius* is remarkable for its beautiful, whitish, trifoliate leaves, each secondary vein being terminated by a well-defined point. Its flowers are a lively green, and come about the month of March.

Helleborus Hybrid (*Lenten Lilies*).— By far the most important group after the true Christmas Rose, and its forms, are the fine varieties raised in gardens, the hardiest of them:—from the bold and free *H. Orientalis*, a native of Greece and Asia Minor, and in some cases crossed with other species. The spotted and variously coloured forms raised in this way are excellent, and, while quite distinct from the true Christmas Rose and its forms, are more vigorous in growth, and coming into flower at the end of winter or dawn of spring, they open well without protection in many parts of the country. They are not nearly so difficult about soil as *H. Niger* and its forms, growing in any free and good soil in many cases without any special making of the soil, liking it deep. They do best in partial shade in the southern countries. Almost too vigorous for the choice parts of the rock-garden, it is easy to place them near its approaches among the shrubs, or in a half-shaded wall approaching. A great many beautiful varieties have been raised in England, and also in France and Switzerland, and these varieties have been given fine names; but, to some extent, they are repetitions of each other, and it is not worth while being very particular as to whether they are named or not, if we have kinds that please us in colour.

HELONIAS BULLATA (*Stud Flower*).—A distinct and handsome marsh perennial, growing 12 inches to 16 inches high, and having handsome purplish-rose flowers arranged in an oval spike. It is suitable for the artificial bog, or for moist ground near a rivulet. In fine sandy and very moist soils it thrives well as a border plant. North America. Syn., *H. latifolia*.

HEMIPHRAGMA HETEROPHYLLA.—A dwarf trailing plant of the Figwort family, bearing inconspicuous flowers, succeeded by bright red berries about the size of small Peas, on slender creeping stems. It is rather tender, and requires a sheltered and well-drained spot in the rock-garden. Himalaya.

HEPATICA. (See **ANEMONE.**)

HERNIARIA.—Dwarf trailing perennial plants, forming a dense turfy mass that remains green throughout the year. There are two or three species, but the most important is *H. glabra*, useful on account of its dwarf compact growth, and is always of a deep green, even in a hot and dry season. They grow in any soil, but the flowers are inconspicuous.

HESPEROCHIRON PUMILUS.—A pretty Californian rock plant, stemless, dwarf in growth, with leaves borne on slender stalks, forming a tuft, the flower bell-shaped, half-inch across, white, varying to a purplish tinge. It grows in marshy ground, and in damp places in the Rocky Mountains and Northern Utah, and is apparently quite hardy, as it thrives in ordinary soil in well-drained parts of the rock-garden. *H. californicus* is a species of somewhat the same habit.

HIERACIUM (*Hawkweed*).—A very extensive genus of Composites, consisting chiefly of perennial herbs with yellow flowers. Some of the yellow alpine and other kinds are valuable in botanical collections, and many of them are beautiful, but the prevalence of yellow flowers of the same type makes them less important, and not a few are too large and coarse for the rock-garden.

HIPPOCREPIS COMOSA (*Horseshoe Vetch*).—A small prostrate British plant, with pretty little deep-yellow flowers, in coronilla-like crowns, the upper petal faintly veined with brown, the pinnate leaves small, and leaflets smooth. It is a capital little plant for the upper ledges of rocks in dry positions, as in such places the shoots will fall down some 18 or 20 inches; easily raised from seed; partial to chalky soils; rather common in the South of England, but not a native of Ireland or Scotland.

HORMINUM PYRENAICUM.—A Pyrenean plant, forming dense tufts of foliage and having purplish-blue flowers, in spikes about 9 inches high, which appear in July or August. It is hardy and of easy culture, but is not a plant of much character from a garden point of view.

HOTTONIA PALUSTRIS (*Water Violet*).—A beautiful British waterplant, which I include here in consequence of having seen it thrive better on soft mud banks than when submerged. The deeply-cut leaves formed quite a deep green and dwarf turf over the mud, and from these arose stems, bearing at intervals whorls of handsome pale-lilac or pink flowers, which might perhaps be more justly called the Water Primrose, as it is nearly allied to the Primulas, and it may be grown either in the water or on a bank of soft wet soil at its margin. It grows from 9 inches to 2 feet high, flowers in early summer, and may be found in abundance near London on the banks of the Lea river, and in many other places, and is pretty freely distributed over England.

HOUSTONIA CÆRULEA (*Bluets*). A delicate North American mountain plant, with many pale sky-blue flowers, fading to white, and with yellowish eyes, crowding on thread-like stems to a height of 1 inch to 2½ inches, from close low cushions of leaves shorter than many mosses, less than half an inch high when fully exposed. It is usually considered somewhat difficult to grow, but this arises chiefly from its minuteness; in level exposed spots it does very well in moist peaty soil, the chief care required being to keep it quite clear of weeds or coarse-growing neighbours. It is suitable for association with the choicest mountain-plants. I have grown this plant well in the open air in London; it withstood the evil influences of showers of soot.

Houstonia purpurea is another good kind; both inhabit open grassy places and among wet rocks.

HUTCHINSIA ALPINA (*Alpine H.*).—A neat little rock-plant, from moist and elevated parts of nearly all the great mountain-chains of Central and Southern Europe, with shining leaves, and pure white flowers, in clusters on stems about 1 inch high. It is quite free in sandy soil, and easily increased by division or by seeds. Planted in an open spot, it becomes a dense mass of white flowers.

HYACINTHUS (*Hyacinth*).—Usually the cultivated Hyacinths are not plants for the rock-garden, but a few species come in gracefully, particularly the Amethyst Hyacinth.

Hyacinthus azureus (*Azure Hyacinth*). —A very dwarf and pretty plant, hardy and amenable to ordinary culture, and one of the earliest as well as the most charming of our early spring flowers. It is a jewel for the rock-garden, arising from close carpets of little plants, that save it from the splashings of the winter rains.

H. amethystinus (*Amethyst Hyacinth*), though nearly related to *H. azureus*, is a charming hardy plant, flowering at a time when there is a dearth of flowers. The mistake with a bulb like this is to have two or three or even a dozen in a clump. Instead of by the dozen it should be grown by the hundred, and no prettier sight can well be imagined than a large sheet of this, with its racemes of amethyst flowers. I find it most precious in a group between rock-shrubs, or arising from carpets of *Cinquefoil Sandwort*, or any creeping rock-plant, and it is as hardy and enduring, good in form, and delicate in colour. S. Europe.

Hyacinthus Orientalis.—This is said to be the parent of all the garden Hyacinths in cultivation. The wild types of the garden and Roman Hyacinths, or at least as near as possible to their original forms, are in cultivation at the present time, but so inferior to the varieties we now grow, that no one would care to have them. The varieties are *albulus* and *provincialis*.

HYDROCHARIS MORSUS-RANÆ (*Frog-bit*).—A pretty native water-plant, having floating leaves and white flowers, and well worth introducing in pools. It may often be gathered from neighbouring ponds in spring, when the plants float again after being submerged in winter.

HYDROCOTYLE (*Pennywort*).— Small creeping plants, usually with round leaves and inconspicuous flowers. There are several kinds grown, their only use being as a surface growth to the artificial bog. The most desirable are *H. moschata* and *microphylla*, two New Zealand species, and *nitidula*, though all of these are somewhat tender. The common *H. vulgaris* is rather too rank a grower.

HYPERICUM (*St John's Wort*).— Handsome shrubs, some dwarf, and occasionally of much beauty for the rock-garden, where the best of the larger ones may be used among the shrubs. They are usually of easy culture in ordinary soils. Some of the perennials are good rock-plants, and the best of these is *H. olympicum*, one of the largest flowered kinds, though not more than 1 foot high. It is known by its very glaucous foliage, and erect single stems, with bright yellow flowers about 2 inches across. It may be propagated easily by cuttings, which should be put in when the shoots are fully ripened, so that the young plants may become well established before winter. *H. nummularium* and *humifusum*, both dwarf trailers, are also desirable, and, owing to their dwarf compact growth, several of the shrubby species are well suited for the rock-garden. Of these, the best are *H. ægyptiacum, balearicum, empetrifolium, Coris, patulum, uralum,* and *oblongifolium.* The last three are larger than the others, but as they droop they have a good effect among the boulders of a large rock-garden, or on banks. *H. Hookerianum, triflorum, aureum, orientale* are among the kinds having some beauty, but the species from warmer countries than ours are apt to disappear after hard winters. *H. Moserianum* is a handsome hybrid kind.

Hypericum reptans is a beautiful dwarf, and graceful trailer, with small leaves, and wiry prostrate branches, each of which bears a single flower at its tip. In proportion to the size of the foliage the flower is very large, as it reaches 1¾ inches in diameter. This is best seen when grown between stones, and allowed to carpet a sloping or perpendicular surface. Himalaya.

Among other kinds worth a place are *H. Budlleyi,* and *H. empetrifolium.*

Hypericum polophyllum.

IBERIDELLA ROTUNDIFOLIA (*Round-leaved I.*).—A distinct plant, rarely more than a few inches high, with pretty, rosy-lilac, sweet-scented flowers in April, May, and June. The leaves are thick, smooth, leathery, and of a glaucous olive-green, and the flowers are produced in short racemes or corymbs, and usually attain a height of from 3 to 6 inches. Flowering with the vernal Gentian, the Bird's-Eye Primrose, the alpine Silene, and the little yellow Aretia, it is admirable for association with such plants. It grows naturally very high on the Alps, but thrives in loamy soil, and is easily raised from seed. A native of the Alps of Switzerland, Savoy, and Austria. It is occasionally found with white flowers in a wild state.

IBERIS (*Candytuft*).—For the rock-garden, these perennial, half-shrubby plants are essential, hardy, of great endurance, and good effect, and they can be grown anywhere, in any soil, and are easily increased. Although dwarf, they are so wiry and enduring, that they might well be used in bold groups between the rock-garden and its surroundings.

Iberis corifolia (*Coris-leaved Candytuft*). —A very dwarf kind, only 3 or 4 inches high when in flower, and covered with small white blooms in May. Few alpine plants are more worthy of general culture. It is probably a small variety of the Evergreen Candytuft, but for garden use it is distinct enough. Southern Europe; easily propagated by seeds, cuttings, or division, and thriving in any soil.

I. correæfolia (*Correa-leaved Candytuft*). —This plant is readily known from any other cultivated kind by its entire and rather large leaves, by its compact head of large white flowers, and by flowering later than the other white kinds. Both the flowers and the corymb are larger than in the other species, and the blooms stand forth more boldly from the smooth dark-green leaves. It is an invaluable hardy plant, and particularly useful in consequence of coming into full beauty about the end of May or beginning of June, when the other kinds are fading away. Of its native country we know nothing; but once Mr Jennings, of the Wellington Nurseries, informed me that it was raised in, and first sent out from, the Botanic Garden at Bury St Edmunds, and it is probably a hybrid. Mr J. G. Baker considers it to come nearest to *I. Pruiti*, of the Nebrode Mountains, in Sicily. Readily increased by cuttings, and also by seed.

Iberis Gibraltarica (*Gibraltar Candytuft*).—This is larger in all its parts than the other cultivated kinds, has oblong spoon-shaped leaves, nearly 2 inches long; the large flowers, often reddish-lilac, being arranged in low close heads, and appearing in spring and early summer. I am doubtful of its hardiness, and should advise its being wintered in pits or frames till sufficiently abundant to be tried in the open air. It should be planted on sunny spots. A native of the South of Spain; increased by seeds and cuttings.

I. Tenoreana (*Tenore's Candytuft*).—A dwarf species, with toothed leaves, which, with the stems, are hairy, and a profusion of white flowers changing to purple. As the commonly cultivated kinds are white, this one will be the more valuable from its purplish hue, added to its neat habit. It, however, has not the perfect hardiness and fine constitution of the white kinds, and is apt to perish on heavy soils in winter; but on light sandy soils it is a good plant. A native of Naples, and easily raised from seed.

I. sempervirens (*Evergreen Candytuft*). —This is the common rock Candytuft of our gardens, as popular as the yellow *Alyssum* and the white *Arabis*. Half-shrubby, dwarf, evergreen, and perfectly hardy, it escaped destruction where many herbaceous plants were destroyed; and as in April and May its neat tufts of dark-green are transformed into masses of snowy white, its presence has been tolerated longer than many other fine old plants. When in good soil, and fully exposed, it forms spreading tufts often more than a foot high, and they last for many years. Like all its relatives, it should be exposed to the full sun rather than shaded, if the best result is sought. A native of Greece, Asia Minor, Dalmatia, and S. Europe, and readily increased by seeds, cuttings, or division.

I. Garrexiana is a variety of the Evergreen *Iberis*, not sufficiently distinct to be worthy of cultivation; in fact, it and several other Iberises prove to be mere varieties, and very slight ones, of *I. sempervirens* when grown side by side.

ALPINE FLOWERS FOR GARDENS

Iberis jucunda.—A beautiful and very dwarf Candytuft, with soft, rosy, lilac-flowers in corymbed clusters, on slender twisted stems, over small sea-green foliage, the plant rarely more than 4 inches or 5 inches high. It is easily raised from seed, and should be cultivated in numbers, so as to form good-sized patches.

INCARVILLEA. — Distinct and beautiful perennials of recent introduction, probably hardy, coming from the high mountains of China, where there are, no doubt, many other beautiful things in Nature's vast storehouse. Though the habit is bold, they may very well find a home on the rock-garden until more plentiful and better known. They, so far as now known, flower in summer, are of easy culture in ordinary soil, and do not seem difficult of increase.

Incarvillea Delavayi.—We owe the introduction of this beautiful plant to the Abbé Delavay, who found it in Yunnan, Western China, at a height of 8,000 feet to 11,000 feet. It has a stout root-stock, with a very short subterranean stem, from which spring the bright green leaves, each a foot or more long. The flower-scape varies from a foot to 2 feet, and bears from two to a dozen or more flowers, 2 inches long and 2 inches wide, rich rose, with a few purple streaks, and a tinge of yellow in the throat.

Mr C. M. Mayor, of Paignton, Devon, sending me a photo of a very fine plant, says: "It was planted in deep, light, ordinary garden-soil, in a sunny spot, the crown covered with sand to a depth of 3 inches. I found that mulching with rotten manure or other moisture-holding material, if in contact with the bases of the frond-like leaves, causes them all to rot off—a rot which quickly spreads to the tuberous root itself."

Incarvillea grandiflora resembles *I. Delavayi* in general characters, differing in its shorter leaves, more rounded leaflets,

Violet Cress (Ionopsidium acaule). Engraved from a photograph by Miss Wolley Dod.

bearing only one or two large flowers, whilst the colour is a deep rose-red.

I. Olgæ is hardy in the southern counties, and has bright green pinnate leaves and, borne upon the upright ends of the branches, panicles of rose-pink tubular flowers, each an inch long and wide. Turkestan.

There are other beautiful species of these not yet introduced or sufficiently tried.

IONOPSIDIUM ACAULE (*Violet Cress*).—This, being an annual plant, is only introduced here in consequence of its peculiar beauty for adorning bare spots on the rock-garden devoted to very minute alpine plants. As it sows itself, the cultivator will have no more trouble with it than with a hardy perennial. It frequently flowers at 1 inch high, and rarely exceeds 2 inches, the small flowers being of a pale violet tinge, and the leaves roundish and compactly arranged. It will flower a couple of months after being sown; and, when sown in spring in the open ground, the self-sown seeds of the summer flowers soon start into growth, and the second crop flowers in autumn, and far into winter. A native of Portugal and Morocco.

IRIS (*Flag*).—Of these wonderfully varied and beautiful plants, the majority are too vigorous for the rock-garden; but a certain number of the dwarf species might well find a home on it, such as the little American crested Iris. Also some of the new cushion Irises may there find conditions that suit them. The various forms of the Dwarf Flag (*I. pumila*), are often very pretty in colour, and are easily grown.

Iris cristata (*Crested I.*).—A dwarf and charming Flag, usually running about with its creeping and rooting stems exposed on the surface, not rising above the ground more than a few inches, having flowers, however, as large as many of the coarser species. It flowers in May; blue with spots of a deeper hue on the outer petals, and a stripe of orange and yellow variegation down the centre of each. The plant is readily distinguished at any season from any other dwarf species by the creeping stems growing well above the ground. Even young tufts push so boldly out of the ground that a top-dressing of an inch of fine soil placed around them cannot fail to help the roots. It loves and flourishes luxuriantly on rich but free and light soil, in a warm position. I have never seen it do so well as in the Glasnevin Botanic Gardens, but have seen it thrive both to the north and south of London. On the rock-garden, it thrives best on level sandy spots. A native of mountainous regions in North America, with all the gem-like loveliness of the choicest Swiss alpine flowers; was introduced by Mr Peter Collinson, so long ago as 1756.

Iris pumila (*Dwarf Crimean I.*).—Often flowering at 4 inches, the dwarf Iris, even in favourite soils, rarely exceeds 10 in height: the stems usually bear one or two deep-violet flowers, large and beautiful in April and May. It thrives in ordinary garden soil, the lighter and deeper the better; the finest specimens I have ever seen were in a deep sandy peat, and they were twice the ordinary size. There are several varieties: yellow, white, light blue, and deep dark violet, respectively known under the names of *I. pumila*, *lutea*, *alba*, *cærulea*, and *atrocærulea*. Each of the varieties is worthy of cultivation, and easily increased by division of the rhizomes.

I. reticulata (*Early Bulbous I.*).—Distinct from other early Irises, and perhaps the most valuable of all, considering its early bloom, violet scent, and rich colour. The root is a tuber; leaves four-angled and rather tall when fully developed; and the flowers, borne on stems 3 to 6 inches high, are of the most brilliant purple, each of the lower segments marked with a deep orange stain. It blooms in early spring, long before any other Iris shows itself, and loves a deep sandy soil and a warm well-drained position. There is no more beautiful plant for a sunny bank on the lower slopes of the rock-garden. Southern Europe, Asia Minor. Increased by division.

I. stylosa (*Algerian Flag*).—A lovely winter-blooming Iris, quite hardy on all warm, dry soils, but its flowers are of delicate texture, and suffer from rough gales. There are several varieties having flowers of lighter or darker shades of soft lilac or lilac-purple, and there is a white form with golden-crested petals. All are beautifully and easily grown in the open air, but it only flowers well in warm

ALPINE FLOWERS FOR GARDENS

sandy soil, and therefore where such soil does not occur naturally, the best place for it is the rock-garden, in well-drained

Iris stylosa (*Algerian Flag*). (Engraved from photograph by Mr S. W. Fitzherbert.)

and warm slopes, where its tufts of grassy leaves will look well throughout the year.

Dwarf Bulbous Iris.—Apart from the above older plants of our gardens of recent years, a number of dwarf bulbous Iris have come into cultivation, for which the rock-garden will often afford a good place. Of these, some of the prettiest are :

Iris Bakeriana.—A charming little hardy Iris about 5 inches high; standards pale blue, falls white with purple spots and a rich black purple lip; flowering in February. It is sweet-scented.

I. Boissieri, lilac dark blue, with yellow blotch, very charming species.

I. Danfordiæ, brilliant yellow, with small greenish spots, very dwarf, early spring flowering, quite hardy.

I. Histrio, blue, streaked yellow and blotched deep purple. Not only one of the hardiest of the Irids, but one of the earliest, being earlier than *I. reticulata.*

I. Histrioides.—A beautiful dwarf Iris; the early flowers are bright ultramarine, with markings on a white ground.

I. orchoides, bright yellow, hardy and free on many soils.

I. Persica (*Persian Iris*).—Light blue, blotched with purple, and lined with orange, early, sweet-scented.

I. Persica purpurea, a most beautiful variety, of a rosy purple colour.

I. Rosenbachiana, short upright leaves, flowers deep violet, very long falls, which are marked blue and yellow.

Iris Sophenensis, beautiful dwarf Iris, in the way of *I. reticulata*, bright blue flowers.

I. Willmottiana.—Lavender blue, white and dark blue spotted, a pretty new Turkestan Iris.

ISOPYRUM THALICTROIDES

(*Meadow-rue I.*).—A graceful little plant allied to the meadow-rues, with pretty white flowers, valuable for its maidenhair-fern-like foliage. It is useful as an elegant ground-plant below rock shrubs as well as for its own sake, is hardy, and easy to grow on any soil. Comes from the Pyrenees and mountainous parts of Greece, Italy, and Carniola, is easily propagated by division or by seed. The leaves rarely rise more than a few inches high, the flower-stems from 10 to 14 inches.

JANKÆA HELDREICHI.—This is the prettiest of the Ramondia family, and is a native of the mountains of South Macedonia, growing in ravines and dells. Owing to failures in its cultivation, it has been considered a miffy plant, dying away in our gardens in spite of the most careful handling. It likes to be moderately moist at the roots and have shade and moisture in the air. The blooms are of a deep and bright blue, somewhat nodding, and shaped like those of a Soldanella. Their beauty is heightened by the silver-grey leaves.

JASIONE(*Sheep's Scabious*).—Dwarf, perennials and annuals of the Bellflower family, interesting, but not of highest importance for the rock-garden, *J. humilis* is a creeping tufted plant, about 6 inches high, bearing small heads of pretty blue flowers in July and August. Though a native of the high Pyrenees, it often succumbs to the damp and frosts of our climate, and it therefore requires a dry well-drained part of the rock-garden, and should have a little protection in winter during severe cold and wet. *J. perennis* is taller, often above 1 foot high, with dense heads of bright blue flowers, from June to August; it is a rock-garden plant, stronger than the preceding, thriving in good light loam, and a native of the mountains of Central and South Europe. These perennial kinds may be propagated best from seed, as they do not divide well. *J. montana* is a neat, hardy annual, with small, pretty bright blue flower-heads in summer. Seed in autumn or spring. A native plant.

JASMINUM (*Jasmine*).—Beautiful shrubs, the hardy ones among the best introduced to our country, and of very wide and precious use. Where any bold rock-gardening is carried out, these should be used, and may be very gracefully used. They are so often the victims of crucifixion against walls, that it will be pleasant to see them showing their native grace of habit.

Jasminum humile (*Indian Yellow Jasmine*).—A handsome kind, hardy, with evergreen foliage, which adds to its value. It flowers freely, and its yellow bloom amidst the deep green foliage is welcome in summer and autumn. Being an Indian plant, it should have a warm aspect and good warm soil. (Syns., *J. revolutum* and *J. wallichianum*.)

J. **nudiflorum** (*Winter Jasmine*).—A lovely Chinese bush, which is happy enough in our northern climate to flower very often in the depth of winter, clustering round cottage walls and shelters, and often more lovely when not too tightly trained. In wet years it will be noticed increasing as freely as twitch at the points of the shoots. It should be planted in different aspects, so as to prolong the bloom.

Jasminum officinale (*White Jasmine*).—The old white Jasmine of our gardens, one of the most charming shrubs ever introduced for warm banks; it is best on rocky or sandy soils. There are several varieties of it, the best being *J. affine*, with flowers larger than those of the ordinary kind. It is almost evergreen, except in exposed places.

It is a native of Persia and the north-western mountains of India, naturalised here and there in Southern Europe.

JEFFERSONIA DIPHYLLA (*Twin-leaf*).—A plant very little grown, and usually regarded as a botanical curiosity; but when planted in sandy peat associated with plants like the *Epimedium*, *Rhexia*, and *Spigelia marilandica*, it becomes a pretty spring flower, as well as interesting from its curiously paired leaves. The flowers are white, with yellow stamens, about an inch across, and freely borne when the plant is in vigorous health. A good plant for peaty and somewhat shady spots on the rock-garden, planted in sandy peat. A native of rich woods in North America. Careful division in winter.

JUNIPERUS (*Savin*).—Often graceful bushes of the great Pine family, clothing the alpine rocks where the tree has no chance from poverty of the rocky soil and exposure. Few evergreen rock shrubs are more useful for a quiet and graceful effect than the common Savin and its forms, and particularly that known in Nurseries as the Tamarix-leaved Savin (*J. tamariscifolia*), for carpeting stony ground, planting on dry banks where little else could grow. Some of the northern dwarf forms of Juniper are grown on rock-gardens under the name of *J. nana*.

KALMIA (*Mountain Laurel*).— Among the loveliest of evergreen shrubs of the northern world. The smaller kinds are of the highest value, and the large one essential for the bold rock-garden, being not only a first-rate evergreen, but the flowers are of great beauty, coming too at a very good time, between the great crowd of spring flowers and the coming of the Roses. If one had only these and half a dozen other groups of shrubs of the northern moors and mountains, a very enduring and graceful rock-garden might be made from them alone. And that almost without trouble in the many parts of our islands where rocks crop out, as in Wales, Ireland, and Scotland. Nor do we want rocks, as they grow like weeds on the peaty moors of England.

Kalmia angustifolia (*Sheep's Laurel*), grows about 1½ feet high, and bears in early June dense clusters of rosy pink flowers. It is a graceful, hardy, and easily grown shrub, excellent for the rock or drier parts of bog-garden. Newfoundland, Hudson Bay, and southward.

K. glauca (*Swamp Laurel*).—A dwarf evergreen shrub with smooth leaves silvery on the lower surface, with purplish flowers. Excellent for the rock-garden among the mountain bushes, and quite free in peat or moor soil. Newfoundland, Hudson Bay, and Alaska.

K. hirsuta (*Hairy Laurel*).—A dwarf evergreen shrub, distinguished from the other kinds by its hairy leaves, and not quite so hardy, being a native of Virginia, and Florida in Pine Barrens.

K. latifolia (*Mountain Laurel*).—This is the finest as it is the commonest in gardens, and should be planted wherever the soil is suitable. Like the Rhododendron and Azalea, the Kalmia is best grown in a moist peaty soil, or one light or sandy. It will not thrive in stiff or chalky soils. Its lovely clusters of pink wax-like flowers open about the end of June, when the bloom of the Rhododendron and Azalea is on the wane, and last for a fortnight or longer. There are varieties of *K. latifolia*, having in some cases larger flowers, and in others, flowers of a deeper colour, the finest being *Maxima*, which is superior in size of flower.

The Myrtle-leaved Kalmia (*K. myrtifolia*) seems to be only a variety of *K. latifolia*, with smaller foliage. The growth is dwarf and compact, and the flowers are almost as large as those of *K. latifolia*. Canada and southwards, in sand and rocky woods.

KERNERA SAXATILIS.—A neat little plant, very like the dwarf Scurvy Grass (*Cochlearia*). It forms a compact tuft of foliage, and in early summer is a dense mass of tiny white blooms. It grows in any soil in an open position in the rock-garden, where it is an attractive plant in spring, and may be freely propagated by seeds. Europe.

LATHYRUS(*Everlasting Pea*).—For the greater part, these perennial trailers are too large for our purpose, if we take the narrow view of the rock-work with small plants only; but in a bolder kind of rock-garden, with its mountain shrubs, the rarer and more beautiful kinds may come in very well. Moreover, the freedom of the shrubby rock-garden allows us to dispense with staking, which is a great gain, as I think these plants never look so well as in their own way of growth. The effect is much better when they fall over rocks or banks. Even the stoutest kind, with its white and prettily coloured forms, is handsomer falling down banks than in any other way. But when we have to deal with Everlasting Peas of such rarity and beauty as the Greek *L. sibthorpii* and the Californian *L. splendens*, we have plants by no means so free, and which may well grace the rock-garden. Some good plants once known by other names are now included in *Lathyrus*. Syn., *Orobus*.

Lathyrus cyaneus (*Blue Bitter Vetch*).—A dwarf vetch-like plant, with large, handsome, bluish flowers among masses of light green leaves, with two or three pairs of leaflets, flowering in spring, the plant growing little more than 6 inches high. I have only observed this plant growing on very cold stiff ground scarcely acceptable to coarse weeds, and there it was quite hardy and flowered regularly, ascend in a zigzag manner to about 1 foot in height, bearing leaves with two or three pairs of leaflets, and rather closely arranged racemes of flowers supported on a footstalk a couple of inches long. The flowers, though small, are beautifully variegated, the upper petal being a fine rose-colour with a network of full purplish-crimson veins, the points of the wings being blue. It is a hardy, easily-grown plant, and

Leiophyllum buxifolium.

so that it is probable it would do much better on light good soils. It comes from the Caucasus, and is best for warm, sheltered, sunny spots. It is sometimes met with under the name of *Platystylis cyaneus*, under which name it was figured by Sweet. Syn., *Orobus cyaneus*.

Lathyrus variegatus (*Variegated Vetch*).—A compact plant, with two firm and opposite keels on its wiry stems, which may be increased by seeds or division. Southern Italy and Corsica.

Lathyrus vernus (*Spring Everlasting Pea*).—From black roots spring rich healthy tufts of leaves, with two or three pairs of shining leaflets, the flower buds showing soon after the leaves, and eventually almost covering the plants with purple and blue flowers with red veins, the keel of the flower tinted with green, and the whole changing to blue. It is no fastidious alpine plant that, when carried to our gardens in the cultivated plains, sickens and dies, but a vigorous native of Southern and Central Europe, well able to make the most of our warm deep sandy loams, growing in almost any soil, and hardy everywhere. It varies a good deal—all the better, of course—the most marked of the known varieties or subspecies being *ruscifolius* and *flaccidus*.

LEDUM (*Labrador Tea*).—The best of the few species of *Ledum* grown in gardens is *L. latifolium*, which represents the genus well. Its usual height is under 2 feet, but sometimes it reaches 3 feet. It is dense and compact, and has small dull green leaves of a rusty brown beneath. During the latter part of May it bears clusters of small white flowers, which being abundant are showy. It is a very old garden plant, and was brought from North America more than a century ago. The Canadian form of it (*Canadense*) is found in some gardens, but does not differ materially from the type. A form called *Globosum* is finer, as the flower-clusters are larger and more globular. *L. palustre* is commoner than *L. latifolium*, smaller in every part, and much inferior. It is dwarf and spreading, and its flowers are white. A native of both North America and Northern Europe. They thrive best in a peaty soil or sandy loam, and are usually included in a collection of so-called American plants, and are charming grouped in the bog-garden, fully exposed.

LEIOPHYLLUM BUXIFOLIUM (*Sand Myrtle*).—A neat and pretty tiny shrub, forming compact bushes from 4 to 6 inches high, and densely covered with pinkish-white flowers in May, the buds of a delicate pink hue.

It is suited for grouping with diminutive shrubs, such as the Partridge Berry and smaller Daphne, thriving in sandy peat. A native of sandy "Pine Barrens" in New Jersey, and often to be had in our Nurseries under the name of *Ledum thymifolium*.

LEONTOPODIUM ALPINUM (*Edelweiss*).—A native of high pastures on many parts of the great continental mountain ranges. The flowers are small, yellowish, the leaves covered with white down, like those of many mountain composite plants, but it is distinguished by a beautiful whorl of oblong leaves, springing star-like from beneath the closely set and inconspicuous flowers, and almost covered with white, dense, short down. It is a hardy perennial, growing from 4 to 8 inches high, and thriving in firm, sandy, or gritty and well-drained soil, in well-exposed spots in the rock-garden. The soil should be poor, as in rich soil it loses its charm, and often perishes through overgrowth. It is best to raise it from seed.

LEUCANTHEMUM ALPINUM (*Alpine Feverfew*).—A very dwarf plant, with small fleshy leaves, deeply cut, and hoary, and not rising more than half an inch above the surface. It bears pure white flowers more than an inch across, and with yellow centres, borne on hoary little stems, from 1 to 3 inches long. It is a rather quaint and pretty plant, and well deserves cultivation on the rock-garden, in bare level places, on poor, sandy, or gravelly soil. Syns., *Chrysanthemum alpinum* and *Pyrethrum alpinum*. Alps of Europe. Division or seed.

LEUCOJUM (*Snowflake*).—Graceful bulbous plants, the taller of which are easily grown plants anywhere, even naturalised in riverside soils; one or two of the smaller ones are very pretty, coming out of tufts of low plants in the spring, particularly the vernal Snowflake.

Leucojum vernum (*Spring Snowflake*).— A dwarf, stout, broad-leaved plant, like a *Galanthus*, but with larger and handsomer flowers, and appearing about a month later than the Snowdrop; fragrant, the segments white, an inch long, and each distinctly marked with a green or yellowish spot near the point, drooping and usually produced singly on stems from 4 to 6 inches high. It is more worthy of cultivation than the Snowdrop, and that is as high praise as we can give to any dwarf spring-flowering plant. It has long been known as a continental plant, and was valued and grown in our gardens, when hardy flowers were more esteemed than they are at present; but its existence as a true native was not known with certainty till recent years ago, when it was found in abundance, on the "Greenstone heights, in the neighbourhood of Britford." It is not by any means a common plant, and those who have it would do well to place it in positions where it is likely to thrive in light, rich, well-drained soil, or in borders, and as, after the plant has flowered, the leaves attain the length of nearly a foot, and are nearly or quite three-quarters of an inch across, a sheltered position, where they may not be torn by winds, will be best. It is apt to dwindle on some cold soils.

Edelweiss (*Leontopodium alpinum*). (Engraved from a photograph by Mr G. S. Symons, Chaddlewood, Plympton.)

LEUCOTHOE.—Beautiful evergreen shrubs of the Heath family, most of them very old garden plants, and common in collections of American plants. There is a striking family likeness between the common kinds, the best-known being *L. acuminata*, which grows from 1½ to 2½ feet high, and has slender arching stems clothed with long pointed leaves. In early summer the stems are profusely wreathed with tiny white bell-shaped flowers, extremely pretty. *L. axillaris* is similar, and so are *L. Catesbœi* and *L. racemosa*, all of which are known in gardens under the generic name *Andromeda*. They are natives of North America, hardy, and thrive in any light soil, preferring peat or leaf-mould. A newer and very beautiful species is *L. Davisiæ*, introduced a few years since from California, and therefore neither so common nor so hardy as the others. It makes a neat little evergreen bush 2 or 3 feet high, and has small leaves on slender stems, which in May are terminated by dense clusters of small white flowers in short erect spikes. It is one of the choicest of evergreen hardy shrubs, is thoroughly deserving of general cultivation, and thrives with Rhododendrons and Azaleas in peat soil.

LEWISIA REDIVIVA (*Bitter Root*). —A singular and pretty plant, allied to the Ice plants, and forming rosettes of leaves, 2 to 3 inches long, on a thick, woody stalk. After the leaves attain their full growth in spring or early summer, beautiful flowers issue from the rosettes, nearly hiding the plant, each blossom 3 to 4 inches across, and consists of eight or twelve shaded pink petals, the centre being nearly white and the tips rose-colour, the whole having a satiny lustre. The flowers open only during sunshine. Native of the west parts of North America, particularly in Washington Territory and Oregon. Should have a warm position in the rock-garden, in dryish soil, or between stones on an earth-mortared wall.

LIBERTIA.—Beautiful plants of the Iris Order, of which some are hardy in peaty and leafy soils. *L. formosa* is beautiful at all seasons, even in the depth of winter, owing to the colour of its foliage, which is as green as the Holly; and it bears spikes of flowers of snowy whiteness like some delicate Orchid. It is dwarf and compact, and has flowers twice as large as the other kinds. They lie close together on the stem, and remind one of the old double white Rocket. *L. ixioides*, a New Zealand plant, is also a handsome evergreen species, with narrow grassy foliage and small white blossoms. *L. magellanica* is also pretty when in flower. All of these thrive in borders of peaty soil, and in the rougher parts of the rock-garden, but they grow slowly on certain loamy soils, living perhaps, but never showing freedom and grace. Increased by seed or by careful division in spring.

LILIUM (*Lily*).—Most of these handsome plants are too large for the rock-garden; a few, however, of the smaller ones may well come into it. And the idea so much urged in this book, that we ought in every case almost to associate the mountain shrubs with the alpine flowers, when carried out, gives us a chance of growing Lilies and other choice bulbs among the shrubs. The shelter and partial shade of the shrub helps the bulbs in various ways, and gives us a good opportunity of growing these beautiful plants in their fine variety of good form and colour. As the manner and descriptions of Lilies are to be found in so many books and lists, there is no need to name them here.

LINARIA (*Toad Flax*).—Annual and perennial plants, rather fine and graceful in form, some, though not many, pretty. Some of the species have not beauty enough for our present purpose, and a close selection of the best only should be made where the aim is beauty.

Linaria alpina (*Alpine Toadflax*).—A true alpine plant, from the Alps and Pyrenees, found on moraines and *débris* of the mountains; allied to the Ivy-leaved Linaria, but quite different in aspect, forming dense, dwarf, smooth and silvery tufts, covered with bluish-violet flowers, with two bosses of intense orange in the centre of the lower division of each. Its habit is spreading, but neat and very dwarf, rarely rising more than a few inches high. On the Alps I have seen it flowering profusely at 1 inch high, the leaves which attain a length of three-quarters of an inch in our gardens being almost rudimentary and scarcely perceptible beneath the flowers, which quite obscure stem and leaves, being larger proportionately than on the cultivated plant. It is usually a biennial; but in favourable spots, both in a wild and cultivated state, becomes perennial. Its duration, however, is not of so much consequence, as it sows itself freely, and is one of the most charming subjects that we can allow to "go wild" in sandy, gritty, and rather moist earth, or in chinks of rockwork. In moist districts it will sometimes even establish itself in the gravel walks. It is readily increased from seed, which should be sown in cold frames, in early spring, or out of doors.

L. antirrhinifolia.—An elegant little rock plant, forming a very neat spreading mass about 6 inches to 8 inches high. It has the advantage of not spreading so rapidly as some of its congeners, flowering incessantly throughout the summer. The flowers are of a bright purple colour. The plant is of the easiest possible culture, and can be highly recommended for the rock-garden.

Linaria crassifolia (*Thick-leaved Toadflax*).—A small and pretty, though not showy species, 3 to 6 inches high, flowering in summer; fine blue, with a yellow throat. A native of Southern Spain, near the town of Chiva. This plant resembles *L. origanifolia*, but the living plants present a marked difference. The rock-garden, walls, ruins, borders, light, sandy soil. Division and seed.

L. Cymbalaria (*Ivy Toadflax*).—This is the wild Ivy-leaved Linaria, that drapes over so many walls so gracefully. It has a white variety. The plant itself would be here, were it not that it usually takes possession of old walls, but it is always one of the most graceful of the plants that adorn them, and it should be encouraged. It occurs on old walls and stony places in many parts of Europe, and is wild in Britain, but probably only naturalized. Any soil suits it, or dry walls without soil. It usually establishes itself. Seed.

L. hepaticæfolia (the *Hepatica-leaved Toadflax*), from Corsica, is also a good alpine plant, but not so attractive as alpina. It is nearly always in flower, in summer and autumn, and masses in a rock-garden are good in effect.

LINNÆA BOREALIS (*Twinflower*).—A fragile trailing evergreen, bearing delicate, fragrant, and gracefully drooping pale pink flowers. This plant is named after Linnæus, with whom it was a favourite. A native of moist mossy woods, in Northern Europe, Asia, and America, and sometimes of cold bogs or rocky high places in Britain, occurring in fir woods in a few places in Scotland and Northern England. It loves a sandy peat and moist soil, and may be grown as a trailer, the shoots being allowed to fall down over the faces of the rocks, or in mossy rocky ground among bushes, on the fringes of the bog garden, or in some half-shady position, in the hardy fernery. It usually enjoys a somewhat shady position, but, if in proper soil, will bear the sun. Readily increased by division.

LINUM (*Flax*).—Annual and biennial plants of much delicate beauty of colour. Some of the dwarfer perennial kinds are most charming flowers in their various shades of blue, and well deserve to be grown in groups. The habit of "dot" planting is against our seeing the best effect of the mountain flaxes.

Linum alpinum (*Alpine Flax*).—A dwarf and quite smooth Flax, growing only from 3 to 8 inches high, and bearing large dark-blue flowers in summer. A charming rock plant, native of the Alps, Pyrenees, and many hilly parts of Europe, thriving well in warm well-drained spots on rockwork, in a mixture of sandy loam and peat. There are several varieties—*alpicola*, *collinum*, and *crystallinum*; *L. austriacum* is intimately related to it.

L. arboreum (*Evergreen Flax*).—This is the neat, glaucous, dwarf, spreading shrub, with many clear large yellow flowers, an inch and a half across, sometimes seen in our gardens under the name of *L. flavum*. Although said to be tender in the colder and drier parts of the country, it thrives well in others in the open air, and in all is well worthy of a place. A native of hilly parts of South-Eastern Europe, Asia Minor, and North Africa; usually propagated by cuttings. It is sometimes grown as a frame and greenhouse plant, but should be tried everywhere in warm spots on dry borders, banks, or rockworks. It begins to bloom in early summer, often flowering for months at a time.

L. campanulatum (*Yellow Herbaceous Flax*).—A herbaceous plant, with yellow flowers in corymbs on stems from 12 to 18 inches high, distinct from anything else in cultivation, and well worthy of a place in a collection of alpine plants. A native of the South of Europe, flowering in summer and flourishing freely in dry soil on the warm sides of banks, and propagated by seeds. *Linum flavum* is said to be different from this by its shorter sepals, and several minor characteristics; but Messrs Grenier and Godron found these very inconstant and differing very much in the French plant. Syn., *L. flavum*.

Linum narbonnense.—A beautiful and distinct sort, bearing during the summer months large, light sky-blue flowers, with violet-blue veins. A fine plant for the lower flanks of the rock-garden, on rich light soils, forming lovely masses of blue, from 15 to 20 inches high. A native of Southern Europe, thriving in any good soil.

L. perenne (*Perennial Flax*).—A plant found in some parts of Britain, particularly in the Eastern countries, but rare, usually growing in dense tufts from 12 to 18 inches high, with bright cobalt-blue flowers more than an inch in diameter, the stamens in some being longer than the styles, in others shorter, the petals overlapping each other at the edges. Mr Syme considered it probable that *L. alpinum* and *L. Leonii* are forms that may be included under *L. perenne*. *L. perenne album* is also an ornamental plant, and there is also a variety with blue flowers variegated with white, known in gardens as *L. Lewisii variegatum*, but this marking is not very conspicuous or constant. *L. sibiricum* and *L. provinciale* are also included under *perenne*. They are all of very easy culture in common garden soil.

L. monogynum (*New Zealand Flax*).—A beautiful kind, with large pure white blossoms, blooming in summer. It grows about 1½ feet high in good light soil, and its neat and slender habit renders it particularly pleasing for the borders of the rock-garden or for pot-culture. It may readily be increased by seed or division; it is hardy in the more temperate parts of England, but in the colder districts is said to require some protection. *L. candidissimum* is a finer and hardier variety. Both are natives of New Zealand.

Linum salsoloides (*Heath Flax*).—A hardy, dwarf, half-shrubby species, somewhat like a dwarf Heath, with the stem twisted at the base, from 3 to 6 inches high, blooming in June and July; white with a purple centre. A native of the South of Europe, this plant is well adapted for the rock-garden, in well-drained sandy soil.

L. viscosum (*Viscid Flax*).—Half-shrubby, slightly branching downy stems; about 1 foot high. Flowering in summer; lilac, with deeper veins, nearly 1 inch across. The rock-garden, in moist sandy loam. Seed and division. Pyrenees.

LIPPIA (*Fog Fruit*).—*L. nodiflora* is a dwarf perennial creeper of the Verbena order, bearing in summer heads of pretty pink blooms. It grows in any situation or soil, and is a good plant for quickly covering bare spaces in the rock-garden. Division. Southern United States, and California.

LITHOSPERMUM (*Gromwell*).—Dwarf, half-shrubby, very beautiful plants of the Forget-me-not order, but unhappily not hardy in our country, except in the best conditions of culture. The warmest part of the rock-garden is the best for them. But they come from the burning rocks and sands of Spain and North Africa, and though they promise much, few survive our hard winters.

Lithospermum Petræum (*Rock Gromwell*).—A neat dwarf shrub, in colour somewhat like a small Lavender bush. Late in May or early in June all the little grey shoots of the dwarf bush begin to show small, oblong, purplish heads, and early in July the plant is in full blossom, the flowers of a fine violet blue, with protruded anthers of a deep orange red, the buds of a reddish-lilac. The flowers are barely more than a quarter of an inch long, and tubular, not at all open, but as every shoot is crested by a densely-packed head of flowers, the effect is pretty and distinct. The best position for this plant is somewhere on a level with the eye, on a well-drained, deep, but rather dryish sandy soil on the sunny side. Dalmatia and Southern Europe; cuttings, or seeds, if they can be obtained.

Lithospermum Prostratum (*Gentian-blue Gromwell*).—A charming little evergreen spreading plant, having lovely blue flowers, with faint reddish-violet stripes, about half an inch across, in profusion where it is well grown. A native of Spain and the South of France, easily propagated by cuttings, and valuable as a rock-plant from its prostrate habit and the fine colour of its flowers—a blue scarcely surpassed by that of the Gentians. It may be planted so as to let its prostrate shoots fall down the sunny face of a rocky nook, or allowed to spread into flat tufts on level spots. In cold or wet soil it should be raised on banks, and planted in sandy earth.

L. purpureo-cæruleum, a British plant, *L. Gastoni* and *L. canescens* are also worthy of culture in large collections, but the tender nature of most of the kinds limits their use in our country.

LLOYDIA SEROTINA.—A small bulbous Liliaceous plant, frequently seen as soon as the snow melts, in flower by the alpine pathways. It is most suitable for botanical collections. Alps.

LOISELEURIA PROCUMBENS.—In a wild state on the Alps, or on mountain moors, this is a wiry trailing shrub, growing quite close to the ground, the plants occasionally forming a rather dense tuft, bearing small reddish flowers in spring, when the snow melts. It is very rarely seen in a thriving state under cultivation, and most of the plants transferred from the mountains to gardens usually perish. This is sometimes owing to the finest plants being selected instead of the younger ones. I never saw it in such perfect health in a garden as in that of the late Mr Borrer, in Sussex, where it flourished in compact masses thrice its usual size, in deep sandy peat. Its true garden home is the rock-garden, and it will seem well worthy of a place to most lovers of rare British plants. On the high Alps tiny plants of it are charming. Syn., *Azalea procumbens*.

LONICERA (*Honeysuckle*).—Given the idea of the shrubby rock-garden, we have here again a fine group of plants usually well-trained, grown and often over-pruned on walls: are themselves rock-shrubs, and will associate and mingle very well with many shrubs that we may use in or near the rock-garden. There are various kinds worth growing, a description of which will be found in "The English Flower-Garden," and other works.

One can hardly go wrong with the Honeysuckle as to kind; the European Honeysuckle, with its beautiful forms, the Japanese, the Chinese (including the Winter Honeysuckle), the American, and the forms we call the Dutch, I can imagine nothing fairer than these grown in their natural forms on rocky banks or among shrubs near the rock-garden.

LUZURIAGA RADICANS. — A small half-hardy evergreen from Chili. In the mildest localities, though even in these, it does not thrive so well as in a cool house. It is worthy of a trial in a cool bed of peat, on the north side of the rock-garden, among the larger alpine shrubs.

LYCHNIS (*Campion*).—These showy perennials are usually too tall for the rock-garden, but a few of the mountain kinds are pretty, and quite fitting for the rock-garden.

Lychnis Alpina (*Alpine L.*).—In a wild state, seldom rising more than a few inches high. "In Britain," says Mr Bentham, "it is only known on the summit of Little Kilrannock, a mountain in Forfarshire," but in 1886, under the safe guidance of the late Mr James Backhouse, I had the pleasure of seeing it abundantly in Cumberland in very lonely and high mountain gorges. We found it on the face of a dry crumbling crag, quite 500 feet long, and of great height, and generally in such positions that extermination is impossible. In some places where the rocks overhung, it was in full health, where a drop of rain could scarcely ever fall upon it; but many plants which had sprung from seeds fallen from these cliffs were growing freely in moist shattered rock. In cultivation it is a pretty, if not a brilliant, plant, and may be grown without difficulty in rather moist sandy soil.

Lychnis Lagascæ (*Rosy L.*).—A lovely dwarf alpine plant, with a profusion of bright, rose-coloured flowers, with white centres when young, each about three-quarters of an inch across, and quite obscuring the small and slightly glaucous leaves. In consequence of its exceeding brilliancy of colour, and slightly spreading, though firm, habit, it is well suited for fissures on the exposed faces of rocks, the colour telling a long way off, while it is also a gem for association with the smallest alpine flowers. It is a native of the sub-alpine region of the North-Western Pyrenees, and was introduced some years ago by the late Mr J. C. Niven, of the Hull Botanic Garden, in whose collection I first had the pleasure of seeing it grown. It is distinct from, and more beautiful than, any other alpine or dwarf *Lychnis*. It flowers in early summer, and is most readily increased by seeds. Syn., *Petrocoptis Lagascæ*.

L. Viscaria (*German Catchfly*).—A British plant, found chiefly in Wales and about Edinburgh, but widely distributed in Europe and Asia. It has long grass-like leaves, and very showy panicles of rosy-red flowers, on stems from 10 to nearly 18 inches high in June. The variety called *splendens* is the most worthy of garden cultivation, being of a brighter colour. *L. v. alba* is a white variety, also worthy of a place; and *L. v. flore pleno*, the double Catchfly, is a fine variety, with more rocket-like blooms. They are excellent plants for the rougher parts of rockwork, and as border-plants on dry soils. Any of the kinds are worthy of being naturalised on dryish slopes, or open banks, on which they seem to form the largest, healthiest, and most enduring tufts. Easily propagated by seed or division.

Lychnis.

Lychnis Haageana, with shaggy stems and bracts, and flowers of a splendid scarlet; *L. flos-Jovis*, a downy plant, with rich purplish flowers; *L. Coronaria*, the handsome Rose Campion; *L. fulgens*, with vermilion-coloured flowers, from Siberia; and the double varieties of *L. diurna* and *vespertina*, although, for the most part, handsome plants, are too large for association with rock-plants.

LYCOPODIUM DENDROIDEUM

(*Ground Pine*).—A club-moss, in habit like a Liliputian Pine-tree, and of all its family by far the most worthy of a place in the rock-garden. The little stems, ascending to a height of 6 to 9 inches, from a creeping root, are much branched, and clothed with small bright, shining green leaves; fruit-cones yellow, long, cylindrical, and, like the stems, erect. A native of moist woods in North America, and high mountains of the Southern United States. I have never seen this plant perfectly grown except in Mr Peek's garden, at Wimbledon, where it flourishes as freely as in its native woods, in a bed of deep sandy peat, fully exposed to the sun. Few plants are more worthy of being established in a deep bed of moist peat in some part of the rock-garden, where its distinct habit will prove attractive at all seasons. It is difficult to increase, and as yet exceedingly rare in this country. In attempting its culture, the chief point is the selection of sound well-rooted plants to begin with; small specimens may retain their verdure after the root has perished, and thus often deceive. Some of our native Club-Mosses are worthy of a place in the marsh-garden.

LYSIMACHIA NUMMULARIA

(*Creeping Jenny*).—Were this native a new plant, and not one found mantling over the ditch-side, we should probably think it worth having, with its long-drooping, flower-laden shoots, whether on points of moist rock or sloping banks. Creepers and trailers we have in abundance, but few which flower so profusely as this, growing in any soil. In moist and deep soil, the shoots will attain a length of nearly 3 feet, flowering the whole of their extent. Rarely or never seeds, but easily increased by division. Flowering in early summer, and often throughout the season, especially in the case of young plants. A native of England, but not of Ireland or Scotland.

Lysimachia nemorum (*Yellow Pimpernel*) is also a slender creeping plant, useful in or near the rock-garden. It is a native of all our counties. The other kinds known in gardens are too large for the rock-garden.

MAIANTHEMUM BIFOLIUM

(*Twin-leaved Lily of the Valley*).—A dwarf perennial, allied to the Lily of the Valley, and a native of our own country. Its habit and relationship make it interesting, and it is easily grown in shady or half-shady spots, and under or near Hollies or other bushes. Syn., *Convallara bifolia*.

MALVASTRUM (*Rock mallow*).—

These are in flower like Mallows, but not quite hardy, being natives of the warmer parts of America. *M. Munroanum* is a dwarf plant with rather small orange-red flowers, and *M. lateritium*, a dwarf native of Buenos Ayres, has brick-red flowers. Sometimes in mild districts these plants thrive in the rock-garden or well-drained borders, in light warm soil.

MAZUS PUMILIO (*Dwarf M*.).—A

distinct little New Zealander, creeping underground, so as rapidly to form wide and dense tufts, yet rarely reaching more than an inch in height. The flowers are on very short stems, so as barely to show above the leaves, are pale violet, with white centres; the leaves with a tendency to lie flat on the surface of the soil. It thrives in pots, cold frames, or in the open air, and is best placed in firm open, bare spots, in free sandy soil in warm positions. It is not showy but is an interesting plant, easily increased by division, flowering in early summer.

MECONOPSIS (*Satin Poppy*).—These are perennials and biennials of the Poppy family, of exquisite beauty of colour and, usually, stately form. Well grown, they are almost taller than the plants that we usually associate with the rock-garden; but they are true mountaineers, and can hardly fail to give distinction to a cool ledge. They mostly come from the Himalayas, or Manchuria, or China, while a yellow one is a native of Britain, and a pretty plant too, often sowing itself in all sorts of places, and looking well everywhere, though it shows no trace of the startling dignity and fine charm of the Indian kinds, which are almost as distinct in leaf as in flower. They are all, we believe, quite hardy, but require attention on account of their biennial duration. As they have to be raised annually from seed, the young seedlings require great care in handling. They are also difficult to please as regards position, and strong vigorous plants are almost impossible, unless in rich, deep, light soil and in the south of England a partially shady situation, where they can have abundance of moisture without its becoming stagnant. The better way in handling seedlings is to grow them in pots during the first winter, planting out early in spring, when the stronger plants may be expected to show flower in July. The smaller ones will go on growing, forming large rosettes which will make robust specimens the following summer Except under the most favourable conditions, a slight protection will be required in wet autumns and winters for the rarer kinds, this being best effected by squares of glass raised a few inches above the crowns. All the species usually flower the second year, and the grower's aim should be to get as much vigour into them in that time as possible.

Meconopsis aculeata is usually a small plant in gardens, but well grown, forming bold pyramids of purple flowers. It is singularly beautiful plant. The leaves are cut up. It is a biennial also, and a native of the Himalayas.

M. cambrica (*Welsh Poppy*).—For the rock-garden, or for the flower bed, the Welsh Poppy is one of the most useful On old crumbling walls wherever it can get hold, its ample Fern-like foliag and abundance of orange-yellow blossom are attractive, and it will grow almost anywhere. Where it can be allowed space in out-of-the-way corners, stony ground or even the edges of gravel paths, it flower freely. Seed.

M. Nepalensis (*Nepal Satin Poppy*).—The commonest Indian species found in gardens, is smaller than *M. Wallichi*, and a pretty fine-foliaged plant. The sof yellow-green leaves form dense rosette which are said in a young state to clos up or fold over as a protection to th tender crowns. The flower-stems var from 3 feet to 5 feet high, bearing noddin blossoms 2 inches to 3 inches in diameter and of a soft yellow. It is also biennial requiring a rich deep soil and partia shade. Nepaul.

M. Wallichi (*Wallich's Satin Poppy*) i the finest of the Poppy-worts in cultiva tion, and a handsome biennial, remarkabl

ALPINE FLOWERS FOR GARDENS

inasmuch as it is one of the few, if not the only, truly blue-flowered Poppy in cultivation at the present time. It grows from 4 feet to 7 feet in height, forming a pyramid, extremely beautiful in full flower, the drooping Poppy blooms of a fine pale blue colour and fine in form. The flowers first open at the top or ends of the branches, continuing until those nearest the main stem have opened. It

Meconopsis aculeata.

forms a rosette of large leaves, 12 inches to 18 inches long, deeply cut, and so brittle that, although well able to stand our winters, they are apt to be damaged by snowfalls. The plants like a moist situation in a deep peaty soil, and partially shaded from the mid-day sun. It is biennial, and to keep up a stock, seed should be sown annually, and this as soon as gathered. The varieties *fusco-purpurea*

and *purpurea* are not so good in colour as the fine blue of the old form.

Meconopsis simplicifolia has a tuft of lance-shaped leaves, 3 to 5 inches long, slightly toothed, and covered with a short, dense, brownish pubescence. The unbranched flower-stalk is about 1 foot high, and bears at its apex a single violet-purple blossom, 2 to 3 inches in diameter.

MEGASEA. (See **SAXIFRAGA**).

MELITTIS MELISSOPHYLLUM (*Balm M.*).—A distinct-looking plant of the Salvia order, with slightly hairy ovate leaves, about 2 inches long, clothing the stem to its apex, and from one to three flowers arranged in the axils of the opposite leaves. The flowers are usually nearly or quite an inch and a half long, and opening at the mouth to a little more than an inch deep. The lower lip is the largest, and is usually stained with a deep purplish rose, except a narrow margin, which is a creamy white. The handsome lip reminds one of the flowers of some of our handsome exotic Orchids rather than those of a labiate plant. It varies a good deal in colour; sometimes the lip has not the handsome stain above alluded to, and sometimes the whole flower is of a reddish-purple hue. *M. grandiflora* of Smith is a variety differing in colour. The plant is distinct, and worthy of a place. It naturally inhabits woods, and even when one finds it on the lower flanks of some great alp, it is seen nestling among the shrubs and low hazel-trees. Woody spots near the rock-garden would suit it, and it grows readily among shrubs. Found in a few localities in Southern England, and widely over Europe and Asia. Seed or division, flowering in May about London.

MENYANTHES TRIFOLIATA (*Buckbeam*).—A beautiful British aquatic herb, with trifoliate leaves, flowering in early summer; corolla white inside, tinged with red outside, beautifully bearded. Common in Europe and North America, and at home by margins of lakes, ponds, and streams, or in the bog garden. Division.

MENZIESIA.—Dwarf shrubs and alpine, admirably suited for rock-gardens, or wherever there is a moist peat soil. They are all of compact growth, and pretty in flower.

Menziesia cærulea is a tiny alpine shrub, native of Scotch mountains, and of northern European mountains. A pretty bush for the rock-garden or for choice beds of dwarf plants, 4 to 6 inches high, with pinkish-lilac flowers, flowering rather late in summer and in autumn. Europe.

M. empetriformis.—A tiny shrub, neat in habit and of much beauty, with rosy-purple bells in clusters on a dwarf heath-like bush, seldom more than 6 inches high. This plant is one of the best for the rock-garden, thriving in a rather moist sandy peat soil. It is cultivated with most success in Nurseries in the neighbourhood of Edinburgh. It flowers in summer, and is sometimes known as *Phyllodoce empetriformis*. America.

See also *Erica* for the plant known in Nurseries as *M. polifolia*.

MERENDERA BULBOCODIUM.—A bulbous plant, very like *Bulbocodium vernum*, but flowering in autumn. The flowers are large and handsome, and of a pale pinkish-lilac. Suitable for the rock-garden and bulb-garden, till plentiful enough to be used in borders. Increased by separation of the new bulbs and by seed. S. Europe.

MERTENSIA (*Smooth Lungwort*).—Graceful plants of the Borage order, of much beauty of colour. One, *virginica*, grown in leafy and peaty soil in a cool place, is one of the most graceful of hardy spring flowering plants.

Mertensia alpina is a pretty alpine kind, and should only be associated with the choicest plants. The leaves are bluish-green; the stem from 6 inches to 10 inches high, and has from one to three terminal drooping clusters of light blue flowers in spring or early summer.

M. dahurica, although of a very slender habit, and liable to be broken by high winds, is perfectly hardy. It grows from 6 inches to 12 inches high, with erect branching stems, and flowers in June, bright azure-blue, in panicles. It is a very pretty plant for the rock-garden, where it should be planted in a sheltered nook in a mixture of peat and loam. It is easily propagated by division or seed. Syn., *pulmonaria dahurica*.

M. maritima (*Oyster Plant*).—A beautiful native plant, and though usually found growing in sea-sand, it is amenable to garden culture. Given a light sandy soil of good depth, and a sunny position where its long and branching succulent flower-stems may spread themselves out, carrying a long succession of turquoise-blue flowers, it is a plant that we may expect to see appearing with renewed vigour year after year. It is much loved of slugs, and is best on an open part of the rock-garden.

M. oblongifolia is another diminutive species, with deep green, fleshy leaves. The stems are 6 inches to 9 inches high, and bear handsome clustered heads of brilliant blue flowers.

M. sibirica.—The peculiar value of this species is that it has the beauty of colour and grace of habit of the old *M. virginica*, and at the same time grows and flowers for a long period in ordinary garden soil. The flowers are small and bell-shaped, and in loose drooping clusters that terminate in graceful arching stems. The colour varies from a delicate pale purple-blue to a rosy pink in the young flowers. It is a hardy perennial, and may be propagated by division.

M. virginica (*Virginia Cowslip*).—A

lovely perennial, distinguished from its allies by the smoothness of all its parts, and by its large leaves, the lower ones being 4 to 6 inches long. The flowering stems are from 10 to 18 inches, suspending blooms of a beautiful purple blue, trumpet-shaped, and about an inch long, from the beginning of April to May or early June, and loves a soil cool and light, and a half-shady position. This fine plant often fails to thrive in stiff soils. It is a native of marshy meadows and by streams from Canada to New Jersey, and also southward and westward, so there can be no doubt of its hardiness, but the mistake is often made of planting it in dry borders, though in parts of our islands, where the rainfall is copious, it may succeed in that way. In the drier parts of our islands, the bog-garden is the place for it.

MIMULUS (*Monkey Flower*).—Of this numerous genus few of the species after the common Musk have come into cultivation to stay. The yellow (*M. lutea*) is naturalised, and a pretty plant for the marsh garden. There are one or two brilliant forms of the copper *Mimulus* which succeed well in like positions, but most of the introduced species are too coarse and short-lived in bloom for the rock-garden: the common Musk, *M. moschatus*, is pretty in moist corners.

Mimulus radicans.—A very pretty and interesting species from the shady ravines of New Zealand. It forms a dense creeping mass of dark green obovate obtuse slightly hairy foliage, stems creeping, with short leafy branches, and flowering freely about the end of May; the flowers are white with a very conspicuous violet blotch, the upper lips small and divided, the lower much larger and three-lobed. It is of the easiest cultivation, growing in mud or on old pieces of wood, so long as it is kept damp. When it is protected by taller growing plants, which retain their foliage during winter, it is perfectly hardy, but when fully exposed to a severe winter it frequently goes off.

MITCHELLA REPENS (*Variegated Partridge Berry*).—One of the pretty woodland plants that accompany the May Flower (*Epigœa*), the tree *Lycopodium*, and the Rattlesnake Plantain (*Goodyera*), in the Pine woods of North America. It is a trailing little evergreen, with roundish shining leaves, the flowers white, sometimes tinged with purple, followed by scarlet berries in autumn. I saw it in Long Island, running about in the Moss, beneath Pine trees, and it occurred to me at the time that it would be a pretty addition to shady parts of our rock-gardens, in which it would thrive under the same conditions as the *Pyrolas*, and the *Linnœa*.

MODIOLA GERANIOIDES (*Geranium-like M.*).—A hardy, tuberous-rooted, trailing Malvaceous plant, 4 or 5 inches high, flowering late in summer; rich rosy-purple, marked with a dark line in the centre, solitary, 1 inch or more across, on long and slender flower-stalks. Easily grown in well-drained sandy soil. Division.

MŒHRINGIA MUSCOSA (*Mossy M.*).—A very dwarf evergreen herb, 2 or 3 inches high, with prostrate, thread-like stems, clothed with very narrow leaves, like those of an *Arenaria*. Flowering in early summer, white, small, solitary. A native of Europe, on the margins of woods, in humid parts of mountains. The rock-garden and borders, in fine, very sandy loam. Division and seed.

MORISIA HYPOGÆA.—A pretty hardy alpine, and one of the most charming re-introductions of recent years. It was first flowered by Mrs Marryat in April, 1834, and is figured in Sweet's "British Flower Garden," second series, tab. 190. The flowers, as large as a shilling, and of a bright yellow, come singly on short stalks, rising very little above the tufted glossy foliage in April and May. It seems to do best in a light gritty soil, and is of easy culture on the rock-garden. It buries its seed-pods in the soil, like some of the Violas.

MUHLENBECKIA.—Graceful free-growing evergreen trailers, useful as coverings for rocks or stumps; natives of New Zealand. The best known, *M. complexa*, is a rapid grower, with long wiry and entangled branches, small leaves, and white waxy flowers inconspicuous. *M. adpressa* is larger, and has heart-shaped leaves, and long racemes of whitish flowers. *M. varia* is a small kind, with fiddle-shaped leaves, and is very distinct from either of the above, it being suited for the rock-garden proper, whereas the larger kind should only be used among shrubs or to clothe bold rocks.

MUSCARI (*Grape Hyacinth*).—Very pretty bulbous flowers, distinct and good in colour and form. They come early in the spring, and are very welcome then. Most of the kinds are pretty, the more so, if in association with Narcissus and the flowers of different colours that come about the same time. They are plants mainly of the East, and, though not difficult about soil, are much happier, and increase more freely in open warm soils. Only the prettiest kinds are fitted for, and in need of the advantage of the rock-garden. Among the shrubs, and associated with the dwarf Narcissi, they come in well.

Among these plants we have more names than real distinctions, but some few are very beautiful, such as *M.* *conicum*, which tells well in groups on the rock-garden. Still, they do not tempt us to grow numbers of them, as we get all the beauty of the family from a few kinds.

MUTISIA.—Remarkable and beautiful South American plants, some almost hardy in the milder parts of our islands. In winter the bush-clad rock-garden offers a good place for them. Some few cultivators have been successful with *M. decurrens*; once or twice *M. ilicifolia* has been grown and flowered very well. *M. Clematis* is the least delicate.

Mutisia ilicifolia.—A very distinct and beautiful plant, is a native of Chili, where it grows over bushes, with thin wiry stems. Every part is covered with a cobweb-like tomentum. The leaves are about 2 inches long, toothed, the texture leathery, and the mid-rib growing beyond the blade, and forming a strong twining tendril. The flowers are 3 inches across, with from eight to twelve ray florets coloured pale pink, or sometimes white with pink tips; the disc is lemon-yellow.

M. decurrens.—The most beautiful of the three garden *Mutisias*. Mr Coleman has grown it well amongst Rhododendrons at Eastnor Castle; Mr Gumbleton, Mr Hooke, Mr Ellacombe, and Kew have also had it in good condition. Most cultivators kill this species by planting it in a hot, sunny place, where it gets baked, and soon sickens. It wants a moist, cool soil, a sunny, airy position, and a few slender Pea sticks to clamber upon. The flowers of this are over 4 inches across, a fine orange with a yellow disc.

M. Clematis.—The first coloured picture of this species ever published in any English work was the plate in the *Garden*, 27th July 1883. It is a tall herbaceous climber, 10 to 20 feet high, with leaves ending in branched tendrils. The plant grows freely, does not die off suddenly like the others, and when properly treated it flowers freely. It is probable that this species would thrive out-of-doors in Devon, South Wales, and South Ireland. It grows as fast as *Cobœa scandens*, and is said to be propagated in the same way, viz., by means of cuttings of the young growth. A native of Peru, and Ecuador, at elevations of from 6,000 to 11,000 feet.

MYOSOTIDIUM NOBILE (*Antarctic Forget-me-not*).—A noble perennial, with very handsome flowers like a Forget-me-not. A native of the Chatham Islands in the Pacific, and frequenting there damp sandy shores, it is, for the most part, difficult to grow in our country, but Mrs Rogers at Burncoose, and various others, have succeeded. The neighbourhood of the sea almost essential, though by the use of frames and care the plant can be grown elsewhere, but I have never seen it well done except in Cornwall. It has a thick root-stock, from which arise the large heart-shaped, shining green leaves. The flowers are borne on an erect stem $1\frac{1}{2}$ feet high; it is leafy all the way up, and is terminated by a loose corymb of flowers, in colour exactly like Forget-me-not, but the shade of blue varies. It has been grown in cool houses with some success, but the thing to do, if one can, is to establish it on a sandy moist part of the rock-garden, anywhere within the influence of the sea, using also, if one may, sand from the beach.

Mrs Roger's plants were raised from seed, and grown in a south border, sea-sand piled up around them.

MYOSOTIS (*Forget-me-not*). — Perennial and biennial plants; some true alpines among them, for the most part of easy culture, and precious for their associations as well as beauty.

If the Forget-me-nots are in moist soil, not too heavy, they not only do not need shade, but are better in the open, the plants sturdier and more free flowering, but the wood and water Forget-me-nots will thrive in partial shade.

Myosotis Alpestris (*Alpine Forget-me-not*).—A British alpine plant, found in one or two places in Scotland and Northern England, and of fine colour and beauty. It forms close tufts of dark-green hairy leaves, healthy plants rising to a height of only about 2 inches, and in April a few flowers of a beautiful blue, with a very small yellowish eye, begin to appear among the leaves, and as the weather gets warmer, the little flower-stems gradually rise, and soon the plants become masses of blue, remaining so all through the early summer. Fortunately, it is very easily raised, and comes quite true from seed. It loves to be pinched in between lumps of millstone grit, and is apt to perish in winter if made to grow too grossly. It is quite distinct from, and much finer than, the dwarf mountain form of the Wood Forget-me-not, often met with on the Alps, the leaves always being in very dense tufts close to the earth, while the smallest specimens of *M. sylvatica* seen on the mountains do not branch below the surface, but are rather slender and erect in habit. It is also a true perennial, while the Wood Forget-me-not usually perishes after blooming. The garden home of the Alpine Forget-me-not is on the most select spots in the rock-garden—where it grows best, perhaps, on ledges with a northern aspect, though it thrives perfectly in open sunny spots; the soil to be moist throughout the warm season Syn., *M. rupicola*.

M. Azorica (*Azorean Forget-me-not*).— This has flowers of an indigo-blue, and rich purple when they first open. It was first brought home by Mr H. C. Watson, author of the "Cybele Britannica," who found it near cascades and on wet rocks, with a north-eastern aspect, in the Westerly Azores. It is a little tender, but so beautiful and distinct from our European blue and yellow-eyed Forget-me-nots that it is worthy of being annually raised, in case old plants should perish during winter, and it is easily increased by seed. It is best raised in autumn, and kept through the winter in dry frames, pits or a greenhouse, or in very early spring in a gentle heat, and planted out about the beginning of May in a somewhat shaded or sheltered position, in light but deep and moist soil, in which it will form spreading tufts.

Myosotis Dissitiflora (*Early Forget-me-not*).—This bears some resemblance to the Wood Forget-me-not; but is much earlier in flower, blooming in January and February, and lasting till early summer. Early in the season, and in poor ground, it sometimes opens with pink flowers; but where the plants are healthy and the ground good, it soon expands into tufts of the loveliest sky-blue. In dry ground it is apt to go off with the droughts of spring or early summer; but when placed in some moist cranny, it continues in flower for a long time, and accompanies the Wood Forget-me-not in its beauty, though it begins to show much earlier. For this treasure to our gardens we are indebted to the late Mr J. Atkins, of Painswick, who found it on the Alps near the Vogelberg, and grew it for several years in his garden, before it was in cultivation elsewhere. From him I obtained it, and soon afterwards it passed into general cultivation, at first under the name of *M. montana*. It is quite easily grown in any cool moist soil, and very easy to increase, by pulling the tufts in pieces.

M. Palustris (*Water Forget-me-not*).—This may be grown easily anywhere by the side of a stream, or pond, or moist place, by merely pricking in bits of the shoots, and perhaps this is the best way in most places, particularly where the ordinary soil is dry. But in many districts the climate and soil are congenial, and in such it is often desirable to have a group or two of a plant so great a favourite with all. I have never seen the flowers so large as among Rhododendrons growing in beds of moist peat soil. It thrives, however, in ordinary soil in many gardens, and grows as far north as the Arctic Circle, and is a native of North America as well as of Europe and Asia. It is essential for the water-side, be it streamlet or pool.

M. Sylvatica (*Wood Forget-me-not*).—A native of woods, mountain pastures, in the north of Europe and Asia, and in the great central chain from the Pyrenees to the Caucasus, and also a British plant, though rare, limited to Scotland and the North of England. In a wild state it is said to be perennial, but in gardens usually proves a biennial, and should be sown every year in early summer. It is a very frequent plant on alpine pastures, always in a more compact form than in gardens.

Myosotis cæspitosa.—A variety of this, called *Rechsteineri*, is a dense and minute creeper from the Lake of Geneva. Useful for moist ledges, where it makes matted tufts of pale green herbage, and in early summer bearing little racemes of turquoise-blue flowers, barely 2 inches from the ground. It is one of the best carpet plants for bulbous things in the rock-garden, and quite a thing to be proud of. As its roots get somewhat bare, top-dressings of loam and leaf-mould mixed with a little sand should be applied.

MYRICA (*Sweet Gale*).—Marsh shrubs worthy of a place where the marsh-garden is carried out, or where there are watery or marshy spots near our rocks. Our native Sweet Gale (*M. Gale*) should be wherever sweet-smelling plants are cared for. It is a wiry bush 2 or 3 feet high, having fragrant leaves. In a moist spot, such as a bog, it spreads by underground shoots and makes a large mass. The North American, *M. cerifera* (Wax Myrtle or Bayberry), *M. Pennsylvanica*, and *M. Californica*, are less common. The last is a good evergreen of dense growth, with fragrant leaves, that keep green through the winter. It is a vigorous plant, especially in light soils, and is quite hardy. The Wax Myrtle is met with in old gardens, where it was planted for its spicy foliage. I find the Gales free and vigorous in stiff poor soils, where few things grow well.

NARCISSUS (*Daffodil*).—Although most of these handsome plants are independent of the rock-garden, and its advantages, and grow freely in the coldest soils, one of the most beautiful things we can do is to keep the dwarfest and choicest of them for growing through mossy dwarf plants on the rock-garden, and also in the grassy places near and among the groups of rock shrubs. I have never seen anything more beautiful in nature or in gardens than grassy banks planted with the smaller and rarer Narcissi in the gardens at Warley Place. The effect is all the more precious, coming so early in the spring. Among the smaller Narcissi, the little *N. minimus*, with its flowers bent into the Moss or short turf, is charming for the rock-garden, as are all the smaller wild kinds, and any choice, new variety may also find a home there. For names and descriptions of the kinds, see the "English Flower Garden."

NARTHECIUM OSSIFRAGUM (*Bog Asphodel*).—A small native plant, in growth somewhat like an Iris, with a spike of small yellow flowers. It is an interesting plant for the marsh-garden, and is of easy culture.

NERTERA DEPRESSA (*Fruiting Duckweed*).—The flowers of this diminutive plant are inconspicuous, but when in fruit it is best compared to a small Duckweed growing on firm earth, and bearing numbers of little oranges! They not only occur on the surface of the tufts, but by pushing the fingers between the small dense leaves, the bright berries are found in profusion hidden among them. It is quite distinct, deserves a place for the pretty fruit, and should be associated with the dwarfest plants in firm and moist soil. New Zealand and the Andes of S. America. Division.

NIEREMBERGIA RIVULARIS (*Water N.*).—Of quite a different type to the other members of its family seen in our gardens, the stems and foliage of this trail along the ground, while from amongst them spring erect open, cup-like flowers of a creamy-white tint, just above the foliage. Sometimes the blossoms are faintly tinged with rose, are usually nearly 2 inches across, with yellow centres, and continue blooming during the summer and autumn months. It is said to abound by the side of the Plate River, but only within high-tide mark, its flowers rising so high among the very dwarf grass that the plant is discerned from a great distance. Rooting much at the base, it is easily increased by division.

NYMPHÆA (*Water Lily*).—Wherever water is associated with the rock-garden (I have shown before it is not often a natural condition), the lovely new Water Lilies may lend great interest, and not a few give fine colour. As they are described in so many books and catalogues, there is no need to enumerate them here. As to culture, however, a word may be said. They are usually starved in pots and baskets. The right way is to put them in the soil of ponds or streams, or, failing this, in the case of artificially made pools, use plenty of loamy soil in the bottom (not less than a foot), and protect from the attentions of water-rats and water-hens, if these are about. Otherwise, few flowers will be seen.

ŒNOTHERA (*Evening Primrose*).—Perennial and biennial plants of showy beauty, some more fitted for borders than for rock-gardens, but the smaller and prostrate kinds of high value. From June onwards, they are at their best, some coming into bloom a second time in late summer. They have large bright yellow or white flowers, freely borne. Although known as Evening Primroses, many of them are open during the day, such as *Œ. linearis, speciosa, taraxacifolia*. Most of them are natives of states west of Mississippi, California, Utah, Missouri, and Texas. All will bloom the first year from seed sown early.

Œnothera cæspitosa.—A dwarf plant, 12 inches high, flowering in May, 4 inches to 5 inches across, white, gradually changing to a delicate rose; as evening approaches, coming well above the jagged leaves, retaining their beauty all night, and emitting a Magnolia-like odour. It is a hardy perennial, and is increased by suckers from the roots, and by cuttings, which root readily. Syn., *Œ. marginata*.

Œnothera Cæspitosa.

Œ. fruticosa (*Sundrops*).—This and its varieties are among the finest of perennials, 1 foot to 3 feet high, with showy yellow blossoms. There are about half a dozen distinct varieties, one of the best being *Youngi*, about 2 feet high, bearing many yellow blossoms. It is one of the best of yellow Evening Primroses for small beds, for edgings, or as a groundwork for other plants, and it goes on flowering even after the first frosts.

Œnothera glauca is a handsome North-American species, allied to *fruticosa*. It is of sub-shrubby growth, becomes bushy, and bears yellow flowers. The variety *Fraseri* is a still finer plant, and where an attractive mass of yellow is desired through the summer, there are few hardy plants of easy cultivation so effective. In a large rock-garden a few plants here and there give good colour, and the plants bloom long.

Œ. Missouriensis (*Missouri Evening Primrose*).—A noble, hardy herbaceous perennial, with prostrate, rather downy stems, entire leaves, their margins and nerves covered with silky down, and with clear yellow flowers, 4 to nearly 5 inches in diameter, borne so freely that the plant covers the ground with its flowers. As the seed is but rarely perfected, it is increased by careful division, or by cuttings made in April. It does not make such a free growth in cold clayey soils as it does in warm light ones, and it is best on the lower flanks of the rock-garden. North America. The blooms open best in the evenings. Syn., *Œ. macrocarpa*.

Œ. speciosa (*Pale Evening Primrose*).—A handsome plant, with many large white flowers, which afterwards change to a delicate rose, in these respects somewhat resembling *Œ. taraxacifolia*, but the plant is erect, with almost shrubby stems. It forms neat tufts, usually from 14 to 18 inches high, is a true perennial, and valuable for borders or the rougher parts of the rock-garden. A native of North America; increased by division, cuttings, or seeds, but not seeding freely in this country, and thriving in well-drained loam.

Œ. taraxacifolia.—One of the most beautiful of our dwarf hardy plants, with rather stout stems, that freely trail over the ground, bearing a profusion of large flowers. The leaves are deeply cut, somewhat like those of the Dandelion, but of a greyish tone; the flowers several inches across, white, changing to pale delicate rose as they become older. The plant is quite perennial, but on some cold soils perishes in winter. Where it does so, it should be raised annually from seed. It will thrive in almost any garden soil, best in one rich and deep, and may be used with the best result as a drooping plant in the rock-garden borders. Plants raised in early spring and pricked over bare surfaces of rose-beds, flower well the first year. A native of Chili, flowering all the summer and autumn, and seldom more than 6 inches above the ground.

OMPHALODES LUCILIÆ.—A seldom seen and charming plant, with very glaucous smooth leaves, in hue resembling those of the Oyster-plant, and with flowers of a light sky-blue, with a faint stain of something akin to the palest lilac. A native of Mount Taurus, doing best in sunny parts of the rock-garden, in free gritty soil. Slugs often destroy it.

Omphalodes verna (*Creeping Forget-me-not*).—Like a Forget-me-not, with handsome deep blue flowers with white throats, in early spring. A native of mountain woods on several of the great continental chains, and precious for the rock and every other kind of garden. Easily increased by division. Tufts of it taken up and gently forced in midwinter form beautiful objects in baskets.

ONONIS ARVENSIS (*Rest-harrow*). —One of the prettiest of our wild plants, and well worthy of cultivation on banks. It is a variable plant, forming dense spreading tufts, clammy to the touch, and covered with pink flowers in summer. There is a white variety even more valuable. No plants can be more readily increased from seed or by division. This plant is distinct from the spiny *Ononis campestris*, which forms stems nearly 2 feet high, sometimes even more.

Ononis rotundifolia (*Round-leaved Rest-harrow*).—This species is easily known by its large and handsome rose-coloured flowers, with the upper petal or standard veined with crimson. It is a distinct and pretty plant, hardy, and easily cultivated, flowering in May and June and through the summer. It attains a height of from 12 to 20 inches, according to soil and position, increasing in height as the season advances. It is suitable for the rougher parts of the rock-garden; comes from the Pyrenees and Alps of Europe, and is easily propagated by seeds or division.

O. Arragonensis is a distinct species from Spain, a recent introduction.

ONOSMA TAURICUM (*Golden Drop*).—A handsome evergreen perennial from 6 inches to 12 inches high, forming a dense tuft, and bearing in summer drooping clusters of clear yellow, almond-scented blossoms. The best place for growing it is the rock-garden, in which provision is made for a good depth of soil, so that the plants may root strongly between the blocks of stone. The soil should be a good sandy loam, mixed with broken grit, and the plant placed between large blocks of stone, near which the roots ramify and are kept cool and moist. The tops of dry walls also suit this very fine rock perennial.

OPHRYS (*Bee Orchis*).—These small terrestrial Orchids are singularly beautiful, and among the most curious of plants. There have been many in cultivation, but being chiefly from South Europe and not hardy, they must have protection, and then can be grown only with great attention. There are, however, a few native species that can be grown. Of these, one of the most singularly beautiful is the Bee Orchis (*O. apifera*). It varies from 6 inches to more than

1 foot in height, with a few glaucous leaves near the ground; the lip of the flower is of a rich velvety brown, with yellow markings, so that it bears a fanciful resemblance to a bee. It is usually considered very difficult to grow, but this is by no means the case, and it may be grown easily in rather warm and dry banks in the rock-garden, planted in a deep little bed of calcareous soil, if that be convenient; if not, loam mixed with broken limestone may be used. It will be found to thrive best if the surface of the soil in which it grows be carpeted with the Lawn Pearlwort, or some other very dwarf plant, and, failing these, with 1 inch or so of cocoa-fibre and sand, to keep the soil somewhat moist and compact about the plants. Flowers in early summer. Other interesting species to cultivate in a collection of hardy Orchids are *O. muscifera, arachnites, aranifera,* and *Trolli.*

OPUNTIA (*Prickly Pear*).—A large group of plants of the Cactus order, mostly American, but often growing far north into many cold as well as dry regions in California, Utah, and Nevada. Like most Cactuses, they might at first be thought too tender for our country, but some kinds have proved hardy, and the country they come from has severe winters. A most interesting series of species and varieties have been introduced by Mr Spath, of Berlin, who writes of them in the *Garden*, as follows:—

"The hardiness of these species, varieties, and natural hybrids, even in the often trying winters of Berlin, is proved beyond all doubt, having stood in the open for several years without protection. As to soil, they are not particular, but they are thankful for slight manuring, which develops sturdy and healthy specimens in a few years. These produce fine large flowers. When, during the month of July, the plants are covered with their conspicuous flowers, varying through all shades, from light yellow to orange and salmon, from a tender rose to deep and brilliant carmine, they present a picture of unrivalled beauty.

The collection of Colorado *Opuntias,* as far as they have flowered and been named here, is as follows:—

"*O. camanchica lutea, c. orbicularis, c. rubra, c. salmonea.* These four varieties have large and thin joints of roundish shape. *O. fragilis, f. cæspitosa, f. tuberiformis. O. Missouriensis, m. erythrostema, m. salmonea. O. pachyarthra flava. O. pachyclada rosea, p. spæthiana. O. rhodantha, r. brevispina, r. flavispina, r. pisciformis, r. schumanniana. O. Schwerini. O. xanthostema, x. elegans, x. fulgens, x. gracilis, x. orbicularis, x. rosea.*"

Some of them have been grown with success in England. On dry slopes on and partly protected under projecting ledges of rock, they are curious, and the flowers often beautiful in colour, but of tropical associations that hardly go well with alpine plants, and so would be better grouped apart, where they might get some winter protection where needed, and all the sun and warmth could be got for them in our climate. Their nomenclature is still far from clear, and it is probable those arid and cold regions have other hardy and handsome kinds worth introducing.

ORCHIS (*Orchid*). — Perennial ground Orchids often beautiful, hardy, being mostly European or natives of cold countries, not difficult to grow. These are essential for the bog-garden. Some of our native Orchis are deserving of a place, but few succeed well with them, because the plants are often transplanted at the wrong season. The usual plan is to do it just when the first or second flower has opened. At this period of growth, the plant is forming a new tuber for the following year, and if in any way injured, it shrinks and dies. If, instead of this, the plants are marked when in flower and allowed to remain until August or September, when the newly-formed tuber will be matured, the risk of transplanting it is considerably lessened.

The following are among the kinds most worthy of culture :—

Orchis foliosa.—One of the finest of the hardy Orchids, from 1 foot to 2 feet or more in height, with long dense spikes of rosy-purple blossoms, spotted with a darker hue. It begins to flower about the middle of May, and continues for a considerable time. It delights in moist sheltered nooks at the base of the rock-garden, or in some similar place, and it should be planted in deep, light soil. Madeira.

O. latifolia (*Marsh Orchis*).—A native kind, 1 foot to 1½ feet high, flowering in early summer purple in long dense spikes. It is easily grown, forming fine tufts in damp, boggy soil in peat or leaf mould. There are several beautiful varieties of this Orchis, the best being *præcox* and *sesquipedalis*; *O. sesquipedalis* grows about 1½ feet high, and the stem for fully a third of its length is furnished with densely-arranged flowers of large size and of a purplish-violet hue.

O. laxiflora is a handsome species, 1 foot to 18 inches high, flowering in May and June, rich purplish-red, in long loose spikes. Native of Jersey and Guernsey, and suited for the rock-garden in a moist spot, or the marsh-garden. Division.

O. maculata (*Spotted Orchis*).—This is usually pretty in the poorest soils, but is a very different plant in a rich one. If well grown in moist and rather stiff garden loam, it will surprise even those who know it well in a wild state. Obtain it at any season, and carefully plant twelve or twenty tubers in a patch in a half-shady and sheltered position in moist loam. It flowers in summer, and is an excellent plant for the bog-garden. The variety *superba* is a much finer plant.

Orchis maculata superba. (Engraved from a photograph sent by Rev. C. Wolley-Dod.)

Other beautiful kinds are *O. papilionacea, purpurea, militaris, mascula, pyramidalis, spectabilis, tephrosanthos,* and *Robertiana,* but all are difficult to establish freely, as they grow in their natural conditions.

ORIGANUM (*Marjoram*). — The common *O. vulgare* is scarcely a garden plant, but another, *O. Dictamnus* (the *Dittany* of Crete), is a pretty little plant, though somewhat tender. During mild winters, however, it survives unprotected. It has mottled foliage, and small, purplish flowers in heads, like the Hop; hence it is sometimes called the Hop plant.

O. Sipyleum is similar, and quite as pretty. If grown in the open, these plants must have a warm spot in the rock-garden in very light, open soil, and then mostly in the south or very mild districts.

ORONTIUM AQUATICUM (*Golden Club*).—A handsome water-side perennial of the Arum family, 12 inches to 18 inches high. The flowers, which are yellow, densely crowded all over the narrow spadix, and which emit a singular odour, are borne early in summer. The plant may be grown on the margins of ponds and fountain-basins, or in the wettest part of the bog-garden. North America, in rivulets and bogs.

OTHONNA CHEIRIFOLIA (*Barbary Ragwort*).—A plant of distinct character; the leaves and shoots quite smooth and glaucous, and the habit spreading, forming silvery tufts from 8 inches to a foot or so high. It flowers sparsely on heavy and cold soil, but on light soils it blooms freely in May, a rich yellow, and is useful for its distinct aspect; propagated by cuttings. N. Africa.

OXALIS (*Wood Sorrel*).—A large group of dwarf, often curious and often pretty, plants, which, so far as they are hardy, may well come into the warm parts of the rock-garden; but, being mostly plants of the Cape and warm countries, few of those known to us are hardy, excepting always the few that are natives of our own country, among which the most graceful is the little native Wood Sorrel. The following are the kinds of proved hardiness in our gardens. In warmer lands than ours some are apt to become troublesome as weeds.

Oxalis Acetosella (*Stubwort, Wood Sorrel*).—The prettiest kind known so far for our gardens is our native Wood Sorrel, which bore in old times the name of "Stubwort"—a name which should be used always. This grows itself in such pretty ways in woody and shady places that in many gardens there will be no need to cultivate it. Where it must be cultivated it will be happy in shady spots in the rock-garden.

O. **Bowieana.**—A robust grower, forming masses of leaves 6 inches to 9 inches high, the flowers rose, in umbels, borne continuously throughout the summer. It is best for warm soils, and in cold ones seldom or never flowers; on well-drained and very sandy ones it does so abundantly. The soil that suits this fine plant being often found on the rock-garden, it would be well to have a seam or two of it there at the foot of a hot rock. Division. Cape of Good Hope.

O. **corniculata rubra** is a form of a native kind, with brown purple leaves that might be encouraged where there are stony banks, for this handsome plant speedily covers the most unpromising surfaces. In gardens, however, it may become a weed. With me, this plant comes up everywhere among stone edgings and also in the joints between stone pavings, and is so far an interloper sowing itself very pretty.

O. **floribunda.**—A free-flowering kind, quite hardy in all soils, and producing, for months in succession, numbers of rose-coloured flowers with dark veins. There is a white-flowered variety as free-flowering and in every way as valuable as the rose-coloured form. Both are very useful for rockwork and for the margins of borders, and are easily increased by division. This appears to be the commonest kind of *Oxalis* in cultivation. It is hardy enough to encourage one to attempt to naturalise it on any rocky place or about ruins. S. America.

O. **lasiandra** is one of the most distinct, with large dark green leaves, and, in early summer, umbels of numerous flowers of a bright rose-colour. Best on warm parts of the rock-garden. Mexico.

OXYTROPIS CAMPESTRIS (*Field O.*).—A dwarf stemless perennial, about 6 inches high, flowering in summer, yellowish, tinged with purple, erect, in a dense spike. Leaves, with many pairs of leaflets, more numerous and much less silky than those of the *Purple O.* Europe, America, and in Scotland. The rock-garden, in sandy loam. Seed and division.

Oxytropis Pyrenaica (*Pyrenean Oxytrope*).—A very dwarf species, with pinnate leaves, clothed with a short silky down. These barely rise above the ground, as the short stems are nearly prostrate, and seldom exceed a few inches in height ; the flowers, borne in heads of from four to fifteen, are of a purplish-lilac. It is not a showy, but withal a useful kind for the parts of the rock-garden devoted to very dwarf plants. A native of the Pyrenees, increased by seed or division, and should be planted on well-exposed and bare spots, in firm, sandy, or gravelly soil.

O. uralensis (*Purple O.*).—An elegant little perennial, resembling *O. campestris* in habit, but more densely clothed with soft silky hairs in every part ; about 6 inches high, flowering in summer, bright purple, in dense round heads. Scotland and various parts of Europe. The rock-garden, in moist sandy loam. Division and seed.

OZOTHAMNUS ROSMARINIFOLIUS.—A neat little evergreen shrub from Tasmania, almost hardy in the south and coast districts, with small, Rosemary-like leaves, and about the end of summer bearing dense clusters of small white flowers. It thrives in any light soil, and should be planted in an open sunny spot or on a warm bank.

PAPAVER (*Poppy*).—Showy perennial, biennial, or annual plants, for the most part too vigorous for the rock-garden, and in no need of its care ; a few kinds are useful, however. There is no difficulty about their culture, any open spot with sand or gritty soil suiting them. As in our country, the plants are apt to wear out too soon ; it is well to sow a little seed here and there on the rock-garden, and leave the plants to grow where sown.

Papaver alpinum (*Alpine Poppy*).—This dwarf and fragile plant has large white flowers, with yellow centres, its leaves cut into fine acute lobes. A native of the higher Alps of Europe, it may sometimes be seen in good condition in our gardens, but it is liable to perish as if not a true perennial. It varies much in colour, there being white, scarlet, and yellow forms in cultivation. The variety *albiflorum* of botanists has white flowers, spotted at the base ; the variety *flaviflorum* has showy orange flowers, grows 3 or 4 inches high, and is hairy. This last variety is also known as *P. pyrenaicum*.

P. nudicaule (*Iceland Poppy*).—A dwarf kind, with deeply cut leaves, and large yellow flowers on naked stems, from 12 to 15 inches high. A native of Siberia and the northern parts of America, and a handsome plant, easily raised from seed, and forming rich masses of cup-like flowers, but, like other dwarf Poppies, does not seem to be permanent, and should be raised from seed annually. There are several varieties.

PARADISIA LILIASTRUM (*St Bruno's Lily*).—When the traveller in early summer first crawls down from the snowy fields of an Alp into the grateful warmth and English meadow-like freshness of a Piedmontese valley, most likely the first flower he notices in the pleasant grass of the valley is a Lily-like blossom, standing about level with the tops of the blades of Grass and Orchises. The blooms, about 2 inches long, so delicately white that they might well pass for emblems of purity, have each division faintly tipped with pale green, and from two to five flowers occur on each stem. It does not grow in close tufts, as in our borders, but one or perhaps two stems spring up here and there all over the meadows, and if it were an English flower, it might be called the Lady of the Meadows. It is easy of culture on ordinary soils. Slight shelter would prove beneficial, and that may readily be afforded by planting it among dwarf shrubs near the rock-garden. It will be found to flourish in British as well as in Alpine grass, and is easily increased by division or by seeds. Syn., *Czackia Liliastrum*.

PARNASSIA (*Grass of Parnassus*).—Mountain pasture and moor perennials, pretty for the bog-garden or for moist spots in the rock-garden, and not difficult to grow in moist peaty soil.

Parnassia palustris.

Parnassia Caroliniana (*Carolina Grass of Parnassus*).—A native of North America, chiefly in mountainous places, on wet banks, and in damp soil. This is much larger than our Parnassia, the stem reaching from 1 to nearly 2 feet high, the flowers from 1 inch to 1½ inches across, the leaves thick and leathery. It is a good plant, succeeding in deep moist soil, and flowering in autumn. *P. asarifolia*, a native of high mountains in Virginia and North Carolina, does not differ much from this, but has the leaves rounded and kidney-shaped, with larger flowers, and requires much the same treatment. Seed or division.

P. palustris (*Grass of Parnassus*).—A well-known native mountain plant, with white flowers 1 inch or more in diameter, growing naturally in bogs, moist heaths, and high wet pastures. Thrives in moist spots in or near the rock-garden, and may also be grown in pots placed half-way in any fountain or other basin devoted to aquatic plants. Plants or seeds may be easily obtained; seeds should be sown in moist spots as soon as gathered.

PAROCHETUS (*Shamrock Pea*).—*P. communis* is a beautiful little creeping perennial, with Clover-like leaves, 2 to 3 inches high, bearing in spring Pea-shaped blossoms of a fine blue. It is of easy culture in warm positions on the rock-garden, and where the climate is too cold to grow it in the open air it may be grown in a cold frame. Division or seed. Nepaul.

PARONYCHIA.—Small-growing creeping plants of slight value. *P. serpyllifolia*, on account of its dense turfy growth, might be made use of for clothing any dry bank where little else would thrive, or for covering any bare space in the rock-garden.

PASSERINA NIVALIS (*Sparrow-wort*).—An interesting dwarf Alpine plant, nearly allied to the Daphne. It grows to about 1 foot in height, and bears *Mezereum*-like blossoms. It is found at high elevations on the Pyrenees.

PELARGONIUM ENDLICHER-IANUM.—This is interesting as the only species that comes so far north as Asia Minor, is hardy and handsome, with rose-coloured flowers, boldly upheld on stems about 18 inches high, the two upper petals being very large. I first saw it in the Jardin des Plantes, at Paris, where it had remained several severe winters in the open air, thus hardy. A sunny nook would suit it well, sheltered from the north. Seed or division.

PENTSTEMON (*Beard Tongue*).—Beautiful perennial plants of the rocky mountains of North-West America and Mexico, little grown in our gardens, though some are of the highest value as rock-plants. The tall kinds grown in our gardens require frequent moving and rich soil, and are useless for the rock-garden. What we should seek are the true rock and mountain kinds, dwarf in habit, and hardy. They are easily grown on warm open soils, and easily increased by cuttings or seeds, but in the northern and midland districts not many are hardy.

The following are some of the best for the rock-garden. Many are excluded, however; some on account of their rarity, and others because they are not hardy.

Pentstemon azureus is a pretty dwarf branching kind, with numerous branches, bearing many blossoms in whorls, clear violet-blue, towards the end of summer, and lasting a long time. California.

P. crassifolius.—Allied to *P. Scouleri*, but the flowers are of a charming light lavender colour, and the plant admirably suited for a dry knoll of the rock-garden; but this knoll must be well exposed to the sun and on a deep mass of bog soil or peat, so that while the situation of the plant is dry, the roots may find what they require. *P. Menziesii* resembles *P. Scouleri*, but has reddish purple flowers.

P. Fendleri.—This is a pretty and distinct species, glaucous, with a long, erect, one-sided raceme of flowers of a very pleasing light purple colour. In height it rarely exceeds 12 inches to 15 inches. It is hardy in ordinary soils, and is one of the most distinct species in cultivation. *P. Wrighti* is a plant of a similar character with magenta-tinted blossoms, and the variety *angustifolius* is likewise a pretty plant. Both are worthy of culture.

Pentstemon heterophyllus.—A dwarf sub-shrubbery kind, its showy flowers, singly or in pairs in the axils of the upper leaves, of a pinky lilac; plants from seed are very liable to vary. Though hardier than many species, it succumbs to severe winters. California.

P. humilis.—A distinct alpine species, rarely exceeding 8 inches in height, forming compact tufts, its large blossoms of a pleasing blue suffused with reddish-purple: it should be planted in the rock-garden in a fully exposed spot in gritty loam and leaf-mould, and during summer the plant should be copiously watered. It blooms in early June, and is a native of the Rocky Mountains.

P. Jeffreyanus.—A showy kind, and the best of the blue-flowered class, its glaucous foliage contrasting finely with its clear blue blossoms borne during the greater part of the summer. It is a handsome dwarf border plant, but not being a good perennial, the stock should be kept up by the aid of seedlings, which will bloom much more vigorously than old plants. North California.

P. lætus is a close ally of *P. azureus* and *P. heterophyllus*, and, like them, is of dwarf branching habit, with blue flowers in raceme-like panicles about 1½ feet high, blooming in July and August. It is a native of California, and is as hardy as most of the species from that region.

P. ovatus, also known as *P. glaucus*, is a fine vigorous plant, 3 to 4 feet high, the flowers small, but in dense masses, in colour varying from intense ultramarine to deep rosy-purple; their brilliant colour, and the handsome form of the plant, combine to give it a special value. It should be considered a biennial, as it usually flowers so vigorously in the second year as to exhaust itself. Mountains of Columbia.

P. procerus is a beautiful little plant, and about the hardiest of all the species, as it takes care of itself in any soil. It is of a creeping habit, sending up from the tufted base numerous flowering stems 6 to 12 inches high. The small flowers are in dense spikes, and being of a fine amethyst-blue, they make it charming for either the border or the rock-garden It seeds

s

abundantly. It is the earliest to blossom of all the *Pentstemons.*

Pentstemon Scouleri is a small semi-shrubby plant of twiggy growth. Its large flowers are of a slaty bluish-purple, and are arranged in short terminal racemes; they are not produced in great abundance, but, combined with the dwarf and compact growth of the plant, they have charms sufficiently distinct to render it worthy of cultivation. *P. Scouleri* may be readily increased in spring by cuttings of the young shoots, since such cuttings strike freely in a little bottom-heat similar to that used for ordinary bedding plants. Syn., *Menziesii.*

PERNETTYA MUCRONATA.—An Evergreen shrub of the Heath family. Though from South America, it is hardy enough for every English garden. Apart from the evergreen foliage, the berries which it bears in autumn are very showy. After an abundant crop of small white blossoms, the berries are the size of small Cherries, and there are varieties with white, rose, pink, crimson, purple-black, and every intermediate shade. There are few more charming dwarf shrubs than *Pernettyas.* They thrive where the soil is peaty, or sandy. Even a heavy soil may be made suitable by a large addition of leaf-mould and sand. For autumn and winter effects they are excellent, and they may often be used among shrubs on the rock-garden.

PETROCALLIS PYRENAICA (*Beauty of the Rocks*).—A "rock beauty!" as it seems, as one sees its fresh green tufts, not more than an inch high, and cushioned amidst the broken rocks. From these stains of light green spring in April innocent-looking flowers, reminding one of miniature "Ladies' Smocks," on stems that rise little more than half an inch over the leaves. When well grown, its faintly-veined pale lilac flowers seem to form a little cushion, so delicate-looking, that people grow it for years without suspecting it to be fragrant; but it breathes a delicious, faint sweetness. Only suited for careful culture, being of a fragile nature, though hardy, it should be planted in sandy fibry loam, in rather level warm spots on the rock-garden, where it could root freely into the moist soil, and yet be near broken rocks and stones, down the buried sides of which it can send its roots, always in a sunny position. I have seen it grown as a border plant in a moist part of Ireland, but in the hands of a very careful cultivator who grew it in very fine soil on a select border, and took up, divided, and carefully replanted the tufts every autumn. It may also be grown in pots plunged in sand in the open air, and in frames in winter; but it becomes drawn and delicate under glass protection. Easily increased by careful division, and also raised from seed. Alps and Pyrenees.

PHILESIA BUXIFOLIA (*Pepino*).—An exquisite dwarf shrub, with large carmine-red *Lapageria*-like bloom (2 inches long), nestling among the sombre evergreen foliage. It is a precious shrub for the rock-garden in the more favourable coast gardens. The best soil is fibrous peat, with a small portion of loam; the plant should have a sunny aspect, but be sheltered from the north. To increase peg down each shoot to the ground, then cover over with peat and leaf-mould. It will root freely from the stems and soon form a nice bush. South America.

PHLOX. Mostly known in our gardens by the tall kinds; the great majority of these are natives to the mountains of North America. The alpine kinds are brilliant in colour and as easily cultivated as any plants can be for the rock-garden; for which no more precious plants have ever been introduced, and they are easily grown on "dry" walls and as edging plants. Coming from a cold northern country like ours, they rival the spring flowers of Europe in brilliancy and fine colour and abundance of their flowers, and help to add a fresh glory to the spring. All thoughts of special soils or fancies may be given up in their case, as they grow like native plants in ordinary soil, and are easily increased by pulling to pieces. Some, perhaps not a few, kinds are not yet introduced, and this is a pity, as nearly every mountain Phlox we know is beautiful and free under cultivation.

The mountain Phloxes are so closely allied that general cultural remarks may suffice. Well-drained ordinary garden soil and sunny exposure are essential. Though hardy, the damp of mild winters is hurtful to some kinds in low-lying places, and as the plants do not seed freely, they must be increased by cuttings. A sharp knife and a careful hand will soon remove the two or three pairs of leaves with their included buds, without damaging either the slender stem or the joint. These should be taken off in July, when the branches are just commencing to harden, and inserted in sandy soil in a frame where they can be shaded from full sunshine, and given the benefit of the night dews by the removal of the lights. They will soon root and become good flowering plants the following season. With large patches, the readiest way is to sprinkle sandy soil over the entire plant, and to work the same gently amongst the branches with the hand. If this be done during the summer or the early autumn, the trailing branches will form roots the following season, and may be planted elsewhere. Most of them are easily increased by careful division of the tufts in autumn or early spring.

There is a good account of the plants, from a botanical point of view, by James Britten, in the *Garden* of 29th September 1877.

Phlox amœna.—A very hardy little Phlox, spreading with rosy flowers in early summer, a native of dry places in the southern states, but so hardy in Britain that I have seen it naturalised on poor clayey banks in a wood. A good rock and wall plant.

P. Carolina is a handsome plant, about 1 ft. high, with slender stems terminated by a cluster of large showy rosy flowers. Syn., *P. ovata*.

P. divaricata (*Wild Blue Phlox*).—Larger than the Creeping Phlox or Moss Pink, attaining a height of about 1 foot, and bearing lilac-purple blossoms. The plant thrives in good garden soil, and flowers in summer. In moist copses and woods, Canada, and southwards. Syn., *P. Canadensis*.

P. pilosa is a pretty plant, 10 or 12 inches high; with flat clusters of purple flowers $\frac{1}{2}$ to $\frac{3}{4}$ inch in diameter, from June to August. It is one of the rarest in gardens, another kind being sold for it. The true plant reminds one of *P. Drummondi*. Another rare species is the true *P. bifida*, an elegant plant, the flowers bluish-purple. Canada and southwards and westwards.

P. reptans (*Creeping Pink*).—With the large flowers and richness of colour of the taller Phloxes, this mantles over borders and rockworks with a soft green about an inch or two high, and sends up stems from 4 to 6 inches high, each producing from five to eight deep purplish-rose flowers. It is by no means fastidious as to soil or situation, but will be found to thrive best in peat or light rich soils.

As it creeps along the ground, and gives off numbers of little rootlets from the joints, it is propagated with the greatest ease and facility. A person with the slightest experience in propagation may convert a tuft of it into a thousand plants in a very short time. It is almost indispensable for the rock-garden, makes very pretty edgings round the margins of beds, and also capital tufts on the front edge of the mixed border. It may also be used in the spring garden and for vase decoration, and is a native of North America, inhabiting damp woods. It is perhaps better known in gardens as P. *stolonifera* and *P. verna*, than by the above name. Mountain woods of Middle States and Virginia.

Phlox setacea is sometimes considered the same as *P. subulata*, but its leaves are longer and farther apart on its trailing stems, the whole plant being less rigid. The flowers are of a charming soft rosy-pink, and have delicate markings at the mouth of the tube. *P. s. violacea* is a handsome Scotch variety, more lax in growth, and with deeper coloured flowers, almost crimson. Both are lovely plants for the rock-garden, where, with roots deep among the fissures, they thrive in sunshine.

Phlox subulata (*Moss Pink*).—A moss-like little Evergreen, with stems from 4 inches to a foot long, but always prostrate, so that the dense matted tufts are seldom more than 6 inches high, except in very favourable rich and moist, but sandy and well-drained soil, where, when the plant is fully exposed, the tufts attain a diameter of several feet, and a height of 1 foot or more. The leaves are awl-shaped or pointed, and very numerous; the flowers of pinkish-purple or rose colour, with a dark centre, so densely produced that the plants are completely hidden by them during the blooming season. It occurs in a wild state on rocky hills and sandy banks in North America, and there are few more valuable plants for the decoration of the spring garden borders or rocks, being at once hardy, dwarf, neat in habit, profuse in bloom, forming gay cushions on the level ground, or pendent sheets from the tops of crags or from chinks on rockwork. It is easily increased by division, forming roots freely at the base of the little stems, and usually thrives in ordinary garden soil, particularly in deep sandy loam. Excessive drought seems to injure it, but it is less likely to suffer when rooted beneath stones. There is a white variety (*P. subulata alba*), known in many gardens as *P. Nelsoni*, which is also a beautiful plant. Besides this, the late Mr Nelson of Aldborough raised a large number of seedlings, varied in hue, which are given names, and may be had in Nurseries.

P. stellaria (*Chickweed Phlox*).—A fragile-looking but hardy kind, very graceful in bloom in spring, the flowers a bluish-white. It is a pretty rock plant, and with me free on "dry" walls. A

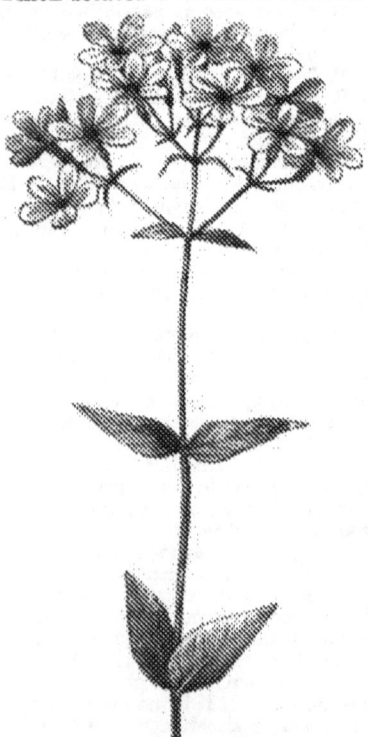

Phlox divaricata.

native of rocky hills in Kentucky and Illinois.

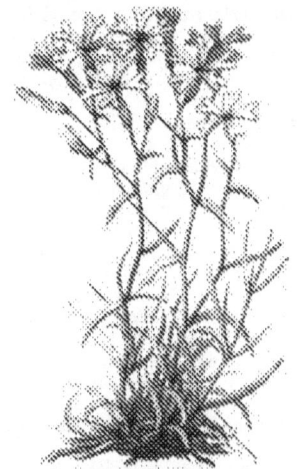

Phlox stellaria.

PHYTEUMA (*Rampion*). — Perennial plants of the Bellflower order, some of them good rock-plants.

Phyteuma comosum (*Rock P.*). — A dwarf distinct alpine plant, with sea-green leaves and flattish heads of flowers very large for the size of the plant; in summer, blue, on very short stalks, in large heads. A plant for the choice rock-garden, in dry sunny spots, in well-drained, very sandy or calcareous soil. I have seen this plant growing from small chinks in arid cliffs, where probably no other plant could exist. What Mr A. W. Clarke says of it is worth following :—

"In winter the plant should be fixed tightly between limestone. A layer of fine broken limestone and strong loam—two parts limestone, one part loam—without any sand, will be a suitable compost. After placing the bottom stone, put a portion of the compost on the stone; then lay on the plant, leaving plenty of room for the root to go down (as it forms a tap root), then add a little more compost on the plant before placing on the other stone. Make these as tight as possible without injury to the roots or crowns of the plant. It should be well looked after in the spring, so that the slugs do not eat all the crowns away. If the slugs get to the plant they will be sure to eat out the centre crowns, then only a few leaves will appear the following year. Top-dress in the autumn and spring with fine, broken limestone, letting it run well between the stones." Alps. Seed.

Phyteuma Sieberi is a neat plant for the rock-garden, requiring a moist sunny situation, and a mixture of leaf-mould, peat, and sand. It forms cushion-like tufts, and in May and June has dark-blue flower heads, on stems 4 to 6 inches long. Division.

P. humile is a dwarf tufted plant for the rock-garden, where it can get a dry sheltered position in winter, and plenty of water in summer. The flowers are blue, and borne in June on stems 6 inches high. Division.

P. Charmeli and **P. Scheuchzeri** are much alike, *P. Scheuchzeri* being dwarfer. It bears pretty blue flowers, on stems from 6 to 12 inches in height, and is evergreen. Seed in autumn.

PIERIS.—Usually rather dwarf, or compact, evergreen shrubs, of much distinction and beauty, natives of China, Japan, and North America, important for the rock-garden, if, as I always urge, we give to the hardy northern and mountain shrub its right place in such gardens. Where, as so often happens in Scotland, Ireland, Wales, and in many districts in England, the natural rock breaks out, and peaty or sandy soil occurs in some places, these bushes are most important, and will be found free in such soils.

The things to be observed are a cool, moist, and not necessarily a peaty soil, always free from lime, as heavy soils can be made to suit them by deep trenching and adding plenty of leaf-mould, with, towards the top, a little peat. The soil in which they grow suits many species of Lilium, which thrive well planted between the shrubs.

Pieris floribunda.—A native of the United States, and forming a compact evergreen bush. The racemes form in October and do not open until the following spring, and carry numerous white flowers. It is a shrub of easy culture.

P. formosa.—In seaside and west-country gardens this is a valuable shrub, the leaves when young of a reddish colour, changing with age to a deep green. The flowers, which are white, borne in a cluster of erect branching racemes, are pendent and almost globular. Himalayas.

P. japonica.—A most graceful evergreen bush, with long clusters of flowers, giving a lace-like effect in the case of well-grown bushes. It is hardy, but slow and poor on loamy soils, thriving on good peat, and should be associated with the choicest evergreens. A precious bush for the rock-garden on peaty or leafy soils.

Other kinds of less importance for the rock-garden are: *P. Mariana* from North America; *P. nitida*, *P. ovalifolia*. Syn., *Andromeda*.

PINGUICULA (*Butterwort*).—Interesting dwarf perennials, natives of Alpine and Arctic bogs or wet rocky slopes.

Pinguicula Grandiflora (*Irish Butterwort*).—Leaves in rosettes, light green, fleshy, and glistening flowers, handsome, two-lipped, spurred like the Horned Violet, more than an inch long, nearly or quite an inch across, of a fine blue. Mr Bentham unites this with the less beautiful *P. vulgaris*, but Mr Syme says: "I cannot conceive how any one who has seen the plants alive can consider them as the same species"; and as *P. grandiflora* has flowers twice as large as *vulgaris*, and is a handsomer plant, it is the kind best worthy of cultivation. It inhabits bogs and wet heaths in the south-west of Ireland, and thrives in moist mossy spots on the northern and shady slopes of the rock-garden or in more open places in moist peat soil. Increased by small green bulbils, which are given off at the base of the rosettes.

Pinguicula Alpina (*Alpine Butterwort*) differs from other kinds in having white flowers, marked more or less with lemon-yellow on the lip, but sometimes tinted with pale pink. It roots firmly, by means of strong woody fibres, and prefers peaty soil, mingled with shale or rough gravel, and shady humid positions, such as is afforded by a rock-garden with a north aspect. A Scottish plant. Ross and Skye.

P. vallisneriæfolia, from the mountains of Spain, differs in its clustered habit of growth. Its leaves are pale yellowish-green, and sometimes almost transparent, occasionally even 7 inches towards the end of the season. The flowers are large, soft lilac colour, with conspicuous white or pale centres. Dripping fissures and ledges of calcareous rocks (frequently in tufa) suit the plant, but it requires free drainage, and continuous moisture.

P. lusitanica, found on the west coast of Scotland, South England, and in Ireland, is small, and has pale lilac flowers. It grows in peaty bogs.

P. vulgaris, a native plant, grows freely in any sunny position in rich moist peat or peaty loam. A small form, with leaves like those of *P. Alpina*, both in form and colour, is found in alpine bogs in the north of England.

PLATYCODON.—*P. grandiflorum*, sometimes called *Campanula grandiflora*, is a handsome perennial, hardy in light dry soils, but impatient of damp and undrained situations, where its thick fleshy roots decay. The flowers are 2 to 3 inches across, deep blue with a slight slaty shade, and in clusters at the end of each branch, and handsome in all forms. Rich loamy soil, good drainage, and an open situation are best. Propagate by seeds, which can be readily procured. The variety *Mariesi* is distinct and good. China and Japan.

PLUMBAGO LARPENTÆ (*Hardy P.*)—A dwarf, herbaceous perennial, once cultivated in greenhouses, but now found to be hardy, and a first-rate plant for rocks or walls. In September nearly covered with flowers in close trusses at the end of the shoots, and of a fine blue, afterwards changing to violet—the calyces being of a reddish violet. The bloom usually lasts till the frosts. I have seen this plant live in cold soils, but it is in all cases best to give it a warm, sandy or other light soil, and a sunny warm position, as under these conditions the "dry" bloom is finer. In consequence of the semi-prostrate habit, it is well suited for planting above the upper edges of vertical stones or tops of walls. A native of China; increased by division of the root.

POLEMONIUM (*Greek Valerian*).—Herbaceous perennials, some pretty dwarf mountain plants among them. The tall kinds are not fitted for the rock-garden.

Polemonium confertum.—A pretty plant, with slender deeply-cut leaves and dense clusters of deep blue flowers on stoutish stems, about 6 inches high. It requires a warm spot in the rock-garden and a well-drained, deep, loamy soil, rather stiff than otherwise. It should be undisturbed for years after planting. Rocky Mountains of North America.

P. humile is a truly alpine plant, with pale-blue flowers on stems a few inches high. In a dry situation and a light sandy soil it is hardy, but on a damp sub-soil is sure to die in winter. North America.

P. reptans is an American alpine plant, its stems creeping, and its slate-blue flowers forming a loose drooping panicle, 6 or 8 inches high. Snails devour it, especially the scaly root-stocks during winter, and must be watched for.

POLYGALA (*Milkwort*).—The hardy Milkworts are neat dwarf perennials, some true Alpine plants among them.

Polygala Chamæbuxus (*Box - leaved Milkwort*) is a little creeping shrub from the Alps of Austria and Switzerland, where it often forms but very small plants. In our gardens, however, on peaty soil and fine sandy loams, it spreads out into compact tufts covered with cream-coloured and yellow flowers. The variety *purpurea* is prettier; the flowers are a bright magenta-purple, with a yellow centre. It succeeds in any sandy, well-drained soil, best in sandy peat. Even when out of flower it is interesting, owing to its dwarf compact habit, bright shining evergreen leaves, and olive-purplish stems.

Polygala calcarea (*Chalk Milkwort*).—A native plant found in the south of England, generally on chalky *débris*, and pretty, usually with blue, but sometimes with pink or whitish flowers, about a quarter of an inch long, in compact racemes; Mr Syme says this has no connecting links with the common Milkwort (*P. vulgaris*). It is known by the flowering shoots rising from rosettes of leaves, and by the leaves on those shoots becoming abruptly smaller and narrower than those below them. It is the handsomest and the easiest to grow of the British species, and does very well in sunny chinks, planted in calcareous soil, forming tufts of violet-blue and white flowers, and blooming in early summer. It should be allowed to sow itself if possible, or the seed may be gathered from wild plants and sown in sandy soil.

P. paucifolia (*Fringed Milkwort*) is a handsome North American perennial, 3 to 4 inches high, with slender prostrate shoots and concealed flowers. From these shoots spring stems, bearing in summer one to three handsome flowers, about three-quarters of an inch long; generally rosy-purple, but sometimes white. It is suited for the rock-garden, in leaf-mould and sand, and for association in half-shady places with *Linnæa borealis*, *Trientalis*, *Mitchella*.

In this enormous genus there are probably handsome hardy plants not yet in cultivation.

POLYGONATUM (*Solomon's Seal*).—Perennials of graceful form not in the ordinary "hard-and-fast rockery," but which come in well among the rock shrubs in the rock-garden in which the mountain shrubs find a home. They thrive in almost any position in good sandy soil, in shady nooks, and under the shade of shrubs. They are increased by seeds or berries, which, sown as soon as gathered in autumn, germinate in early spring; the creeping root-stocks may also be divided to any extent

Polygonatum biflorum, from the wooded hillsides of Canada, of graceful growth, the arching stems 1 foot to 3 feet in height, the small flower stems jointed near the base of the flowers, which are greenish white, two or three together in the axils of the leaves.

P. japonicum.—A distinct species, native of Japan, hardy in this country, flowering in early April, growing about 2 feet in height, the leaves of a firm leathery texture, the flowers white, tinged purplish.

P. latifolium (*Broad-leaved Solomon's Seal*). — A robust plant, the stems being from 2½ feet to 4 feet high, arching, the leaves bright green; flowers large, two to five in a bunch in July. *P. latifolium* var. *commutatum* differs from the above in being glabrous throughout, with a flower-stem 2 feet to 7 feet in height; large white flowers, three to ten in a bunch. North America.

P. multiflorum (*Solomon's Seal*).—A graceful perennial, from 2 feet to 4 feet high, glaucous green; the flowers large, nearly white, one to five in a bunch. It is a free-growing species, of which there are several garden varieties, a double-flowered one, and one in which the leaves are variegated. *P. Broteri* is a variety with much larger flowers; *P. bracteatum*, a form in which the bracts at the base of the flowers are well developed.

P. oppositifolium.—From the temperate regions of the Himalayas, and hardy. It will doubtless do best in a sheltered spot, but even in the open it has given me no trouble, and it is a good plant for shady spots on the rock-garden, the habit graceful, 2 feet to 3 feet in height, leaves glossy green; the flowers, white, marked with reddish lines and dots, are borne in bunches of from six to ten in the axils on both sides in late summer. The fruit is red when ripe.

Polygonatum punctatum.—A beautiful kind from the temperate Himalayas, where it is found at altitudes of 7,000 feet to 11,000 feet, and hardy in our gardens; about 2 feet in height, the stem angular, with hard leathery leaves, flowers white, with lilac dots, two or three in a bunch, in late summer.

P. roseum (*Rosy Solomon's Seal*).—A handsome little plant, allied to *P. erticillatum*. It was first sent to Kew, by Bunge, and varies much in the length and breadth of its leaves, also in the size of its flowers, 2 feet to 3 feet in height, the leaves in whorls of three or more; the flowers in pairs in the axils of the leaves, clear rose-coloured, are pretty amongst the narrow green foliage. North Asia.

POLYGONUM (*Knotweed*).—A very large genus, mostly herbaceous, and some climbing perennials, but few in their right place on the rock-garden, and those not of highest value.

Polygonum affine, one of the *Bistorta* group, is a pretty alpine feature in the Himalayas, where it grows on the wet river banks and meadows, and hangs in rosy clumps from moist precipices. In cultivation it is 6 to 8 inches high, with rosy-red flowers in dense spikes in September and October.

P. Brunonis is similar, and as desirable; the flowers, of a pale rose or flesh colour, borne in dense erect spikes nearly 18 inches high, and continuing more or less through the summer.

P. sphærostachyum.— A beautiful dwarf Knotweed, bearing spikes of deep crimson flowers. A native of the mountains of India, and with more merit as a choice rock plant than any so far as known in gardens.

P. vaccinifolium (*Rock Knotweed*).— Although it comes of rather a weedy race, this is a neat trailing plant, scrambling freely over stones, and producing many bright-rose spikes of flowers in summer and autumn. It comes from 11,000 to 13,000 feet on the Himalayas, which may perhaps have had much to do in refining its character and making it so unlike the Knotweeds that garnish the slime of our ditches. Easily increased by division or cuttings, and thrives in common garden soil. Suited for banks, and the less important parts of the alpine garden.

PONTEDERIA CORDATA (*Pickerel Weed*).—A handsome hardy water plant, forming thick tufts of arrow-shaped, long-stalked leaves from 1½ feet to more than 2 feet high, crowned with blue flower-spikes. *P. angustifolia* has narrower leaves; both should be planted in shallow pools or by the margins of ponds. Multiplied by division of the tufts at any season. North America.

POTENTILLA (*Cinquefoil*).—In these herbaceous or evergreen herbs, we have a family known in our gardens mainly by its large and freer kinds, chiefly hybrids. These are far too free for the rock-garden, and would soon overrun it. Among *Cinquefoils*, however, are some of the most beautiful and easily-grown rock plants, good in colour and valuable for their tufted and good habit for many situations. It is a very large genus, and what we have to guard against for the rock-garden is kinds too vigorous or without distinct beauty.

Potentilla ambigua, from the Himalayas, is a dwarf compact creeper, with, in summer, large clear yellow blossoms on a dense carpet of foliage; perfectly hardy, requiring only a good deep well-drained soil in an open position in the rock-garden.

P. alba (*White Cinquefoil*).—A pretty species, with the leaves in five stalkless leaflets, green and smooth above, and quite silvery, with dense silky down, on the lower sides. It is a very dwarf kind, and not rampant in habit, with white strawberry-like flowers, nearly an inch across, with a dark orange ring at the base. A native of the Alps and Pyrenees, of the easiest culture in ordinary soil,

flowering in early summer, and easily increased by division.

Potentilla argentea (*Silvery Cinquefoil*).—As the name would imply, this plant is covered over with silvery down; it is of a creeping habit, not exceeding 6 inches in height; and though scarcely definite enough in its argent character to give it a status in the gaudy ranks of the flower-garden, it is yet a very desirable plant to place as a variety among dark-leaved plants in a rockery.

P. aurea (*Golden Cinquefoil*).—A dwarf kind, about 2 inches high, with palmate leaves, margined with silvery hairs. The flowers large, yellow, spotted with orange at the base, and borne in a loose panicle from May to July. Suitable either for rockwork or the open ground in the full sun. Increased by division or by seed. Mountains of Central and South Europe.

P. nivea.—Dwarf, with whitish leaves snow-white underneath. The flowers yellow on slender stems, about 2 inches high, in summer. Thriving in the rock-garden in open soil. Seed. Division. Arctic regions of Europe and Asia, and Alps of Europe.

P. splendens.—A species with a woody, branching root-stock and short stems, forming a turfy carpet about 2 inches high, composed of three (rarely four or five) leaflets, which are green and glistening on the upper surface, and covered with silvery down underneath. The flowers a good white, borne singly on long stems from May to July. Pyrenees.

P. alpestris (*Alpine Cinquefoil*).—A native plant, closely allied to the spring *Potentilla* (*P. verna*), but with flower-stems more erect, forming tufts nearly a foot high when well grown, the leaves a shining green, the flowers of a bright yellow, about an inch across. Well worthy of a place on the rock-garden, it matters little how cold the spot, and will enjoy a moist deep soil. *P. verna* is also worthy of a place in the garden, and is of the easiest culture. It is not a very common plant, but is found in a good many parts of the country on rocks and dry banks.

Potentilla calabra (*Calabrian Cinquefoil*).—A silvery kind, particularly on the under sides of the leaves; the shoots prostrate, with lemon-yellow flowers about three-fourths of an inch across. It is chiefly valuable from the hue of its leaves; it flowers in May and June, and flourishes freely in sandy soil. It is worthy of a place in the rock-garden, and wherever dwarf Potentillas are grown. S. Europe.

Potentilla nitida.

P. nitida (*Shining Cinquefoil*).—A pretty little plant, about 2 inches high, with silky-silvery leaves; the flowers of a delicate rose, the green sepals showing between the petals. This native of the Alps is well worthy of a place in the choice rock-garden, and is of the easiest culture and increase. There are several varieties pretty in colour.

P. pyrenaica (*Pyrenean Cinquefoil*).— A dwarf but vigorous kind, with large yellow flowers, the petals round, full and over-lapping. A native of high valleys in the Pyrenees, easily increased by division or seeds, and thriving without any particular attention.

P. fruticosa (*Shrubby Cinquefoil*).—A pretty neat bush, 2 to 4 feet high, bearing in summer clusters of showy yellow flowers. It is suited for dry banks among rock shrubs.

PRATIA ANGULATA.—A pretty plant for the rock-garden, creeping over the soil like the Fruiting Duckweed; the flowers white, and like a dwarf Lobelia, numerous in autumn, giving place to violet-coloured berries about the size of peas. It is hardy. New Zealand. Syn., *Lobelia littoralis*.

PRIMULA (*Primrose*). — Alpine, mountain, pasture, marsh, or waterside dwarf perennials, of the greatest interest and much beauty, inhabiting all the great northern continents and the mountains of India in numbers sometimes enough to impart their own lovely colour to the landscape in mountain ground. Coming as they do from an immense variety of situations in mountain ground, their culture is of more complexity than that of most alpine plants, though not especially those of marshy ground. Among the best of them is our native Primrose, which in our northern woods is perhaps more beautiful than any one known kind. In nature many of these plants are deeply covered by snow for a long season, and thus enjoy a rest, which they cannot have in this country, where, in our open, green winters, the growth goes on, and the plants become more stalky than they do in nature. It is necessary, therefore, now and then to divide and top dress in the spring, in order to keep them in health. In the case of the high alpine kinds, in our dry summers, it is necessary to see that they are kept moist. In the southern parts of our country these kinds should be grown on the north and west sides of the rock-garden. Some of the fine Indian kinds thrive in ordinary soils, especially in the north and in moist districts, and some, like the Indian rosy Primrose (*P. rosea*), and the Japan Primroses, may be grown almost at the water's edge. The kinds we describe here are those of which we have some knowledge in cultivation, or have seen on the mountains of Europe. In the vast mountain ranges of India and Asia, probably the number of species is not even known yet.

As to aspect on the rock-garden, Mr W. A. Clarke, in "Alpine Plants," says:

"*P. calycina* should have a north-east aspect, well-drained position, rough loam and limestone, two parts each. *P. Clusiana*, south-west aspect; peat, loam, and sand—two parts loam, one peat, one sand. *P. frondosa*, south aspect; good strong loam, with a little sand. *P. glutinosa*, shady place or north aspect; peat, loam, and sand; *P. involucrata*, north aspect.

"*P. minima* will do in a sunny place if it can be well watered in summer. In a partially shaded place it grows well, but does not flower so prettily.

"*P. nivalis.*—A partially shaded place in deep, peaty loam suits this species well. *P. sikkimensis.*—Plant on the north side of a bog in good loam and leaf-mould."

To some extent, the question of aspect depends on where we are—north, south, or west. The many forms of the *Auricula* are varieties of one alpine *Primula*, and have the same needs as to moisture and aspect. In some districts the natural conditions of open ground suit them admirably; in other southern and dry districts we cannot grow them unless on cool shady borders, if at all.

Frequently, in addition to their high and cool alpine home-conditions, the Primroses grow wedged in between rocks without apparent nourishment, but the roots deep in the chinks where such moisture as exists can alter them very little. I remember in the Maritime Alps an enormous tuft of *Primula Allioni* in the seams of a great bare cliff, hundreds of feet above our heads; and, therefore, in our rock-gardens it is well to use pieces of grit or stone to protect the plants, and do a double good in keeping the moisture in the ground and also other and coarser plants away from these often very small alpine Primroses. We may frequently wedge them in between lumps of grit or sandstone. The marsh-loving kinds will not want this attention. The many natural hybrids, tender, or doubtful species, are left out of the following selection of the Primroses in cultivation, or observed in a wild state in Europe.

Primula Allioni (*Allionis Primrose*).—A bright richly coloured kind, blooming in March or April, the flowers about an inch in diameter, of a fine rosy purple colour, with white centre, and borne on very short stems. This charming Primrose is, unfortunately, not one of the easiest to cultivate as though loving moisture at the roots, it is susceptible to much moisture on the leaves, especially during the winter. For this reason, it succeeds best when planted sideways between stones, *i.e.* with its roots in an almost horizontal position, so that water can drain off from the leaves. A form of *P. Allioni* is found in the Tyrol, and is known to botanists as *P. tirolensis*, but the difference between the two forms is slight.

P. calycina.—From the Alps of Lombardy; is a dwarf Primrose of easy culture in the rock-garden. It has umbels of from three to five rosy-purple flowers springing from a short stalk in May or June. It thrives in a heavy soil and shaded from the sun.

P. amœna.—Allied to our wild Primrose, but distinct purple flowers coming out before the snow has left. In leaf it is not unlike *P. denticulata*, and the fact that it possesses the vigour of that plant, and also has much larger flowers, makes it welcome. It is so much earlier than the common Primrose that, while that species is in flower, *amœna* has finished blooming, and sent up almost the same kind of strong tuft of leaves which the common Primrose does after its flowers are faded. A sheltered and slightly shaded position will tend to the health of the plant. It is readily propagated by division of the root, and is a native of the Caucasus. The umbel is many-flowered, the blooms larger than those of *P. denticulata*, borne about 6 or 7 inches high; the leaves woolly beneath and toothed.

Primula auricula (*Auricula*). — The parent of the *Auricula* of which Parkinson, writing more than two hundred years ago, enumerates twenty-one varieties, and says there were many more; and in 1792 the Catalogue of Maddock, the florist, named nearly five hundred sorts. In our own time these have come to be almost forgotten as florists' flowers. *P. auricula* lives in a wild state on the high mountain ranges of Central Europe and the Caucasian Chain, and is one of the many Primulas which rival the Gentians, Pinks, and Forget-me-nots, in making the flora of Alpine fields so beautiful. Possessing a vigorous constitution, and sporting into a goodly number of varieties when raised from seed, it attracted early attention from lovers of flowers; its more striking variations were perpetuated and classified, and thus it became a "florists' flower." I do not desire to approach the subject from the florists' point of view, believing that to be a narrow and to some extent a base one; so much so, indeed, that I cannot regret that their practices and laws about the flower have taken but weakly root. To lay down mechanical rules to guide our appreciation of flowers must for ever be the shallowest of vanities. But, without seeking to conform or select them according to mechanical rules, we may preserve and enjoy all their most attractive deviations from the wild forms of the species.

The varieties of cultivated *Auriculas* may be roughly thrown into two classes: First, self-coloured varieties, with the outer and larger portion of the flower of one colour or shaded, the centre or eye being white or yellow, and the flowers and other parts usually smooth and not powdery; second, those with flowers and stems thickly covered with a white powdery matter, or "paste." The handsomest of the not-powdery kinds, known by the name of "alpines," to distinguish them from the florists' varieties, are the hardiest. The florists' favourites are always readily distinguished by the dense mealy matter with which the parts of the flower are covered. They are divided by florists into four sections: green-edged, grey-edged, white-edged, and selfs. In the green-edged varieties, the gorge or throat of the flower is usually yellow or yellowish; then comes a ring varying in width of white powdery matter, surrounded by another of some dark colour, and beyond this a green edge, which is sometimes half an inch in width. The outer portion of the flower is really and palpably a monstrous development of the petal into a leaf-like substance, identical in texture with that of the leaves. The "grey-edged" have also the margin of a green leafy texture, but so thickly covered with powder that this is not distinctly seen. This, too, is the case with the "white-edged," the differences being in the thickness and hue of the "paste," or powdery matter. In fact, the terms green-edged, grey-edged, and white-edged, are simply used to express slight differences between flowers all having an abnormal development of the petals into leafy texture. It is a curious fact that between the white and the grey the line of demarcation is imaginary, and both these classes occasionally produce green-edged flowers. The "selfs" are really distinct, in having the outer and larger portion of the corolla of the ordinary texture, a ring of powdery matter surrounding the eye.

The enumeration and classification of such slight differences merely tend to throw obstacles in the way of the flower being generally grown and enjoyed in gardens. By all means let the florists maintain them, but those who merely want to embellish their gardens with some of the prettier varieties, need not trouble themselves with named sorts at all. One fact concerning the florists' kinds should, however, be borne in mind, —they are the most delicate and difficult to cultivate. The curious developments of powdery matter, green margins, etc.,

ALPINE FLOWERS FOR GARDENS

have a tendency to enfeeble the constitution of the plant. They are, in fact, variations that, occurring in Nature, would have little or no chance of surviving in the struggle for life. The grower will do well to select the free sorts — good varieties of the border kinds.

Their culture is simple: light vegetable soil and plenty of moisture during the growing season being the essentials. In many districts the moisture of our climate suits the *Auricula* to perfection, and in such may be seen great tufts of it grown without attention. In others, it must be protected against excessive drought by putting stones round the plants, and cocoa-fibre and leaf-mould are also useful as a surfacing. In a plant so much degraded by florists from its natural form and colour as this Primrose, it is well to return to the natural colour and some very fine yellow-flowered kinds have been raised by Mr Moon and others, more beautiful than the florists' kinds.

Auriculas are easily propagated by division in spring or autumn—best in early autumn. They are also easily raised from seed, which ripens in July, the common practice being to sow it in the following January in a gentle heat. It should be sown in pans thinly. The plants need not be disturbed till they are big enough to prick into a bed of fine rich and light soil, on a half-shady border in the open air. It is a most desirable practice to raise seedlings, as in this way we may obtain many beautiful varieties. When a good variety is noticed among the seedlings, it should be marked and placed under conditions best calculated to ensure its rapid increase, and propagated by division.

Primula capitata.—One of the finest of Primroses, in autumn bearing dense heads of flowers of the deepest purple, which as regards depth is variable, and is shown to advantage by the white mealy powder in which the flowers are enveloped. It is not so vigorous as *P. denticulata*, though hardy, and it cannot be termed a good perennial, as it is apt to go off after flowering well. It is therefore advisable to raise seedlings. This is easy, as the plant seeds freely in most seasons,

and the seedlings flower in the second year. An open position with a north aspect in good loamy soil well watered in dry weather suits it best. India.

Primula carniolica is a native of Northern Italy and the Tyrol, the flowers, bluish-purple or lilac, with a white centre. The leaves are oblong, about 2½ inches long, very smooth, and arranged in a rosette. A variety, *multiceps*, has larger flowers. The position of *P. carniolica* should be a half-shady one, and it should be planted sideways on sloping or perpendicular rocks.

P. cortusoides (*Cortusa-like Primrose*).— This is entirely distinct in appearance from any of the species commonly grown, the leaves being large and soft, not nestling firmly on the ground like many of the European species, but on stalks 2 to 4 inches in length; the deep rosy clusters of flowers on stalks from 6 to 10 inches high. In consequence of its taller and freer habit, the plant is liable to be disfigured if placed in an exposed spot, therefore it should have shelter in a sunny nook, surrounded by low shrubs, or in any position where it will not be exposed to cutting winds. The soil should be light and rich, with a surfacing of cocoa-fibre or leaf-mould. It is one of the most beautiful and easily-raised Primroses, readily increased from seed. Siberia.

P. denticulata (*Denticulated Primrose*). —A Himalayan Primrose, with neat dense umbels of many small lilac flowers, on stalks from 8 inches to a foot high, springing from leaves, hairy on both sides, and densely so beneath. It is often grown in pots, but is hardy in deep light loam with a dry bottom, selecting a spot sheltered on the coldest sides. Division or by seeds. Although hardy, the leaves are injured by the first sharp frosts, so that it is well to keep it in well-drained warm positions. It is a variable plant, and some of its more distinct forms have received garden names, of which the principal are mentioned below. It is paler in colour than any of its varieties, and its foliage and flower-stalks are not mealy. *P. pulcherrima* is a great improvement on the species. It grows from 10 to 12 inches high, and has a more globular flower-truss, which is of a deep lilac colour. The stalks are olive-green, and, like the leaves, are slightly mealy. *P. Henryi* is a very strong-growing variety, but does not otherwise differ from *P. pulcherrima.* It is a handsome plant, often 2 feet across, and in Ireland it reaches even larger dimensions. *P. cashmeriana* is the finest variety. The flowers are of a lovely dark lilac, closely set together in almost a perfect globe on stalks over 1 foot high. They last from March till May. The foliage is beautiful, and, like the stalk, is of a bright pale green, thickly powdered with meal. They all prefer a cool situation, with a clear sky overhead, and delight in an abundance of moisture during warm summers.

Primula erosa (*Himalayan Primrose*). —Sometimes grown under the name of *P. Fortunei*, with shining leaves, quite smooth, and sometimes quite powdery, which, with its smoothness, distinguishes it at a glance from *P. denticulata*. The purplish blossoms with yellow eyes in flattish heads expand in early spring, and are borne on stems usually mealy. Drs Hooker and Thompson noticed it blooming at great elevations among the snow on the Himalayas, and, as might be expected from this, it is quite hardy in this country, and the way to enjoy its beauty is to place it in a sunny but sheltered nook on the rock-garden, in sandy loam, lightened with peat and leaf-mould, and with the drainage perfect. It should never be allowed to suffer from drought in summer.

P. farinosa (*Bird's-Eye Primrose*).— Slender powdery stems, from 3 to 12 inches high, springing from rosettes of musk-scented leaves, with their under sides clothed with a silvery-looking meal, bear the graceful lilac-purple flowers of the Bird's-Eye Primula. No sweeter flower holds its head up to kiss the breeze that rustles over the bogs and mountain pastures of Northern England. To find it inlaid over moist parts of the great hill-sides on an early summer morning as one ascends the Helvellyn range for the first time, is, to a lover of our wild flowers, a pleasure long remembered. In the Alps of Dauphiny the valleys are coloured with its flowers, and where the bottom of the valley only is moist, a river, as it were, of this Primrose in bloom runs through it. I have mostly seen it in very moist spots where running water spreads out all over the surface, still, however, continuing to flow; but it is also found under different conditions. A moist, deep, and well-drained crevice, filled with peaty soil or fibry sandy loam, will suit it to perfection. It is easy to cultivate in pots, the chief want, whether in pots or in the open, being abundance of water in summer, and where this does not fall naturally, it ought to be supplied artificially. When planted on the rock-garden in the drier districts, it would be well to cover the soil with cocoa-fibre or leaf-mould, which would protect the surface from evaporation; broken bits of sandstone would also do. It varies a little in the colour of the flower, there being pink, rose, and deep crimson shades.

P. farinosa acaulis is a diminutive variety of the preceding. The flowers are not freely upheld on stems like those of the common wild form, but nestle down in the very hearts of the leaves, and both flowers and leaves being very small, when a number of plants are grown together on one sod, or in one pan, they form a little cushion of leaves and flowers not more than half an inch high. The same positions will suit as have been recommended for the Bird's-Eye Primula, but being so very dwarf, it ought to have more care. If any weeds or coarse plants were allowed to vegetate over or near it, it would of course suffer.

Primula glutinosa (*Glutinous Primrose*). —A distinct little Primrose, and growing abundantly in peaty soil at elevations of 7,000 or 8,000 feet on mountains near Gastein and Salzburg, in the Tyrol, and in Lower Austria. The leaves are nearly strap-shaped, but winding towards the top, where they are somewhat pointed and regularly toothed. The stem is as long again as the leaves, growing from 3 to 5 inches high, bearing from 1 to 5 blossoms, purplish-mauve, with the divisions rather deeply cleft. Grow in moist peaty soil.

ALPINE FLOWERS FOR GARDENS

Primula integrifolia (*Entire-Leaved Primrose*).—A most diminutive Primrose, recognized by its smooth, shining leaves, lying quite close to the ground, and in spring, when in bloom, by its handsome rose flowers, with the lobes deeply divided, one to three flowers being borne on a dwarf stem, but little above the leaves, and these flowers are often large enough to obscure the plant that bears them. It is common on the higher parts of the Pyrenees, and I met with it in abundance in North Italy. Scores of plants sometimes grew together in a sod, like daisies, wherever there was a little bank or slope not covered by grass; and it was also plentiful in the grass, growing in a sandy loam. There should be no difficulty in growing this plant on flat exposed parts of rocks, the soil moist and free, but firm. The best way would be to try and form a wide tuft of it, by dotting from six to a dozen plants over one spot, and, if in a dry district, scattering a little cocoa-fibre mixed with sand between them. This, or stones, will help till the plants become established. It flowers in early summer, and is increased by division and by seeds. *P. Candolleana* is another name for this, and *P. glaucescens* is a variety of it.

P. latifolia (*Broad-leaved Primrose*).— A handsome and fragrant Primrose, with from two to twenty violet flowers in a head, borne on a stem about twice as long as the leaves. This is less viscid, larger, and more robust than the better known *P. viscosa* of the Alps, the leaves sometimes attaining a length of 4 inches and a breadth of nearly 2 inches. It grows to a height of from 4 to 8 inches, flowers in early summer, comes from the Pyrenees, the Alps of Dauphiny, and various mountain chains in Southern Europe, and in a pure air will thrive on sunny slopes in sandy peat, with plenty of moisture during the dry season, and perfect drainage in winter. It will bear frequent division; and may also be well and easily grown in cold frames or pits.

P. longiflora (*Long-flowered Primrose*). —Related to our Bird's-Eye Primrose, distinct from it, and larger than those of the best varieties of that species, the lilac tube of the flower being more than 1 inch long. It is not difficult to cultivate, and the treatment for *Primula farinosa* will suit it. In colour it is deeper than the Bird's-Eye Primrose. Austria.

Primula marginata (*Margined Primrose*).—Distinguished by the silvery margin on its greyish, smooth leaves, caused by a dense bed of white dust which lies exactly on the edge of the leaf; and by its sweet, soft, violet-rose flowers, in April and May. I have grown this plant well in the open air in London, and in parts of the country favourable to alpine plants it will prove almost as free as the common *Auricula*. Even when not in flower, the plant is pretty, from the hue of the margin and surfaces of the leaves. Our wet and green winters are doubtless the cause of this and other kinds becoming lanky in the stems after being more than a year or so in one spot. When the stems become long, and emit roots above the surface, it is a good plan to divide the plants, and insert each portion firmly down to the leaves. This will be all the more beneficial in dry districts, where the little roots that issue from the stems would be more likely to perish. It is a charming plant where it thrives freely. In the open ground a few bits of broken rock, placed around each plant, or among the plants, if they are planted in groups or tufts, will do good by preventing evaporation, and also acting as a protection to the plant, which rarely exceeds a few inches in height. A native of the Alps of Dauphiny, and various ranges in the south of Europe, but not of the Pyrenees. Division.

P. minima (*Fairy Primrose*). — With very small leaves, prostrate, but the flowers make up for the diminutive leaves, being nearly an inch across, and quite covering the minute rosette from which they spring. It is a native of the Alps of Austria, and flowers in early summer, the stem rarely bearing more than one, but occasionally two flowers, rose-coloured, or sometimes white. Bare spots are the best places for it, the soil to be sandy peat and loam; it is suited for association with the very dwarfest alpine plant. It may be propagated by division or by seed, and comes from the mountains of southern Europe.

Primula Floerkiana is like the Fairy Primrose, probably only a variety of it, and in the flowers only differing by bearing two, three, or more, instead of a single bloom. There is also a difference in the leaves, which in *P. minima* are nearly square at the ends, but in *P. Floerkiana* are roundish there, and notched for a short distance down the sides. It is a native of Austria, and will be found to enjoy the same conditions as the preceding. Of both it is desirable to establish widespreading patches on firm bare spots, scattering ½ inch of silver sand between the plants to keep the ground cool.

P. Munroi (*Munro's Primrose*).—This has not the brilliancy or dwarfness of the Primulas of the high Alps, nor the vigour of our own wild kinds, but it is distinct, and is of the easiest culture in any moist soil. It grows at high elevations on the mountains of Northern India, near water, and bears creamy-white flowers, with a yellowish eye, more than an inch across on stems 5 to 7 inches high, springing from smooth green leaves a couple of inches long. The flowers are sweet, and it highly merits culture in the bog garden, and flowers from March to May. *P. involucrata* is an allied kind, from the same regions, somewhat smaller, thriving under the same conditions.

P. nivea (*Snowy Primrose*).—A dwarf species, freely bearing trusses of lovely white flowers, quite distinct in aspect from any other in cultivation, happily easy of culture, and may be grown in pots or in the open ground. If in pots, it should be frequently divided; for it has a tendency, in common with other choice Primulas, to get somewhat naked about the base of the shoots, and, as these protrude rootlets, the whole plant is likely to go off if not taken up and divided into as many pieces as possible. Every shoot will form a plant, inasmuch as each is usually furnished with little rootlets, which take hold of fresh soil immediately. In a wild state the natural moisture and the accumulating *débris* of the mountain enable them to use those exposed rootlets, and thrive; but in cultivation I have found it best to divide such fine *Primulas* as this, and plant them down to the leaves when their stems have grown much above the soil. The ground would also be the better of being covered with an inch or so of cocoa-fibre. In moist and cool districts there would be less trouble, but, in all, care should be taken to give the Snowy Primrose what it deserves—a select place, a light free soil, and plenty of water during the summer. It flowers in April and May, is a native of the Alps, and is by some supposed to be a variety of *P. viscosa*.

Primula officinalis (*Cowslip*). — The Cowslip of our meadows is worthy of a place in gardens; but the many handsome kinds that have sprung from it are more valuable from a garden point of view.

Polyanthuses for rich colour surpass all other flowers of our gardens in spring. At one time the *Polyanthus* was highly esteemed as a florists' flower, but nearly all the choice old kinds are now lost, and florists who really pay the flower any attention are few. In consequence, however, of the great facility with which varieties are raised from seed, nobody need be without handsome kinds, and raising them will prove interesting amusement.

P. Parryi.—A pretty rocky mountain Primrose, bearing about a dozen large, purple, yellow-eyed flowers, nearly 1 inch across in summer on stems about 1 foot high. Though an alpine plant, and growing on the margins of streams near the snow-line, where its roots are bathed in ice-cold water, it has succeeded in Britain in moist, loamy soil mingled with peat; it is hardy, and requires shade from extreme heat rather than protection from cold. North-West America.

P. suaveolens of Bertolini is a variety of the Cowslip, found in many parts of the Continent, and not sufficiently distinct or ornamental to merit cultivation.

P. elatior is the true as distinguished from the common Oxlip. It is not an ornamental species, the flowers being of a pale buff-yellow, and it is readily distinguished by its funnel, and not saucer-shaped corolla, which is also destitute of the bosses which are present in the Primrose and Cowslip. It is found in woods and meadows on clayey soils in the eastern counties of England, particularly in Essex, Suffolk, and Cambridgeshire.

Primula palinuri (*Large-Leaved Primrose*).—This is distinct from other cultivated Primroses, inasmuch as it seems to grow all to leaf and stem, whereas many of the other kinds often hide their leaves with flowers. In April the yellow flowers appear in a bunch at the top of a powdery stem, emit a cowslip-like perfume, and are pretty, though they rarely fulfil the promise of the vigorous-looking plant. I have seen it flourish in rich light soil as a border-plant in various parts of these islands, and established plants are easily increased by division. Southern Italy.

P. purpurea (*Purple Primrose*).— A handsome Primrose, from elevations of 12,000 feet or more on the Himalayas, and allied to *P. denticulata*, though finer; the flowers, of an exquisite purple, are larger, in heads about 3 inches across. Sheltered and warm positions, but not very shady, will best suit it, the soil being a light deep sandy loam and decomposed leaf-mould. I have never seen it thrive so well as when planted in nooks at the base of rocks which sheltered it, where it enjoyed more heat than if exposed.

P. Scotica (*Scotch Bird's-Eye Primrose*). —This, one of the most lovely of its family, is a near ally of the Bird's-Eye Primrose. Its rich purple flowers, with large yellowish eye, open in the end of April, supported on stems from ½ an inch to 1 inch high, growing an inch or two taller as the season advances. It is said by some botanists to be simply a variety of the Bird's-Eye Primrose, but the seedlings show no tendency to approach the larger and looser *P. farinosa*, and Mr Boswell Syme, who has carefully observed the living plant both in a wild state and cultivated in his own garden, declares it to be "perfectly distinct." The leaves are powdery on the under side, broadest near the middle, shorter, and less indented than those of *P. farinosa*, which are broadest near the end; and the whole plant is about large enough to associate with a dwarf moss or lichen. A native of the counties of Sutherland and Caithness, and of the Orkney Isles, growing in damp pastures. The best place for it is on some spot where it would have perfect drainage, and not be injured by strong-growing plants shading it. The soil should be a friable loam, mixed with sandy peat or a little cocoa-fibre, and made firm; a few pieces of broken porous rock should be placed firmly in the ground around it, so as to show half their size above the surface, prevent evaporation, and also act as a guard to the little plant. If a coating of dwarf moss is spread over the earth after a time, I should not remove it, believing the plant to enjoy such a carpet. Although so small, it is, when in health, vigorous, and seeds freely, the self-sown seedlings having often formed with me good plants on the mossy surface of the ground. I have grown it in the open air near London; but, as a rule, it is best for all who do not try it in a pure atmosphere to grow it in well-drained pots or pans, using the same kind of soil, and protecting the plants in a cool shallow frame in winter, placing the pots out of doors in summer, plunged in coal-ashes or sand. In all cases the plant should be abundantly watered in dry weather. Easily propagated by seeds, which should be sown soon after they are ripe in shallow pans of sandy peat or fibrous loam mixed with cocoa-fibre, and placed in an open pit or shallow cold frame.

Primula sikkimensis (*Sikkim Cowslip*). —One of the most remarkable of Primroses; when well grown, it throws up strong flower-stems from 15 inches to 2 feet high, bearing many bell-shaped, pale-yellow flowers, without a spot of any other colour, the pedicel mealy, the blooms of an agreeable perfume. Some of the stems bear a head of more than five dozen buds and flowers, and each flower is nearly 1 inch long and more than ½ inch across. It is hardy, and loves deep well-drained and moist ground; near water, or in deep boggy places, suit it best; begins to flower in May, and remains in flower for many weeks. It is said to be the pride of all the Primroses of the mountains of India, inhabiting wet boggy localities, at elevations of from 12,000 to 17,000 feet, and covering acres of ground with its yellow flowers. Propagated by division, as it rarely or never matures its seeds in this country. It is well to raise it from good seed now and then, as it is apt to disappear in some soils.

Primula Stuartii (*Stuart's Primrose*).—A noble and vigorous yellow Primrose, a native of the mountains of Northern India, to some parts of which, according to Royel, it gives a rich yellow glow. It grows about 16 inches high, has leaves nearly a foot long, mealy below, smooth above; the umbels being many-flowered. Like *P. denticulata* and the purple Primrose, the place most suitable for this is some perfectly drained and sheltered spot; if convenient, plant it against the base of rocks, which will shelter it from cutting winds, though, when sufficiently plentiful, this precaution may be dispensed with. A light deep soil, never allowed to get dry or arid in summer, will suit it well.

P. viscosa (*Viscid Primrose*).—This is the lovely little Primrose that travellers who visit the Alps in early summer see opening its clear rosy-purple flowers with white eyes at various altitudes: sometimes, in crossing a high pass, it comes into view, plant, flower, and all, not bigger than a shilling, but still bravely flowering—indeed, nearly all flower; while on sunny slopes and in the valleys it may be seen nearly as large as the *Auricula*. It may be grown in any position in light, peaty, or spongy loam, with about one-half its bulk of fine sand, provided its roots are kept moist during the dry season. A native of the Alps and Pyrenees; easily increased by division, and may also be raised from seed. Varieties are sometimes found with white flowers, but rarely. The handsome purple Primroses known in gardens under the name of *P. ciliata* and *P. ciliata purpurea* are varieties of this, the last said to be a hybrid between it and an *Auricula*. Syn., *P. villosa*.

P. vulgaris (*Common Primrose*).—The Gentians and dwarf *Primulas* do not do more for the Alps than this for the hedge-banks, groves, open woods, and borders of fields and streams of the British Isles.

The forms of the plant most precious for the garden are the beautiful old double kinds. No sweeter or prettier flowers ever warmed into beauty under a northern sun than their richly and delicately tinted little rosettes. The best known and most distinctly marked kinds are the double lilac, double purple, double sulphur, double white, double crimson, and double red.

The double kinds, more delicate and slower-growing than the single ones, require more care, and in their case the development of healthy foliage after the flowering season should be the object of those who wish to succeed with them. Shelter and partial shade are the two conditions chiefly necessary to secure this. Open woods, copses, and half-shady places are the favourite haunts of the Primrose in a wild state. In them, in addition to the shade, it enjoys shelter not merely from tall objects around, but also from the long grass and other herbaceous plants growing in close proximity; and we should also take into account the moisture consequent upon such companionship, and let these facts guide us in the culture of the double kinds. As will be readily seen, a plant exposed to the full sun on a naked border would be under a different condition to one in a thin wood; the excessive evaporation and searing away of the leaves by the wind would be sufficient to account for the failure of the exposed plant. It is therefore desirable, in the case of the beautiful double Primroses, to plant them in shaded and sheltered positions, using light rich vegetable soil, and, if convenient, keeping the earth from being too rapidly dried up by spreading cocoa-fibre or leaf-mould on it in summer.

They are increased by division of the roots, and to take them up in order to divide these is the only disturbance they should suffer. The double Primroses well grown, and the same kinds barely existing, are such very different objects, that nobody will begrudge giving them the trifling attention necessary to their perfect development. Occasionally they may be seen flourishing by some cottage or old country garden, where they find a home more congenial than the bare fashionable flower-garden of our own day, and they are well worthy of a place on the cooler sides of the rock-garden or among the mountain shrubs near it.

Primula rosea (*Rosy Indian Primrose*). —A brightly-hued Primrose, from 6 to 8 inches high, the flowers in umbels of from 6 to 9 blooms, on a rather stout stem, rosy carmine in colour, with a yellow throat. The leaves are very smooth, about 4 inches long, and serrated at the margin. It is a charming plant for a bog garden, and thrives in any damp, light soil. I have seen it flourish in a sunny bog-bed even better than in a shady one, but it will not endure a dry, sunny position. In Scotland it grows apace in ordinary garden borders, owing to greater rainfall. The plants are easily grown from seed or increased by division of the root-stock.

P. rosea grandiflora.—Of this variety the flowers are more robust, and borne on taller and stouter stems; the colour a deeper carmine-crimson.

P. frondosa.—A member of the mealy section of *Primula*, this is the best, most vigorous, and the freest bloomer, growing with great vigour and freedom where *P. farinosa* is a failure. Growing 9 inches high, the plant when seen in a colony is very pretty, and in quite open spots will come into flower earlier than many species of the genus. It is a fine plant and truly perennial. The best place is the rock-garden, and here on a level spot, rather low down, and afforded some protection by higher rocks from mid-day sun, the plant will form a pretty picture for a long time. When sown as soon as ripe, the plant may be largely increased by seeds, the seedlings to be grown in colonies, and the soil chiefly loam, with small broken rock intermingled, and a coating of small stones on the surface.

P. Sieboldi (*Siebold's Primrose*).—Though this handsome Primrose has been considered a variety of *P. cortusoides*, it is distinct for the size of its flowers, the breadth of its foliage, the creeping character of its root, its exclusively vernal habit, its pseudo-lobed or grooved seed-vessel, and the roundish flattened form of its seed.

Since its introduction from Japan, numerous beautiful varieties have been raised, some of the most distinct being *Clarkiæflora, Lilacina marginata, Fimbriata oculata, Vincæflora, Cærulea alba,* Mauve Beauty, Lavender Queen, *laciniata,* and *maxima*. These possess a fine diversity of colour, and some have the petals fringed. One of the chief merits of these *Primulas* is that they bloom early, flowering about the month of April when flowering plants are rare; and another is, that they are free bloomers, throwing up successive flower-stems, and lasting a long time in perfection. The best soil for them is light and rich, consisting of fibry loam, leaf-mould, pulverised manure, and some grit to keep it open. They are impatient of excessive moisture, and when put in open ground should be planted in well-drained soil, or in raised positions in the rock-garden. The roots creep just below the surface, and form eyes from which any variety can be easily propagated. *P. Sieboldi* is a perennial, which loses its leaves in autumn and winter, when it goes to rest, and breaks up again early in spring.

Primula japonica.—One of the handsomest of Primroses, a good perennial, and is not at all tender. It is a first-rate border plant, and in moist shady spots of deep rich loam it grows vigorously, throwing up flower-stems 2 feet or more high, and unfolding tier after tier of its crimson blossoms for several weeks in succession. It may be grown in the rock-garden as well as in the border, and is an excellent wild-garden plant, thriving almost anywhere, and sowing itself freely. It is said to be rabbit-proof. There is a white form, a pale pink, and a rose form, but the best is the original rich crimson. In raising *P. japonica* from seed, it should be borne in mind that the seed remains some time dormant, unless it is sown as soon as it is gathered, and that it must on no account be sown in heat. A cool frame is the proper place for the seed-pan, and till the seed has germinated, care must be taken to prevent or keep down the growth of Moss and Liverwort on the soil. This Primrose is grown finely at Enys, in Cornwall, along the margin of a pond.

Primula prolifera.—This, better known under the name of *P. imperialis*, is a tall Indian Primrose, allied to *P. japonica*, but with yellow flowers arranged in whorls. It is, perhaps, too tender for the north of England, but in sheltered places in Cornwall it grows to a height of about 3 feet. Peaty soil seems to suit it best.

P. Poissoni.—A Chinese Primrose, found in the mountains of Yunan, and hardy. In Messrs Veitch's Nurseries, at Exeter, it withstood even the severe winter of 1894 without protection, and it is handsome and easy to cultivate, thriving in a moist situation. The flowers are bright rose, with a slight flush of mauve, and have a yellow centre. They are fully the size of a shilling, and are arranged in verticillate tiers of eight or twelve blossoms, each after the style of *P. japonica*, but the tiers are a little further apart than in the last-named variety, showing often 2 inches or more of stem between the tiers. It grows about 12 inches high. The leaves are pale glaucous green, about 5 inches or 6 inches long and 2 inches wide, smooth, the midrib widened towards the base of the leaf and of a pink colour.

P. Wulfeniana.—An excellent rock Primrose, preferring calcareous soil, the flowers large, deep purple, in umbels of about five flowers each, and is one of the easiest to grow, planted in a slanting position.

P. luteola.—One of the handsomest of the yellow Primroses, and a fine plant when well grown. The flower-stems are sometimes 1½ to 2 feet high, though usually under 1 foot in height. They sometimes become fasciated, and thus carry a huge cluster of flowers 4 to 6 inches across. These flowers are like those of a *Polyanthus* or an *Auricula*, but they are borne in more compact heads. It likes a moist situation in full exposure, and thrives in rich borders of rather moist soil, or on the lower banks of the rock-garden.

P. spectabilis.—A native of the Tyrol, growing about 6 inches high, and bearing umbels of about seven or eight rosy purple flowers. The leaves are smooth and have the margin entire and horny. It is a good rock-garden plant of easy culture.

Primula clusiana.—The variety is a native of the calcareous rocks of the Eastern Alps, the flowers large, rosy crimson with white centre, and borne in large umbels on a stem about 9 inches high. It thrives in chalk-soil. In addition to the above, there are known in cultivation : *P. alpina* (Siberia), *angustifolia* (N. America), *apennina* (Piedmont), *Arctotis* (Europe), *assimilis* (Europe), *auricula*, (Europe), *Balbisii* (Europe), *Berninæ* (Switzerland), *biflora* (Switzerland), *ciliata* (Europe), *columnae* (Europe), *commutata*. (Europe), *coronata* (Tyrol), *cottia* (Alps), *decipiens* (Alps), *deorum* (Bulgaria), *digenea* (Europe), *dinyana* (Switzerland), *discolor* (N. Italy), *Dumoulinii* (Alps), *Facchinii* (N. Italy), *flagellicaulis* (Europe), *flœrpkeana* (Alps of S. Europe), *floribunda* (Himalaya), *Forbesii* (China), *Forsteri* (Tyrol), *gambeliana* (Himalaya), *Goebelii* (Tyrol), *grandis* (Caucasus), *Heerii* (Switzerland), *heterodonta* (China), *hirsuta* (Europe), *Huteri* (Tyrol), *imperalis* (Java), *juribella* (S. Tyrol), *Kaufmanniana* (Turkestan), *Kolbiana* (N Italy), *minutissima* (Himalaya), *mistassinica* (N. America), *mollis* (Himalaya), *muretiana* (Switzerland), *obovata* (Venetian Alps), *Obristii* (N. Italy), *obtusifolia* (India), *œnensis* (S. Tyrol and Italian Alps), *pedemontana* (Piedmont), *Peyritschii* (Tyrol), *prolifera* (Himalaya), *pubescens* (Europe), *pumila* (Tyrol), *Reidii* (Himalaya), *rhætica* (Switzerland), *Rusbyi* (New Mexico), *Salisii* (Switzerland), *Sendtneri* (Tyrol), *sibirica* (Asia and Arctic America), *similis* (Tyrol), *spectabilis* (Tyrol), *Steinii* (Tyrol), *Sturii* (Styria), *suffrutescens* (California), *Tyrolensis* (Tyrol), *variabilis* (Europe), *venusta* (Styria), *verticillata* (Arabia), *vochinensis* (Carinthia).

PRUNELLA GRANDIFLORA

(*Self-heal*).—A handsome and vigorous

plant, distinguished by its large flowers from the common British Self-heal, which is unworthy of cultivation. There is a white as well as a purple variety, both handsome plants, that thrive in almost any ground, but prefer a moist and free soil and a position somewhat shaded. They are apt to go off in winter on the London clay, at least on the level ground. A native of continental Europe; flowering in summer, but this and other kinds are only of secondary use in the rock-garden and among shrubs on banks.

PULMONARIA (*Lungwort*).— These plants are more fitted for borders than for the rock-garden. The beautiful plant for many years known as *P. virginica* is now *Mertensia*.

PUSCHKINIA SCILLOIDES (*Striped Squill*).—A fascinating little plant, and the most delicately beautiful among early mountain flowers. The flowers white, striped, and tinged with blue, the small prostrate leaves concave; easily grown, it does not last long in flower, but few spring flowers do. The best position for this is on low banks, in the rock-garden, or in positions where its flowers may be seen somewhat beneath the eye, associated with dwarf *Primulas* and other diminutive spring flowers. A native of the Caucasus, flowering in spring, easily increased by division of the root, and flourishing best in very sandy light soil.

PYROLA (*Wintergreen*). — Dwarf evergreen herbs, inhabiting mountain woods or copses, moors, and wet places among sand dunes. They are not difficult to cultivate in moist peat or sand, associated with the right sort of plants as to stature and wants.

Pyrola rotundifolia (*Larger Wintergreen*).—A native plant, inhabiting woods, bushy, and reedy places; with leathery leaves, and handsome drooping racemes of white fragrant flowers, ½ inch across, ten to twenty flowers, on a stem from 6 inches to a foot high. *Pyrola rotundifolia*, var. *arenaria*, is another very graceful plant, found on sea-shores, and differing in being dwarfer, deep green, and smooth. Both are beautiful plants for shady mossy flanks of rock in free vegetable soil, and flourish more readily in cultivation than any species of their family. In America there are varieties of this plant with flesh-coloured and reddish flowers, none of which are in cultivation with us, and several of the American kinds seem to me well worthy of being brought over.

Pyrola uniflora, media, minor, and *secunda,* are also interesting plants, of which the first, a very rare one in our Flora, is the prettiest. *P. elliptica,* a native of North America, is also in our gardens, though rare.

PYXIDANTHERA BARBULATA (*Bearded P.*).—A curious and minute American plant, plentiful in sandy dry "pine barrens" from New Jersey to North Carolina. It is an evergreen shrub, yet smaller than many mosses; the leaves narrow, awl-pointed, and densely crowded; the flowers are placed singly, and are stalkless, but very numerous, rose-coloured in bud, white when open. The effect of the rosy buds and five-cleft white flowers on the dense dwarf cushions is singularly pretty. Generally found in low, but not wet, places, and usually on little mounds, it is a gem for the rock-garden, on which it should be planted in pure sand and vegetable mould, fully exposed. Flowers in early summer; increased by division.

RAMONDIA (*Rosette Mullien*).— Dwarf plants found on steep and somewhat shady rocks, and, according to Ramond, exclusively in valleys leading from north to south; having leaves in rosettes spreading very close to the ground, blistered, deeply wrinkled, and densely covered with short hairs—quite shaggy beneath and on the leaf-stalk. The shady side of rocks or moist depressions, or the shade of evergreen bushes, suits them best in any free soil. I have seen them succeed well as edgings to beds of evergreen bushes in peat soil. They are increased by division only and the whole should be moist always. They may be increased from the leaves, breaking off the leaf close to the plant, and pegging the foot-stalk into sandy peat, keeping the soil meanwhile moist and the leaves fresh by covering with a bell-glass.

Ramondia pyrenaica form rosettes of leaves, deeply wrinkled, and covered with brown, shaggy hairs on the under surface and the lower parts of the leaf-stalk. The

Ramondia pyrenaica.

when the rosettes are clustered together, and then it must be done with care, owing to the closely-nestling character of the leaves and the few roots. To raise them from seed we should take care that the flowers are fertilised; with good seed growth is quick, and flowering plants may be had in two years. A mixture of peat and plenty of sand, with sandstone the size of Cobnuts, forms a capital compost, leaves spread out close upon the soil, and the flower-stalks emerge from beneath the leafage in the month of June or earlier. Usually there are three flowers to each stem, though on strong plants as many as five are found, each having a diameter of 1 inch or rather more, purplish-violet in colour, and having a rich orange eye or centre. There is a white variety, and there is more than one white-flowered kind, one a pure and spotless flower.

R. Nataliæ is a rare plant from Servia, having light purple flowers with orange stamens, and *R. serbica* has large, handsome foliage, and violet-purple flowers.

RANUNCULUS (*Buttercup*).—These are alpine, northern pasture, water and waterside plants, many of the perennial and mountain kinds, from their boldness, hardiness, and beauty, admirably suited for the rock-garden. Although as interesting as any of the great families of rock plants, they are not nearly so difficult to grow and keep, if care be taken to prevent them being overrun by coarser plants.

early spring, as they often eat out the crowns before they are fairly above ground, and the flowers are lost for the season. A little rough grit will do much to prevent this occurring; if placed over the crowns the fine must be taken out, only using the rough grit."

Ranunculus amplexicaulis (*Lady Buttercup*).—A beautiful plant, with large white flowers having yellow centres, one to five blooms being borne on a stem, which is clasped by smooth sea-green leaves, which set off its snowy bouquet of flowers. I know no more graceful plant

Lady Buttercup (*Ranunculus amplexicaulis*). (Engraved from a photograph.)

Mr W. A. Clark, in "Alpine Plants," rightly attaches importance to top-dressing some of the higher alpine species, and says "that great care must be taken to top-dress or replant just after flowering, as the plants work out of the ground, and this can be done before the hot weather begins. If left without top-dressing, they will no doubt shrivel up with the sun, as the roots will have been left all exposed. A sharp look-out for snails is essential in the

for the rock-garden. A native of the Alps, Pyrenees, thriving in light, rich loam, usually growing 7 inches to 10 inches high, flowering in gardens in April or May, and increased by seed or division. It is worthy of the best positions, and is very pretty grouped in a free way.

R. aconitifolius (*Fair Maids of France*). This white-flowered Crowfoot, which grows from 8 inches to a yard high in moist parts of valleys and woods in the Alps and Pyrenees, is too large for cultivation in the rock-garden among the choicer and smaller things; but its double variety is a beautiful old border flower.

The flowers are not large, but are white and double, and resemble a miniature double white Camellia. A rich, moist soil will be found to suit it best on the shady side of the rock-garden, and among bog-loving shrubs.

Ranunculus alpestris (*Alpine Buttercup*).—A diminutive species, from 1 inch to 3 inches or 4 inches high, and forming neat tufts, each stem bearing from one to three white flowers in April. The leaves are of a dark glossy green, roundish-heart-shaped, and deeply divided. It is a native of most of the great mountain ranges of Europe, in moist, rocky places on the higher pastures, and one of the best plants for the rock-garden. It is not difficult to grow in moist, sandy, or gritty soil, in positions exposed to the sun and moist in summer.

R. Traunfellneri seems to be a diminutive of the preceding, the whole plant, even as we have observed it in cultivation, being not more than 1 inch high. The same treatment will suit it; but, being smaller, it will require a little more care in selecting some firm spot fully exposed to the sun and air, but kept moist with a surfacing of grit, sand, or small stones, till the plant grows into a little spreading tuft.

R. bilobus is another form from S. Tyrol.

R. anemonoides, a native of the Alps of Styria and the Southern Tyrol, is a handsome species, with bluish-green leaves; flowers large, with numerous divisions, of a greenish-white on the inside and pink on the outside, appearing before the leaves, and very early. It does best in the rock-garden in a cool place, and in moist, porous soil.

R. bullatus (*Marigold Buttercup*).—A dwarf stout perennial, easy to cultivate, with showy double flowers, the blossoms as large as those of the double Marsh Marigold. The plant thrives in heavy soil. Division of the roots.

R. crenatus.—A native of granitic mountains in Styria, with roundish leaves, the flowers large, white, two or three together at the extremity of stem, 3 inches or 4 inches high in April or May. It does well in the rock-garden in gritty or open soil.

Ranunculus glacialis (*Arctic Buttercup*).—A well-named plant, as it is an inhabitant of very high places on the Alps, and may often be seen in flower near the snow and in the Arctic regions. The flowers are large, white-tinted, of a dull purplish-rose on the outside; the calyx with shaggy brownish hairs, the leaves smooth, deeply cut, and of a dark green. It will thrive in a cool spot in deep, gritty soil, moist during the warm months. I have seen it thriving with its roots below stones. On the Alps it blooms in early summer; in our gardens somewhat earlier. It is easily raised from seed, and in its native habitat spreads about freely. This is the plant which Mr Ruskin met with high up among the icy rocks, near the margins of the snowy solitude of the Alps, and which pleased him so much there. It is often washed down by the rock streams, and found in the river flats.

R. gramineus (*Grassy Buttercup*).—A graceful plant, which may well represent on the rock-garden the beauty of some of the taller kinds that are too vigorous for it. Easily known by its Grass-like leaves, 6 inches to 12 inches high. The flowers in May are yellow. There is a double variety, but it is seldom seen. Southern Europe. Division. An easily-grown plant.

R. Lyallii (*Rockwood Lily*).—Dr Hooker calls this plant the "most noble species of the genius"—"the Water Lily of the shepherds." Indeed, even in the dried specimens, of which there are many in the Kew herbarium, the resemblance to our common white Water Lily is striking. The plant is said to grow in moist places in the Southern Alps, the Wurumui Mountains, in the glacier regions of the Forbes River, near Otago, and elsewhere in the Middle Island of New Zealand, at heights of from 1000 feet to 5000 feet above the sea. In habit it seems almost identical with our Marsh Marigold, but it is twice or thrice larger. The leaves are circular, 12 inches to 15 inches in diameter peltate, as in the *Nelumbium*, the flowers borne in panicles; each flower of the purest waxy-white colour, 3 inches to 4 inches across. To raise a stock it has been recommended that the seed be sown in well-drained pans or boxes filled with peat and coarse grit in equal parts, stood in a cool place on the north side of a wall, watered well, and covered with a sheet of glass.

To English growers, the most interesting experience is that of Mr Bartholomew, Park House, Reading, who has grown this plant well. His plant was on the north side of a summer-house, in 2 feet of soil, chiefly peat, which was liberally watered all through the summer. When it died down in the autumn, a little cocoa-nut fibre was placed over the crowns, and, with a view to saving the plant as far as possible from alternate freezing and thawing, a sheet of glass, raised on bricks, was placed over it. It flowered freely and ripened seed at Reading. It also bloomed for three years in succession in a Nursery at Aberdeen, the seedlings having been raised there.

Ranunculus montanus (*Mountain Buttercup*).—A dwarf compact plant, with tufts of deep green, glossy leaves, covered in spring with many yellow flowers, somewhat larger than those of our common Buttercup. Although like the Buttercups in colour, it is unlike in its dwarf, close habit, usually flowering at 3 inches high, and, though growing freely enough, not spreading about with the coarse vigour of many of its fellows. It is a native of alpine pastures on the principal great mountain-chains of Europe, growing freely in moist, sandy soil, and should be planted so as to form spreading tufts, as it represents in a modest way the beauty of yellow kinds too vigorous for the rock-garden. Readily increased by seed or division.

R. Parnassifolius (*Parnassia-Leaved Buttercup*).—Distinct, with beautiful white flowers, from one to a dozen or more being borne on each stem, which grows from 3 inches to 8 inches high, and is somewhat velvety, and of a purplish hue. The leaves are of a dark brownish-green, sometimes woolly along the margins and nerves. It is rare in gardens, though abundant in many parts of the Alps on calcareous soils. No plant is more worthy of culture in the rock-garden in sandy, well-drained loam. There is a variety with narrow leaves.

Ranunculus pyrenæus (*Pyrenean Buttercup*).—A slender-leaved plant, 6 inches to 10 inches high, and from the Alps, as well as the Pyrenees, where it abounds. *R. plantagineus* from the Piedmont, and *R. bupleurifolius*, usually found in moist valleys in the Pyrenees at a much lower altitude, are varieties of the species. All have white flowers, and are of easy culture.

R. rutæfolius, syn. **callianthemum** (*Rue Buttercup*).—This, with deeply divided leaves, reminding one somewhat of those of a very dwarf Columbine, and white flowers with orange centres about an inch across, on stems from 3 inches to 6 inches high, bears from one to three flowers, sometimes rose-tinted on the outside. A native of high and cool parts of the granitic continental ranges; increased by seed or division.

R. Seguieri (*Seguir's Buttercup*).— Like the Glacier Buttercup, about 6 inches high, with three-parted leaves, though distinct. Usually the flowers are solitary, and rarely as many as two or three on each stem. The flowers are white, with distinctly rounded petals. Native of the calcareous Alps of Provence, Dauphiny, and Carniola.

R. Thora (*Venom Buttercup*).—The roots of this, like small Dahlia tubers, and said to be poisonous, were formerly used by the Swiss hunters to poison their darts. It is yellow-flowered, with very smooth leaves. *R. Thora*, distributed through Switzerland, the Carpathian, and other mountain chains on rocks and in pastures near the snow-line, thrives in gritty loam.

RAPHIOLEPIS OVATA. — (*Japanese Hawthorn*).—A Japanese evergreen shrub, hardy in the southern counties at least, with thick dark evergreen leaves and large white and sweet-scented flowers, borne in clusters at the ends of the young branches. It is a low spreading bush, and should not be crowded with other shrubs. Some of the other species, such as *R. indica* and *R. salicifolia*, both from China, are not hardy enough for the open air.

RHAMNUS ALPINA.—Among these shrubs there is one tiny thing which is of some value for the rock-garden, as it spreads its small shining leaves over the rocks, clasping them close; the flowers are the most unattractive imaginable, but we have so many ugly ill-placed stones in a rock-garden, that anything which throws a veil over them we may have a place for.

R. Perieri is a dwarf form of the evergreen *Rhamnus*, useful for the rock-garden, where evergreen effects are sought.

RHEXIA VIRGINICA (*Meadow Beauty*).—An American plant of the *Melastoma* order, hardy, forming little bushes, 6 to 12 inches high; the stems square, with wing-like angles; the flowers rosy purple, in summer and early autumn. A native of North America, from a considerable distance north of New York to Virginia, and westward to Illinois and the Mississippi, usually in sandy swamps. It is very rare, indeed, to see it well grown in this country, though no plant is better for the bog-garden. The only place I noticed this plant invariably doing well was in Osborn's old Nursery, at Fulham, in beds of moist sandy peat. Deep, sandy, boggy soil, with moisture at all times, will suit it best. Careful division. There are other kinds, natives of Eastern North America, but probably tender, owing to their more southern habitats; whereas this kind, proved to be hardy in our climate, grows as far north as Maine.

RHODIOLA.—Plants of the *Crassula* family, resembling some of the larger Stonecrops. They have fleshy leaves and heads of small flowers, which are not, however, very attractive.

RHODODENDRON.—This noble family of shrubs, which we see so often massed in not very pretty ways, has great claims on the rock gardener, for many of the species are true mountain plants, like those of the Alps of Europe, America, India, and China. In the first part of this book there is a striking instance of the use of the Rhododendron in natural rock ground, and the many parts of our country, where such ground occurs, afford beautiful opportunities for like effects, even when we are dealing with the ordinary stout-growing kinds. But on the mountains of Asia and China, as well as Europe, there are dwarfer and more alpine kinds, which may be used even in the smaller sort of rock-garden. The main precaution to take in all cultivation of Rhododendrons in choice gardens is not to have anything to do with the usual grafting on *ponticum*, because, if we plant in any bold way, and do not continually watch the suckers, the shoots of *R. ponticum* will come up and kill the kinds we want. So always, in rock-gardens at least, insist on having plants from layers, and most kinds are easily increased in this way.

Rhododendron ferrugineum and **hirsutum**, each bearing the name of "Alpine Rose," and which often terminate the woody vegetation on the great mountain chains of Europe, are easily had in our Nurseries, and well suited for the rock-garden in open peat soil. *R. Wilsonianum, myrtifolium, amœnum, hybridum, dauricumatrovirens, Gowenianum, odoratum,* and *Torlonianum,* are also dwarf kinds, which may be used in the bush rock-garden—the last two very sweetly scented. In some soils the alpine kinds are not easily established, owing in part to our often very snowless winters. Place among flat stones in cool ground where possible.

RHODORA CANADENSIS (*Canadian Rhodora*).—An early flowering shrub, allied to the Rhododendron.

Being a native of the swamps of Canada, it is very hardy, thriving in a moist light soil, though it prefers peat. In very early spring it bears clusters of rosy-purple flowers before the leaves unfold. It is a thin bush, 2 to 4 ft. high, and may find a place among the shrubs near the alpine garden.

RHODOTHAMNUS CHAMÆ-CISTUS (*Thyme-Leaved R.*).— A small Rhododendron-like plant, rising scarcely a span high, and thickly clothed with small fleshy leaves, ciliated at the edge, and with exquisite flowers, of purple, bearing three or four together in early summer. This plant is very rarely seen thriving in gardens, and for its successful cultivation requires to be planted in limestone fissures, in peat, loam, and sand in about equal proportions. A native of calcareous rocks in the Tyrol, and one of the most precious of dwarf rock-shrubs for association with tiny alpine bushes.

RODGERSIA PODOPHYLLA.—A handsome leaved plant of the Saxifrage family. The leaves measure 1 ft. or more across, on erect stalks from 2 ft. to 4 ft. high, and are cleft into five broad divisions. They are of a bronzy-green hue, distinct from any other hardy plant. The flowers, on tall branching spikes, are inconspicuous. It likes a peaty soil and a shady situation, and is easily propagated by cutting the stoloniferous root-stock, from one of which as many as twenty plants can be made in one year. It is a native of Japan, and hardy in our climate, and a striking plant among shrubs near the rock-garden.

ROMANZOFFIA SITCHENSIS (*Sitcha Water-leaf*).—A very dwarf alpine plant of the Rockfoil order, a few inches high; white flowers, May. Suitable for select part of the rock-garden.

ROMNEYA COULTERI (*Bush Poppy*).— If, as I urge, we associate the choicer shrubs with the rock-garden, this lovely half-shrubby plant may come in a queen-flower, even among the fairest. It is hardy and enduring on good soils, and grows rapidly with me on rich loam. Where the winter is feared, the best protection for it is a mulch over the roots of some light and porous material. Pine-needles form the best covering, or rough cocoa-nut fibre. A point in starting is to get healthy plants in pots, planting in spring and not disturbing the roots much. It may be increased by cuttings and seed.

ROSA (*Rose*).—Given the shrubby rock-garden we have an opening for wild Roses (or the dwarfest of them) with the mountain shrubs. Not a few Roses are mountain and alpine plants, such as the Pyrenæan, Scotch and Gallica Roses, any of which might well grace the rock-garden. Among natural rocks or banks, any wild Rose might be grown with advantage.

ROSMARINUS OFFICINALIS (*Rosemary*).—A grey aromatic bush of the stony hill-sides of Southern Europe, often grown on cottage walls with us, but I never like it so well as a group on a hot and poor sandy or rocky bank in the southern countries, or in the milder sea-shore gardens.

RUBUS (*Brambles*).—These, which run everywhere in Britain and stop our progress in the woods, are not wholly without interest for the rock-garden, though many of them are too large for it. A few of the smaller kinds, such as *R. arcticus* (which grows a few inches high and bears numerous rosy-pink blossoms), the Cloud-berry, *R. Chamæmorus* (also dwarf and with white blossoms), the Dewberry (*R. Cæsius*), and *R. saxatilis*, are pretty for the rock-garden in moist soil.

RUSCUS (*Butcher's Broom*).—Wiry half-shrubby plants, often neglected, but having some good qualities, even for the rock-garden or shady places near. The hardy kinds may be planted under the shade of trees. Propagate by division of the roots. The *R. aculeatus* (Common Butcher's Broom) is a native of copses and woods, bearing bright red berries where the two sexes are present. This dense, much-branched Evergreen rarely grows more than 2 ft. high, and its thick, white, twining roots strike deep into the ground. The Alexandrian Laurel (*R. racemosus*) is a graceful plant, with glossy dark green leaves, and is one of the best plants for partial shade, and thrives best on free leafy, or peaty soil. *R. Hypophyllum*, a very dwarf kind, and *R. Hylpoglossum* are also in cultivation, and of easy culture in ordinary soil.

RUTA (*Rue*). — *R. albiflora* is a graceful autumn-flowering plant, about 2 ft. high, with leaves resembling those of the common Rue, but more glaucous and finely divided. The small white blossoms, borne in large drooping panicles, last until the frosts. In some localities it is hardy, but should have slight protection in severe weather. It is also known as *Bœnning-hausenia albiflora*, and is a native of Nepaul. Another pretty plant is the Padua Rue (*R. patavina*), 4 to 6 in. high, with small golden-yellow flowers of the same odour as the common Rue, which I saw used with pretty effect in the Belvedere Garden in Vienna.

SAGINA GLABRA (*Lawn Pearl-wort*).—A plant known from being much talked of a few years since as a substitute for lawn-grass, and though it has not answered the expectations formed of it in that way, it is a minute alpine plant, welcome for forming carpets as smooth as velvet, dotted with many small white flowers, the light, fresh green, moss-like carpet being starred with them in early summer. It is useful in forming carpets of the freshest and closest verdure beneath taller, but small and rare bulbs, or other plants, which it may be desired to place to the best advantage. It is multiplied by pulling the tufts into small pieces, and replanting them at a few inches apart; they soon meet and form a carpet. Although it does not generally form a good turf, yet it is possible, by selecting a rather deep, sandy soil, and by keeping it clean and well rolled, to make a close turf of it; but this is rarely worth attempting, except on a small scale, and when it begins to perish in flakes here and there, it should be taken up and replanted.

SALIX (*Willow*). — Among the Willows there are certain dwarf kinds which, though without the floral beauty characteristic of the Alpine flower, may yet be useful here and there in the rock-garden and in the marsh-garden, among them being the Netted Willow (*S. reticulata*), the Thyme-Leaved Willow, the woolley Willow (*S. lanata*), and *S. herbacea*, any other dwarf mountain or Arctic Willow, all of the easiest culture and increase.

SANGUINARIA CANADENSIS (*Bloodroot*).—A distinct North American plant with thick underground stems, from which spring large greyish leaves, cut into wavy or toothed lobes, and full of an orange-red and acrid juice. The stems from 4 to 8 inches high, each bear a solitary and handsome white flower in March. It grows best in moist places and in rich soil, but, like many other plants, it has a dislike to certain soils, and is not always easy to establish; the most likely places being peaty or leafy hollows.

SANTOLINA INCANA (*Hoary S.*).— A small silvery shrub, with numerous branches and narrow leaves, covered with dense white down, the flowers rather small, pale greenish-yellow, growing readily in ordinary soil, and may be useful on the rock-garden. It is considered a variety of the better-known *S. Chamæcyparissus*, the Lavender Cotton. This, and its other variety, *squarrosa*, are suitable for banks, but forming spreading silvery bushes, 2 feet high, in suitable soil, are not suited for intimate association with very dwarf alpine plants.

Other species of *Santolina* are suited for like purposes, *S. pectinata* and *S. viridis*, forming bushes somewhat like the Lavender Cotton. *Santolina alpina* is of more alpine habit, forming dense mats quite close to the ground, from which spring yellow button-like flowers on long slender stems. It grows in any soil, and may be used on the less important parts of the rock-garden. Cuttings of the shrubby species strike readily, and *S. alpina* is easily increased by division.

SAPONARIA (*Soapwort*).—Perennial herbs and alpine plants or annuals belonging to the Pink family.

Saponaria Boissieri is a dwarf plant of quick and free growth, somewhat tufted in character, and spreading out into good-sized plants. It bears freely bright pink flowers.

S. cæspitosa is a neat little alpine perennial from the higher regions of the Pyrenees, flowering in August, but in the lowlands its rose-coloured blossoms appear towards the end of June. It forms rosettes of leaves, the flowers, in a thick cluster, are on short, stout stems. This graceful little plant is valuable for the rock-garden. A sandy soil suits it best, and it endures our winters.

S. lutea, from Savoy and Piedmont, has yellow flowers and a woolly calyx. The leaves are narrow, and not unlike those of the Alpine Catchfly.

S. ocymoides.—A beautiful trailing rock-plant, with prostrate stems and many rosy flowers. It is easily raised from seed or from cuttings, thrives in almost any soil, and is one of the best plants we have for clothing the arid spots, particularly where a drooping plant is desired. Although it grows freely in poor soil when it is planted with the view of allowing it to fall freely over the face of the rock, it will do much better by giving it a deep, loamy soil.

SARRACENIA (*Pitcher Plant*).—Growing naturally in turfy bogs in North America and Canada, these very curious perennials, with hollow pitcher-shaped leaves, are hardy so far as temperature is concerned, and we have seen the Trumpet Leaf (*S. flava*), and the Huntsman's Cup (*S. purpurea*), growing on spongy peat and sphagnum in Great Britain and Ireland. One point very essential to their success in the open air in this country is good shelter. In North America these and many other beautiful bog-plants are sheltered all through the winter by deep snow, which alike preserves leaves and root from the sudden extremes so often fatal to their leafage here at home during winter and early spring. *S. purpurea* and *S. flava* may be planted out in May or June on sods of peat or fibrous loam, either in a bog-bed or on the sunny margins of either pond or stream, and if these succeed, other kinds may with more confidence be tried. At Glasnevin, *S. purpurea* has lived outside, in a spongy bog near the ornamental water there, for many years, and also at Newry and elsewhere. All through the summer full sunshine is an advantage, and there should be plenty of moisture around the mossy sod on which it is planted. On the approach of winter a wire cylinder may be placed round the plants, and on the advent of frost a top covering of dry leaves or bracken fern may be placed lightly around the leaves, so as to protect them, to check evaporation, and to prevent harm from bright early morning sunshine after dry and frosty nights. With some simple attention and shelter of this kind from November to March, these plants may be grown in the open air with success, and prove of much interest.

SAXIFRAGA (*Rockfoil*). — Dwarf tufted perennial herbs of the Alps and higher mountains, frequent in northern and cold countries. Many of them are quite hardy and give with simple culture, beautiful effects, even in the neighbourhood of smoky towns. They fall into different sections or groups, offering a striking diversity of colour, even when out of flower, in their delicate foliage often freshest in autumn and winter.

In the Arctic Circle, in the highest Alpine regions, on the arid mountains of Southern and Eastern Europe, and Northern Africa, and throughout Northern Asia, they are found in many interesting forms. For the purposes of cultivation some rough division is convenient, as Saxifrages are very different in aspect and uses. There is the Mossy or *Hypnoides* section, of which there are many kinds, and their Moss-like tufts of foliage, so freshly green, especially in autumn and winter, when most plants decay, and their countless white flowers in spring make them precious. They are admirable for the fresh green hue with which they clothe rocks and banks in winter. They are indeed the most valuable winter "greens," in the Alpine flora.

Next to these we may place the silvery group. These have their greyish leathery leaves margined with dots of white, so as to give to the whole a silvery character. This group is represented by such kinds as *S. Aizoon* and the great pyramidal-flowering *S. Cotyledon* of the Alps. Considering the freedom with which they grow in all cool climates, even on level ground, and their beauty of flower and foliage, they are perhaps the most precious group of Alpine flowers we possess, and all can grow them. The London Pride section is another. The plants of this section thrive with ordinary care, in lowland gardens, and soon naturalise themselves in lowland copses. But the most brilliant, so far as flower is concerned, are found in the purple Saxifrage (*S. oppositifolia*) group and its near allies. Here we have tufts of splendid colour in spring with perfect hardiness. The large leathery-leaved group, of which the Siberian *S. Crassifolia* is best known, is important; they thrive in ordinary soil and on the level ground. Such of the smaller and rarer alpine species as require any particular attention should be planted in moist, sandy loam, mingled with grit and broken stone, the soil made firm.

Very dwarf and rather slow-growing kinds, like *S. cæsia* and *S. aretioides*, should be surrounded by half-buried pieces of stone, to prevent their being trampled on or overrun. Stone will also help to preserve the ground in a moist healthy condition in the dry season, when the plants are most likely to suffer. Very dry winds in spring sometimes have a bad effect when such precautions are not taken. The broad-leaved Indian Rockfoils (*Megasea*) are among the most easily grown, increased, and enduring of hardy plants. Where we seek for evergreen effects in winter, there is nothing to equal them, and their flowers have much beauty in spring.

In this large family, as in others, a first consideration should be whether we look at the plants from the artistic or the collector's point of view. If we wish to get good effects, I say the artistic way is the right one. By treating the rock-garden as a book or herbarium, we cannot get the broad and simple effects that are necessary for a good result. We want the charm of the most distinct things, but for effect a few kinds from each group will give us a better result than a large number. The dotting of a great number of species is against good effect, but here, as in all cases, individual taste should have its way, and it may be interesting to study a section by fully representing it, and to make most of the kinds we prefer.

The Rockfoils are a numerous family, with so many forms that it would take a book to describe them, as Mr Correvon of Geneva has described them fully in various articles written for the *Garden* in 1891. I once saw nearly seventy kinds of the mossy Saxifrages in the late Mr Borrer's garden at Henfield, in Sussex; but as regards effect, half a dozen of these will give us all we require.

The great Indian Rockfoils, syn. *Megasea*, have been in our gardens for many years, but in not one place out of twenty do we ever see them made a right use of; they are thrown into borders without thought as to their habits, often as single plants, and are soon overshadowed by other things; and in such ways we never get any expression of their beauty. Yet, if we took a little trouble, and grouped them in effective ways, they would go on for years, giving fine evergreen foliage at all times of the year, and, in the case of some, showy flowers on tall stems. Half the trouble that a gardener gives every year to some evanescent plant that will only show for a few weeks in summer, if given to the placing of these properly, would afford us a good result for years. In addition to the wild kinds, a number of fine forms have been raised in gardens of late years. Some thought should be given to the placing of these things, their mountain character telling us that they ought to be in open banks, borders, or bluffy places exposed to the sun, and not buried among heaps of tall herbaceous vegetation. They are easily grown and propagated, and a little thought in placing them in sufficiently visible masses is the only thing they call for; the fact that they will endure and thrive under almost any conditions should not prevent us from showing how good they are in effect when held together, either as carpets, bold edgings, or large picturesque groups on banks or rocks. The following is a selection of the best of the kinds in cultivation.

Saxifraga aizoides (*Yellow Mountain Rockfoil*).—A native plant, abundant in

Scotland, the north of England, and some parts of Ireland, in wet places, by the sides of mountain rills, and often descending along their course, into the low country, bearing at the end of summer or autumn bright yellow flowers, half an inch across, and dotted with red towards the base. Although a moisture-loving mountain plant, it is quite easy to grow in lowland gardens, doing best in moist ground. Wherever a small streamlet is introduced to the rock-garden or its neighbourhood, it may be planted so as to form spreading masses, as it does on its native mountains. Division, or by seed. When the leaves are sparsely ciliated, it is, according to Mr Syme, the *S. autumnalis* of Linnæus.

Saxifraga aizoon (*Aizoon Rockfoil*).—Not a showy kind, having a greenish-white bloom, but it spangles over many a low mountain-crest and high alp-flank in Europe and America with its silvery rosettes, and in our gardens these form firm and roundish silvery tufts in any common soil. Plants of it established two or three years form grey-silvery tufts, a foot or more in diameter, and about 6 inches high. As to its culture, nothing can be easier; it grows as freely as any native plant, and best when exposed to the full sun. Easily increased by division. There are several varieties.

S. Andrewsii (*Andrew's Rockfoil*).—This British plant is considered by some botanists to be a garden hybrid, and with pretty good reason, judging by the leaves and flowers; but nothing more has been ascertained about its history. Mr Andrews found it first in Ireland, but it has not since been discovered. Among the green-leaved kinds there is no better. Its flowers are large, but I never could see any good seed on it. The leaves are long, firm in texture, and with a membranous margin; the prettily spotted flowers being larger than those of *S. umbrosa*, and the petals dotted with red, which, with other slight characters, points to the probability of its being a hybrid between a London Pride and one of the Continental group of encrusted Saxifrages. It does quite freely on any soil, merely requiring to be replanted occasionally when it spreads into very large tufts.

Saxifraga aretioides (*Aretia Rockfoil*).—A gem of the encrusted section, forming cushions of little silvery rosettes, almost as small and dense as those of *Androsace helvetica*, and about half an inch high. It has rich yellow flowers in April, on stems a little more than an inch high. The stems and stem-leaves are densely clothed with short glandular hairs like those of a *Drosera*. It is not difficult to grow, but requires a moist and well-drained soil, and being so dwarf, must be guarded from overrunning by coarser neighbours. Pyrenees; increased by seed and careful division.

S. aspera (*Rough Rockfoil*).—A small grey, tufted, prostrate plant, with ciliated leaves, with few flowers, rather large, of a dull white colour, on stems about 3 inches high. *S. bryoides* is considered a variety of this, and forms a densely tufted diminutive plant, with pale yellow flowers, the rosettes of leaves being almost globular, and the plant not forming stolons or runners like the preceding. Both are natives of the Pyrenees; *S. bryoides* in the most elevated regions. Both are easy of cultivation, growing freely in the open air, even in London, but rarely flowering there.

S. biflora (*Two-flowered Rockfoil*).—A beautiful dwarf kind, allied to the British species, *S. oppositifolia*, but larger, and distinguished by producing two or three flowers together, and by having its leaves thinly scattered, and not packed on the stems like those of that species. It is also a much larger plant, and has larger flowers, rose-coloured at first, changing to violet. I found it in abundance on fields of grit and shattered rock, in the neighbourhood of glaciers on the high Alps, in company with *Campanula cenisia*; and just without the margins of the vast fields of snow, under which, even in June, lay numberless plants waiting for an opportunity to open when the snow had thawed. It grew entirely in loose grit, so that, with a little care, masses of the branched imbedded stems and long fine roots could be taken up, entire.

It grows freely in gritty or sandy soil, in well-drained positions in rich light loam, may be increased by division, cuttings, or seed.

Saxifraga Burseriana (*Early Rockfoil*). —This lovely early-flowering Rockfoil is a native of the snowy regions of Europe and of Central and Northern Asia. It is dwarf, and forms spreading tufts of glaucous or greyish-green foliage. The flowers are large, pure white, with yellow anthers, and borne singly or two together on a bright purplish rose-coloured stem in January and February. It soon forms good-sized tufts, preferring a dry, sunny situation and calcareous soil. There are two or three distinct forms of this species which differ chiefly in habit of growth, one being much more tufted than the others.

S. cæsia (*Silvery Rockfoil*). — This resembles an *Androsace* in the dwarfness of its tufts. I have met with it on the Alps, in minute tufts, staining the rocks and stones like a silvery moss, and on level ground, where it had some depth of soil, spreading into little cushions from 2 to 6 inches across. It bears pretty white flowers, about the third of an inch in diameter, on thread-like smooth stems, 1 to 3 inches high. A native of the high Alps and Pyrenees, it thrives in our gardens in firm sandy soil, fully exposed, and kept moist in summer. It may be also grown well in pots or pans in cold frames near the glass; but, being very minute, no matter where it is placed, the first consideration should be to keep it distinct from all coarse neighbours, and even the smallest weeds will injure it if allowed to grow. Flowers in summer, and is increased by seeds or careful division.

S. ceratophylla (*Horn-leaved Rockfoil*).—A fine species of the mossy section, with dark highly-divided leaves, stiff and smooth, with horny points; the flowers pure white, and borne in loose panicles in early summer, the calyces and stamens covered with clammy juice. It quickly forms strong tufts in any good garden soil, and is well adapted for covering rocky ground of any description, either as wide level tufts on the flat portions or pendent sheets from the brows of rocks. Seed or division.

Saxifraga cordifolia. — (*Great Heart-Leaved Rockfoil*). — Entirely different in aspect to the ordinary dwarf section of Saxifrages, with very ample leaves, roundish-heart-shaped, on long and thick stalks, toothed; flowers a clear rose, arranged in dense masses, half concealed among the great leaves in early spring. *S. crassifolia* is allied to this. They often thrive in any soil, and are hardy; but it is well to encourage their early-flowering habit by placing them in sunny positions, where the fine flowers may be induced to open well. They are perhaps more worthy of association with the larger spring flowers and with herbaceous plants than with alpine plants. They may also be used with fine effect on rough rock, or on rocky margins to streams or water, their fine, evergreen, glossy foliage being quite distinct. They may, in fact, be called fine-leaved plants of the rocks. A native of Siberian mountains. *S. ligulata* (*Megasea ciliata*) is a somewhat tender species, and only succeeds out of doors in mild and warm parts of this country. Some good varieties of these great-leaved Rockfoils have been raised of recent years.

S. cotyledon (*Pyramidal Rockfoil*).— This embellishes, with its great silvery rosettes and pyramids of white flowers, many parts of the mountain ranges of Europe, from the Pyrenees to Lapland, and is easily known by its rather broad leaves, margined with encrusted pores and its handsome bloom. The rosettes of the pyramidal Saxifrage differ a good deal in size, and, when grown in tufts, they are for the most part much smaller, from being crowded than from single rosettes. The flower-stem varies from 6 to 30 inches high, and about London, in common soil, will often attain a height of 20 inches, and in cultivation usually attains a greater size than on its native rocks; though in rich soil, at the base of rocky slopes in a Piedmontese valley, I have seen single rosettes as large as I have ever seen them in gardens. The plant is hardy, and second to none as an ornament of the rock-garden, thriving in common soil. Nothing can be easier to propagate by division, or cultivate without any particular attention. It is sometimes known as *S. Pyramidalis*, though some consider this at least a variety, having a more erect habit, narrower leaves, and somewhat larger flowers.

Saxifraga cymbalaria (*Golden Rockfoil*). —Quite distinct in aspect from any of the family, and one of the most useful of all, being a continuous bloomer. I have had little tufts of it, which, in early spring, formed masses of bright yellow flowers set on light green, glossy, small ivy-like leaves, the whole not more than 3 inches high. These, instead of falling into the sere and yellow leaf, and fading away into seediness, kept still growing taller, still rising, and still keeping the same little rounded pyramid of golden flowers until autumn, when

Saxifraga geum (*Kidney-Leaved London Pride*).—Like the London Pride in habit and flowers, but with the leaves roundish, heart-shaped at the base, on long stalks, and with scattered hairs on the surfaces; flowers about a quarter of an inch across, and usually with reddish spots. A native of various parts of Europe, useful for the same purposes and cultivated with the same ease as the London Pride; will grow freely in woods or borders, particularly in moist districts. *Saxifraga hirsuta* comes near this, and is probably a variety.

S. granulata (*Meadow Rockfoil*).—A lowland plant, with several small scaly bulbs in a crown at the root, and common in meadows and banks in England,

Saxifraga cordifolia (*Broad-Leaved Rockfoil*).

they were about 12 inches high. It is an annual or biennial plant, which sows itself abundantly, is useful for moist spots, growing freely on the level ground.

S. diapensiodes.—One of the best of the dwarf Rockfoils, and also one of the smallest. I have grown it very well in an open bed in London, and it would flourish equally well everywhere if kept free from weeds, and in a well-exposed spot; the soil should be very firm and well-drained, though kept moist in summer. The flowers are of a good white, three to five on a stem, rarely exceeding 2 inches high; the leaves packed into such dense cylindrical rosettes that old plants feel quite hard to the hand. A native of the Alps of Switzerland, Dauphiny, and the Pyrenees.

with numerous white flowers, $\frac{3}{4}$ inch across. I should not name it here, were it not for its handsome double form, *S. granulata fl. pl.*, which is often grown in cottage gardens in Surrey. It is very useful in the spring garden as a border-plant, or on rougher parts of rockwood. Mr Bentham considers that the small bulb-bearing *S. cernua* of Ben Lawers may be a variety of the Meadow Saxifrage. As a garden-plant, *S. cernua*, however, is a mere curiosity, though it may be acceptable in botanical collections.

S. hirculus (*Yellow Marsh Rockfoil*).—A remarkable species, with a bright yellow flower on each stem, or sometimes two or three, $\frac{3}{4}$ inch across, and quite different in aspect from any other cultivated kind. A native of wet moors in various parts of England, not difficult to cultivate in moist soil, and thriving best under conditions as near as possible to those of the places where it is found wild. It is best suited for a moist spot near a streamlet of the rock-garden, or for the bog-garden.

Saxifraga hypnoides (*Mossy Saxifrage*). —A very variable plant in its stems, leaves, and flowers, but usually forming mossy tufts of the freshest green, abundant the healthiest tufts in shade, and flowering in early summer. Nothing can be easier to grow or increase by division. Under this species may be grouped *S. hirta*, *S. affinis*, *S. incurvifolia*, *S. platypetala*, and *S. decipiens*, all showing differences which some think sufficient to mark them as species. They all thrive with the same freedom as the Mossy Saxifrage, suffering only from drought or very drying winds.

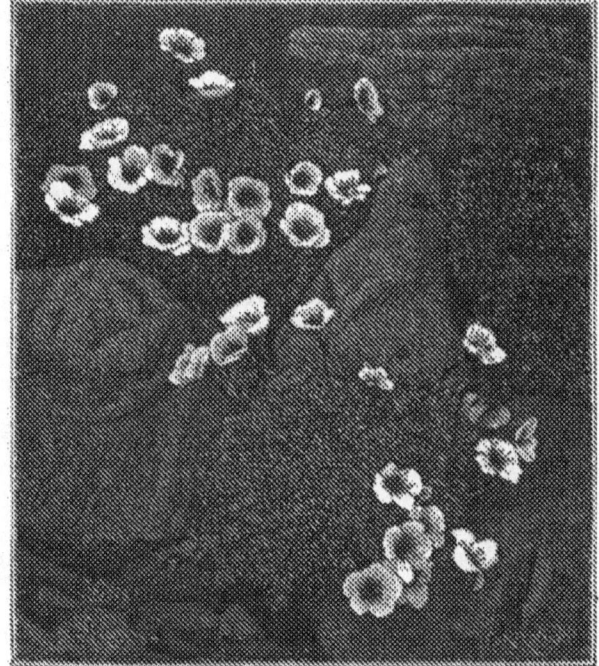

Saxifraga Juniperina (*The Juniper-Leaved Saxifrage*).

on the mountains of Great Britain and Ireland, and common in gardens. In cultivation it attains greater vigour than in a wild state, and no plant is more useful for forming carpets of the most refreshing green in winter and almost in any soil. It thrives either on raised or level ground, in half-shady places or fully exposed to the sun, forming

Saxifraga Juniperina (*Juniper Rock-foil*).—One of the most distinct kinds in cultivation, having spine-pointed leaves, densely set in cushioned masses, looking, if one may so speak, like Juniper-bushes compressed into the size of small round pin-cushions. The flowers are yellow, arranged in spikes on a leafy stem, and appear in summer. It thrives in moist sandy, firm soil, and is well worthy of a place in the rock-garden. A native of the Caucasus. Seed and careful division.

Saxifraga longifolia (*Queen Rockfoil*). —The single rosettes of this are often 6, 7, and 8 inches in diameter. I have indeed measured one more than a foot in diameter. It may well be termed the Queen of the silvery section of Saxifrages, and is so beautifully marked that it is attractive at all seasons, while in early summer it pushes up foxbrush-like columns of flowers from a foot to 2 feet long, the stem covered with short, stiff, gland-tipped hairs, and bearing many pure white flowers.

It is a native of the higher parts of the Pyrenees: hardy in this country; not difficult of culture, and may be grown in various ways. In some perpendicular chink in the face of a rock into which it can root deeply, it is very striking when the long outer leaves of the rosette spread away from the densely packed centre. It may also be grown on the face of an old wall, beginning with a very small plant, which should be carefully packed into a chink with a little soil. Here the stiff leaves will, when they roll out, adhere firmly to the wall, eventually forming a large silver star on its surface. It will thrive on a raised bed, surrounded by a few stones to prevent evaporation and to guard it from injury. It is propagated by seeds, which it produces freely. In gathering them it should be observed that they ripen gradually from the bottom of the stem upwards, so that the seed-vessels there should be cut off first, leaving the unripe capsules to mature, and visiting the plant every day or two to collect them as they ripen successively.

S. lingulata is by some authors united with the preceding, from which it chiefly differs by having smaller flowers, by the leaves and stems being smooth and not glandular, by its shorter stems, and by the leaves in the rosette being shorter and very much fewer in number than in the Long-Leaved Saxifrage. It is also a charming rock-plant, and will succeed with the same treatment and in the same positions as the preceding. *S. crustata* is considered a small variety of the long-leaved Saxifrage with the encrusted pores thickly set along the margins; being several times smaller, it will require more care in planting, and to be associated with dwarfer plants.

Saxifraga Lantoscana (*Foxbrush Rockfoil*).—A beautiful species of the encrusted-leaved section, and a native of Val Lantosque in the Maritime Alps. It reminds one of *S. cotyledon*, but is smaller, the leaves narrower and more crowded in the rosette, and the flower-spike, which is not borne erect, but slightly drooping, is more densely furnished with white flowers. It should be grown in a well exposed position, in a gritty soil well-drained. It remains long in flower, and is one of the best rock-garden plants.

S. Maweana (*Maw's Rockfoil*) is a handsome species of the *cæspitosa* section, larger than any other as regards both foliage and flowers. The latter, about the size of a shilling, form dense white masses in early summer. After flowering, this species forms buds on the stems, which remain dormant till the following spring. Similar, but finer, is a new kind called *S. Wallacei*, which is far more robust, and far earlier, and freer as regards flowering, but which does not develop buds during summer. It is a good plant for the border or the rock-garden.

S. mutata.—A yellow-flowered species, bearing considerable likeness to *S. lingulata* and having the flower - panicle about 18 ins. high. It is rare in cultivation, owing to the fact that it not infrequently exhausts all its vigour in producing blooms, and it rarely matures seeds in this country; and, further, it does not produce offsets, like most of this section. It is a native of the Alps, but limited in its distribution. An allied species, *S. florulenta*, is a beautiful plant of the Maritime Alps, difficult of cultivation in this country.

S. oppositifolia (*Purple Rockfoil*).—A bright little mountaineer, distinct in colour and in habit. The moment the snow melts, its tiny herbage

ALPINE FLOWERS FOR GARDENS

glows into solid sheets of purplish rose-colour; the flowers solitary, on short erect little stems, and often hiding the leaves, which are small, and densely crowded. In a wild state on the higher mountains of Britain and the Continent, in which it has to submit to the struggle for life, it usually forms rather straggling little tufts; but on exposed parts of the rock-garden, in deep and moist loam, it forms rounded cushions fringing over the sides of rocks. Propagated by division, and flowering in early spring. Old plants should be divided. There are the following varieties in cultivation: *S. opp. major*, rosy pink, large; *S. opp. pallida*, pale pink, large; *S. opp. alba*, white.

Saxifraga peltata (*Great Californian Rockfoil*).—A remarkably distinct species, found on the banks of streams in California, well known and a Rockfoil of large size, the hairy flower-stems, which are of an almost purplish-red colour, sometimes attaining a height of more than 3 feet, and terminating in a large umbel of white flowers, with bright rose-coloured anthers. The leaves resemble an inverted parasol in shape, and are large and dark green. They do not appear until after the plant comes into flower. This kind should be grown in a rich, deep, spongy soil, also in a half-shaded position, sheltered from cold, drying winds. It is multiplied by division of the rhizomes and also by seed, and is effective in the dark parts of the bog-garden.

S. retusa (*Purple-Leaved Rockfoil*).—A purplish species, closely allied to our own *S. oppositifolia*, but, in addition to the different character of the leaves, distinguished by the flowers having distinct stalks, and being borne two or three together on their little branches. The small, opposite, leathery leaves are closely packed in four ranks on the stems, which form dense prostrate tufts. A native of the Alps and Pyrenees, flowering in early summer, may be cultivated in the same way as *S. oppositifolia*, and well merits a place in the rock-garden.

S. Rocheliana (*Rochel's Rockfoil*).—A compact and dwarf kind, forming dense silvery rosettes of tongue-shaped white-margined leaves, and with large white flowers on sturdy little stems in spring. I know no more exquisite plant for the rock-garden, or for small rocky or raised borders. Any free, good, moist, loamy soil will suit it, and I have seen it thriving very well on borders in London. It should be exposed to the full sun, and associated with the choicest alpine plants. A native of Austria; increased by seeds or careful division.

Saxifraga sancta.—A native of Mount Athos, at an altitude of 6000 feet. A dwarf species, forming closely-set tufts of foliage, composed of numerous leafy branches of a dull green colour, the leaves pointed, flowers bright yellow, in panicles of two to five blooms.

S. sarmentosa (*Creeping Rockfoil*).—A well-known old plant, with roundish leaves, mottled above, red beneath, with numbers of creeping, long, and slender runners, producing young plants strawberry fashion. Striking in leaf, it is also pretty in bloom, and growing freely in the dry air of a sitting-room, may be seen suspended in cottage windows. It perhaps is most at home running free on banks or rocks, in the cool greenhouse or conservatory; however, it lives in the open air in mild parts of England, and, where this is the case, may be used in graceful association with Ferns and other creeping plants. A native of China, flowering in summer. Closely allied to *S. sarmentosa* is the delicate dodder-like Saxifrage, *S. cuscutæformis*, so called from having thread-like runners like the stems of a dodder, and distinguished by having much smaller leaves, and the petals more equal in size than those of *sarmentosa*, in which the two outer ones are much larger than the others. It will serve for the same purposes as the Creeping Saxifrage, but, being much more delicate and fragile in habit, will require a little more care. The plants grown in gardens as *S. japonica* and *S. tricolor* are considered varieties of the Creeping Saxifrage.

S. tenella.—A very handsome prostrate plant, forming tufts of delicate fine-leaved branches, 4 or 5 inches high, which root as they grow. The flowers, which appear in summer, are numerous, whitish-yellow, arranged in a loose panicle. Similar in growth are *S. aspera, S. bryoides, S. sedoides, S. Seguieri, S. Stelleriana,* and *S. tricuspidata,* all of which are suitable for clothing the bare parts of the rock-garden and slopes, but require moist soil and cool positions. Division in spring or the end of summer.

Saxifraga umbrosa (*London Pride*).—This much cultivated plant grows abundantly on the mountains round Killarney, though it was much grown in our gardens before it was recognised as a native of Ireland. It is needless to describe the appearance of such a familiar plant. It is useful in shady places, fringes of cascades, &c. There are several varieties, as, for example, *S. punctata* and *Serratifolia,* which are distinct enough when grown side by side, and submit to the same culture.

It is believed that the preceding are among those best worth growing. The following is a list of the other species or reputed species believed to be in cultivation now in this country. Those most worthy of culture are marked by an asterisk.

S. adscendens
ajugæfolia
ambigua
androsacea
aquatica
atropurpurea
*Bucklandii
bulbifera
calcarata
*capillaris
condensata
*contraversa
cochleata
*crustata
cuneifolia
*daurica
elatior
elongella
erosa
exarata
flavescens
geranioides
*Gibraltarica
glacialis

S. globifera
Gmelini
*Guthrieana
hieraciifolia
*Icelandica
infundibulum
*intacta
*intermedia
lætevirens
lævigata
*lævis
leptophylla
*marginata
*media
Mollyi
multicaulis
*muscoide
*nervosa
nivalis
Ohioensis
Orientalis
*palmata
Parnassica
*pectinata

S. pedata
pedatifida
petræa
planifolia
pulchella
purpurascens
pygmæa
*recta
recurva
reniformis
Rhei
*rosularis
rotundifolia
rupestris
Schraderi
Sibirica

S. spathulata
Sponhemica
*Stansfieldii
stellaris
stenophylla
*Sternbergii
*tenella
thysanodes
tricuspidata
trifida
trifurcata
trilobata
villosa
virginiensis
Webbiana

SCABIOSA (*Pincushion Flower*).—Annual, biennial, and perennial plants, some dwarf and pretty for the rock-garden.

Scabiosa caucasica (*Caucasian Scabious*).—A handsome plant, flowering from early summer to late autumn, a true perennial on warm soils, but often perishing on cool soils. It forms dense tufts, which yield many blue flower-heads, each usually from 3 to 4 inches in diameter, on long foot-stalks. There is a white variety. Caucasus. Division and seed.

S. graminifolia (*Grass-leaved S.*).—A graceful *Scabious* about a foot high, with pale blue flowers and silvery white leaves; it is very useful for the rock-garden. Southern Europe. June to October. Division and seed.

S. pterocephala (*Wing-headed S.*) is a very dwarf-tufted hardy perennial, rarely exceeding 4 inches or 6 inches in height, even when in flower; flower-heads pale purple in summer. Greece. Division. Syns., *S. Parnassi* and *Pterocephalus Parnassi.*

S. Webbiana is another useful species for the rock-garden or border, forming neat little masses of hoary leaves. Its creamy yellow flowers, borne on long stalks, are pretty from July to August. Division.

All the rock *Scabious* are best in light and well-drained soils.

SCHIVERECKIA PODOLICA.—This small hardy alpine of the Crucifer family is nearly allied to *Alyssum*. It has hoary foliage, and produces, in early summer, a profusion of small white blossoms. It is suited for the rock-garden or the margins of borders, and will grow well in any ordinary soil, but is not of the first merit. South Russia.

SCHIZOCODON SOLDANEL-LOIDES.—The introduction of this pretty mountain plant is due to Captain Torrens, who, in 1891, found the plants growing beside sulphur springs in the mountains of Japan, and, after carrying them hundreds of miles, succeeded at last in bringing home three or four living plants. The flowers of the *Schizocodon* are like those of a large *Soldanella*, prettily fringed, deep rose in the centre, passing into blush or almost white towards the edges, and deserves a good place in the rock-garden, in moist gritty soil.

SCILLA.—Beautiful early flowering bulbous plants, charming in colour, and hardy, and so free that they do not need the comforts of the rock-garden, but the colour is so good and the habit so dwarf, that they may be often used with good effect to come through groups of dwarf rock plants, such as the mossy Rockfoils and the Sandworts. Only the dwarfer kinds, however, are fitted for this purpose, some kinds being too vigorous, and these are omitted here.

Scilla amœna (*Tyrolese Squill*).—A distinct, early-spring flowering kind, opening soon after *S. sibirica*, and readily known from any of its relatives by the large yellowish ovary in the centre of the dark indigo-blue flowers. The leaves, usually about ½ inch across, attain a height of about 1 foot, and are easily injured by cold or wind, so that a sheltered position is that best suited to its wants. Tyrol;

increased from seeds or by separation of the bulbs.

Scilla bifolia (*Early Squill*).—A precious kind, bearing in the dawn of spring, indeed often in winter, masses of dark blue flowers, four to six on a spike, and forming handsome tufts from 6 to 10 inches high, according to the soil and the warmth and shelter of the spot. It thrives well in almost any position, in ordinary garden soil, the lighter the better. Although it blooms earlier than *S. sibirica*, it does not withstand cold wintry and spring rains and storms nearly so well as that species, and therefore it would be well to place some tufts of it in warm sunny spots, either on the rock-garden or sheltered borders. Southern and Central Europe. This species varies very much, and, in consequence, has gone under many names; the best form being *taurica*. The name *S. præcox*, which occurs so often in gardens, and in Nurserymen's Catalogues, does not really belong to a distinct species, and, when best applied, refers to the variety of *S. bifolia*, which usually flowers somewhat earlier than the common form.

S. Italica (*Italian Squill*).—A native not only of Italy but of Southern France and Southern Europe generally. This Squill, with its pale blue flowers, intensely blue stamens, and fragrance, is one of the most distinct, from 5 to 10 inches high, the leaves somewhat shorter; the flowers small, spreading in short racemes, in May. It is perfectly hardy, living in almost any soil, but thriving best in sandy and warm ones. Increased by division, which had better be performed only every three or four years, when the bulbs should be planted in fresh positions. It is worthy of a sheltered sunny spot, particularly as it does not seem to thrive so freely in this country as some of the other kinds.

S. Sibirica (*Siberian Squill*).—A brilliant early flower, perfectly hardy in this country, and, like most other bulbs, thrives best in a good sandy loam. It is needless to disturb the tufts except every two or three years for the sake of dividing them when they grow vigorously. It comes in flower in early spring a little later than *S. bifolia*, but withstands the storms better

than that plant, and remains much longer in bloom.
Of other cultivated Squills, the British ones, S. *verna* and S. *autumnalis*, are worthy of cultivation in collections; the plant usually sold by the Dutch and by our seedsmen as S. *hyacinthoides* is generally S. *campanulata*, and occasionally S. *patula*. The true S. *hyacinthoides* of Southern Europe is scarcely worthy of cultivation; S. *cernua* is not sufficiently distinct from S. *patula*, and one or two southern species allied to S. *peruviana* have not been proved sufficiently hardy for general cultivation.

SCIRPUS (*Bulrush*).—Sedge-like plants, useful for fringing the margins of ponds, which too often present a bare hard line. There are native species that might be transplanted, and the best are S. *triqueter*, S. *atro-virens*, and S. *lacustris*. The true Bulrush is 3 to 8 feet high, and is effective on the margins of ponds or streams, associated with other tall aquatic plants.

SCUTELLARIA (*Skull-cap*). — Perennials of the Sage order, some of interest for the rock-garden. All the kinds may be grown in open loam, the low-growing kinds submit readily to division of the root-stock, and, if need be, the plants are increased by cuttings of the young shoots, by seeds.
Scutellaria alpina (*Alpine Scull-cap*).—A spreading plant, vigorous but neat in habit, and pretty in flower. The pubescent stems are prostrate, but so abundantly produced that they rise into a full round tuft, a foot high or more in the centre, and falling low to the sides; the flowers in terminal heads, purplish, or with the lower lip white or yellow. The form with the upper lip purplish, and lower pure white, is pretty. The variety *lutea* (S. *lupulina*) is an ornamental kind, with yellow flowers. Increased by division, and flowering freely in summer. Alps of Europe.
S. macrantha.—A native of Eastern Asia, has purplish-blue flowers, the blossoms 1¼ inch long. The plant attains to a foot or more high, and may figure in the rock-garden among the more free-growing plants. The plant possesses a firm, woody root-stock, and is hardy.
Scutellaria indica is of dwarf growth, with creeping stems, the flowers blue or bluish lilac, and, though small when compared with those of *macrantha*, it is still worth growing among rock plants.
Other kinds in cultivation are *Orientalis*, *altaica*, *parvula*, *grandiflora*, though, for the most part, these are not frequently seen beyond the limits of botanic gardens.

Scutellaria indica.

SEDUM (*Stonecrop*).—Usually dwarf spreading rock perennials, with thick succulent leaves, which enable them to endure drought in the most arid places. They are often pretty in effect in Nature, but, owing to the dotting and labelling system in gardens, we lose more than half their beauty. In a great number of species are many similar in effect, and no need, therefore, to grow all, as they are not all equally valuable from a garden point of view. In the poorer parts of the rock-garden they are useful, and if we cannot find room for them in it, they do very well on the gravel paths near. They are, perhaps, of all plants, the easiest to cultivate and increase, the smaller species being protected from coarse-growing plants, and so placed that they will not be overrun.

Sedum acre (*Stonecrop*).—Growing on walls, thatched houses, rocks, and sandy places in almost all parts of Britain, this little plant, with its small, thick, bright green leaves and brilliant yellow flowers, is as well known as the common Houseleek. Sheets of it in bloom look gay, and it may well be used with dwarf alpine plants in forming carpets of living mosaic-work in gardens. The fact that it runs wild on comparatively new brick walls near London does away with the necessity of speaking of its cultivation or propagation. There is a variegated or yellow-tipped variety, *S. acre variegatum*; the tips of the shoots of this become of a yellow hue in early spring, so that the tufts look showy at that season.

S. album (*White Stonecrop*).—A British plant, with crowded fleshy leaves of a brownish green, and in summer a profusion of white or pinkish flowers in elegant corymbs. Like the common Stonecrop, this occurs on old roofs and rocky places in many parts of Europe, and may be cultivated with the same facility. It is worthy of a place on walls or ruins, in places where it does not occur naturally, and also on the margins of the pathways or the less important surfaces of the rock-garden.

S. anacampseros (*Evergreen Orpine*).— A species easily recognised by its very obtuse and entire glaucous leaves, closely arranged in pyramidal rosettes on the prostrate branches that do not flower. The rose-coloured flowers are in corymbs, not very ornamental, but the distinct aspect of the plant will secure it a place on the rock-garden, or among very dwarf border-plants. A native of the Alps, Pyrenees, and mountains of Dauphiny, flowering in summer, easily propagated by division, and thriving in any soil.

S. brevifolium (*Mealy Stonecrop*).—One of the most fragile of alpine plants, with pinkish, mealy leaves. A native of the Southern Pyrenees and Corsica, in dry places, it is somewhat too delicate for general planting in the open air; but it may be grown on sunny rocks. *S. farino-* *sum* resembles this, but, so far as my experience goes, it is tender.

Sedum dasyphyllum (*Stonecrop*).—A pretty species, glaucous, or bluish; its leaves smooth, very thick and fat, and very densely packed; flowers of a dull white, tinged with rose, the neat habit of the plant, when not in flower, will always make it a favourite in collections of dwarf plants. It occurs abundantly on rocks, old walls, and humid stony places, in Southern and South-Western Europe, and is found in some places in the south of England. Although hardy on walls and rocks, it has not the vigour and constitution of many of the other Stonecrops, and it is desirable to establish it on an old wall or dry stony part of the rock-garden, so as to secure a stock in case the plant perishes in winter on low ground.

S. Ewersii (*Ewers's Stonecrop*).—A distinct, and diminutive species, with smooth, broad leaves, and purplish flowers in corymbs, the whole plant of a pleasing silvery hue and rather delicate appearance, but hardy, easily increased by division, and flowering in summer. Altai Mountains; of easy culture and increase by division, at any season.

S. glaucum (*Glaucous Stonecrop*).—A minute kind, greyish, forming dense spreading tufts, densely clothed with fat leaves and rather inconspicuous flowers. The neat habit of the plant has made it popular in gardens of late years as a minute surfacing plant. On the rock-garden it may be used in any spot that is to spare, either to form a turf under other plants or for its own sake. Various other *Sedums* are very nearly allied to this, and all are probably but forms of one kind. Hungary.

S. kamtschaticum (*Orange Stonecrop*).— A broad-leaved kind, with dark orange-yellow flowers. It is a prostrate plant, hardy, succeeding in almost any soil, and flowering in summer. Highly suitable for the rougher parts of the rock-garden, where it will take care of itself.

S. populifolium (*Shrubby Stonecrop*).— Distinct from all its race, and forming a small, much-branched shrub, from 6 to 10 inches high, with flat leaves, and whitish flowers with red anthers. Not an ornamental plant, but being so different in habit to the other members of the family, it is worthy of a place in large and botanical collections. It grows in any soil, blooms rather late in summer, and comes from Siberia.

Sedum pulchellum (*American Stonecrop*).—A dwarf species, with purplish flowers arranged in several spreading branchlets, bird's-foot fashion. It is abundant in North America, and at present very rarely seen in our gardens, though far more worthy of cultivation than many commonly grown, flowering in summer, growing in ordinary soil, and easily increased by division.

S. rupestre (*Rock Stonecrop*).— A glaucous densely-tufted plant, with numerous spreading shoots, these shoots generally rooting at the base and erect at the apex. It has rather loose corymbs of yellow flowers, and is frequently grown in gardens. There are several varieties or sub-species, notably the British *S. elegans* and the green-leaved *S. Forsterianum*. A native of Britain and various parts of Europe, and of the easiest culture.

S. Sieboldii (*Siebold's Stonecrop*).—An elegant species, with roundish leaves, of a glaucous hue, in whorls of three on the numerous stems that in autumn bear the soft rosy flowers in small round bouquets. At first the ascending stems form neat tufts, but as they lengthen, they bend outwards with the weight of the flowers at the points, making the plant a graceful one for small baskets or vases. It is hardy, and merits a place on the rock-garden, especially where its graceful habit may be seen to advantage—that is to say, where its branches may fall without touching the earth; but except in favoured places, it does not make such a strong and satisfactory growth as most of the other Stonecrops. Easily propagated by division. In late autumn the leaves often assume a lovely rosy-coral hue. There is a variegated variety, not so good as the ordinary form. Japan.

S. spectabile (*Showy Stonecrop*).—This is one of the finest autumn-flowering plants introduced of late years—distinct, hardy, fine when its delicate rose-coloured flowers, in very large heads, are in bloom, and pretty long before it flowers, from its dense bush of glaucous leaves. It begins to push up its fleshy glaucous shoots in the dawn of spring, keeps growing on all through the early summer, opens its flowers in early autumn, and continues in full perfection till the end of that season. The plant is one of the easiest to propagate and grow, and forms round, sturdy, bush-like tufts of vegetation, 18 inches or more high when well established. Japan.

Sedum spurium (*Purple Stonecrop*).—Several kinds of *Sedum*, with large, flat leaves, occur in our gardens, of which this is much the best, its rosy-purple corymbs of flowers being handsome compared to the dull whitish flowers of allied kinds. A native of the Caucasus; well suited for forming edgings, the margin of a mixed border, or the rock-garden. It is of the easiest culture and propagation, and blooms late in summer, and often through the autumn. The variety *atrosanguineum* is more showy.

The preceding are the most distinct kinds in cultivation. The pretty *S. cœruleum* is an annual, and *S. carneum variegatum* not hardy enough to stand our winters. Several *Sedums* with a monstrous development of stem, or what in botanical language is called *fasciation*, are in our gardens *S. monstrosum*, *cristatum*, and *reflexum monstrosum*, to wit. The following is an enumeration of other species, or reputed species, now in cultivation in this country, the most desirable being marked with an asterisk. They are almost, without exception, of the easiest culture and rapid increase in ordinary soil.

S. aizoides	S. *cyaneum
Aizoon	dentatum
albescens	denticulatum
altaicum	*elegans
anglicum	elongatum
angulatum	Fabaria
arboreum	*farinosum
asiaticum	Forsterianum
aureum	grandifolium
Beyrichianum	*hispanicum
Brauni	hispidum
*corsicum	ibericum
cruciatum	involucratum
*cruentum	Jacquini

S. libanoticum
  littoreum
 *Lydium
 *Maximowiczii
  Middendorfianum
  maximum
 *monregalense
 *multiceps
  neglectum
  ochroleucum
  orientale
  pallens
 *pallidum
  Pittoni
  pruinosum
  pulchrum

S. reflexum
 *sexangulare
 *sexfidum
  spathulifolium
 *speciosum
  stellatum
  Stephani
  telephioides
  Telephium
  teretifolium
  ternatum
  triangulare
 *Verloti
  villosum
  virens
  Wallichianum

SELAGINELLA.—A few graceful mossy kinds of this large family of trailing plants are valuable for clothing shady spots in the rock-garden. These kinds are *S. denticulata*, *S. helvetica*, and *S. rupestris*, plants of a delicate green, mossy growth. *S. Kraussiana*, generally known in plant-houses as *S. denticulata*, is also hardy in many places, and in Ireland grows and thrives better than any of the kinds mentioned. All these plants require a well-drained peaty soil and shaded and sheltered place.

SEMPERVIVUM (*Houseleek*).—Dwarf perennial succulent plants of striking form and variety, inhabiting, like the Stonecrops, hot sandy and rocky places. They are very useful for the rock-garden, and of the easiest culture and increase. Some are beautiful in flower, but perhaps their best quality for the rock-garden is to give us dwarf relief in pretty greens and greys at all times. The late Mr Jordan in his very interesting garden at Lyons accumulated an immense number of forms of the various species from many localities, but from the point of view of the rock-garden a few types of this family will give us all the effect we can desire. Much the best way, however, is to increase the kinds that strike us as most pleasing in colour for our purpose. Of all plants they are perhaps the most easy to cultivate and increase, growing in any soil, the poorer the better perhaps and bearing division at any time. The little offsets will grow freely. Apart from all cultivation and increase, however, we should consider in this, as in so many other cases, the stature of the plants, and only associate them with dwarf plants, and give them full exposure in open sunny places. These are among the plants which grow on the surface of the stone itself, as we see the common kind grow on the roofs of sheds and houses. The others may also be established by putting a piece of stiff clay moistened and dabbed in the face of the stone pressing in the little offshoots of the Stonecrop, which will soon take hold and find their own living on the faces of stones.

Sempervivum arachnoideum (*Cobweb Houseleek*).—One of the most singular of alpine plants, its tiny rosettes of fleshy leaves being covered at the top with a thick white down. Widely distributed over the Alps and Pyrenees, this plant is quite hardy in our gardens; thriving in sunny arid spots, forming sheets of whitish rosettes, which look as if fine-spinning spiders had been at work upon them, and sending up rose-coloured flowers in summer. About London it sometimes suffers from the sparrows plundering the "down." It is easily increased by division, and thrives in sandy loam.

S. ciliatum (*Fringed Houseleek*).— The margins of the leaves of this species are edged with transparent hair-like bodies, the leaves are barred lengthways with brown and deep-green stripes, flowers freely in summer, in close corymbs of many fine yellow flowers, each scarcely ½ inch across. It ought to be placed in some dry spot under a ledge of rock, and might be tried with advantage on the top of an old wall. A native of the Canary islands; easily increased by division or cuttings.

Sempervivum montanum (*Mountain Houseleek*).—A dark-green kind, smaller than the common Houseleek, with an almost geometrical arrangement of leaves, forming neat rosettes, from which spring dull rosy flowers in summer; grows in any soil, is easily propagated. When masses of it are in flower, they are visited by great numbers of bees. Alps.

S. soboliferum (*Hen-and-Chicken Houseleek*).—Growing in dense tufts, and throwing off little round offsets so freely that these are pushed clear above the tufts, and lie rootless, small, brownish-green balls on the surface. The full-grown rosettes are of a light-green, and of a chocolate-brown at the tips of the under side of the leaves, for nearly one-third of their length. The small leaves of the young rosettes all turning inward, they appear of a purplish-brown colour. The rosettes are usually not more than 1½ inch in diameter, but I have seen them in France more than 3 inches; however, whether they were the rosettes of a form larger naturally than the common one, or the result of a higher culture, I cannot say. The plant is well suited for forming wide tufts on banks beneath the eye. It grows freely in any soil.

S. tectorum (*Common Houseleek*).—A native of rocky places, in the mountain ranges of Europe and Asia, and which, having been cultivated for ages on housetops and old walls, is well known. It is needless to describe the culture of a plant which thrives on bare stones, slates, and in the most arid places. It varies somewhat, a glaucous form called *rusticum* being one of the most distinct.

S. calcareum (*Glaucous Houseleek*).—The *Sempervivum* now common in cultivation, under the garden name of *S. californicum*, is really only the French *S. calcareum*, and no finer Houseleek has been introduced. Planted singly, the rosettes attain a diameter of nearly 6 inches, and as the leaves are of a glaucous tone, distinctly tipped at the points with chocolate, it is useful. It is admirable for the rock-garden, is easily increased by division, and thrives in any soil.

In addition to the preceding, which are among the most distinct Houseleeks, there are a great number of species, or so-called species, wild in Europe, which are cultivated in Botanic Gardens. In the following list the more ornamental kinds are marked with an asterisk.

acuminatum	juratum
* anomalum	Mettenianum
* arenarium	molle
assimile	Neilreichii
Braunii	* piliferum
canescens	Pomelli
Cotyledon	* Requienii
dioicum	ruthenicum
* Funckii	* sediforme
* glaucum	stenopetalum
* globiferum	urbicum
grandiflorum	velutinum
* Heuffelli	villosum

The under-mentioned kinds I first saw in cultivation in the Jardin des Plantes, at Paris. They are mostly sorts desirable for cultivation.

affine	fimbriatum
albidum	* Pseudo-arachnoi-
barbatulum	deum
* Boutignianum	Schleani
Comollii	* Verloti
Dœllianum	violaceum
Fauconetti	

SENECIO (*Ragwort*).—An immense family of groundsel-like plants, many of them far too large for our purpose; but some dwarf, silvery, and pretty, as rock-garden plants. There are nearly a thousand kinds, a number of which are not introduced. Any of the dwarf grey kinds may be used with good effect on the rock-garden.

Senecio argenteus (*Silvery Groundsel*).—A sturdy, minute, hoary plant; the leaves quite silvery. The plant is not more than 2 inches high; it withstands any weather, and will live everywhere in sandy soil in well-drained borders.

S. uniflorus (*One-flowered Groundsel*).—A silvery species, growing little more than an inch high, but scarcely equal to the preceding, and not so easily grown. The flowers are poor, and should be removed, as tending to weaken the plant. A native of Switzerland, and perfectly hardy. *S. incanus* is another pretty dwarf alpine kind, and there is also *S. alpinus* and *S. carniolicus* of like use and culture. Increased by seed and division.

SHEFFIELDIA REPENS.—A hardy little New Zealand creeper, with small leaves, small slender stems, and tiny white flowers in summer. It is interesting for the rock-garden, and grows in any good well-drained soil.

SHORTIA.—*S. galacifolia* is an interesting and beautiful plant. First discovered over a hundred years ago by Michaux in the mountains of North Carolina, and rediscovered in 1877, it was found growing with *Galax aphylla*, and forms runners like that plant, being propagated by this means. The plant is of tufted habit, the flowers reminding one of those of a *Soldanella*, but large, with cut edges to the segments, like a frill, so to say, and pure white, passing to rose as they get older. There is much beauty, too, in the leaves, which are of rather oval shape, deep green tinged with brownish-crimson, changing in winter to quite a crimson, when it forms a bright bit of colour in the rock-garden. A correspondent writing in the *Garden* says : "The cultural directions given in Catalogues to keep the plant in a shady situation and grow it in sphagnum and peat, deprive us of its chief charm — *i.e.* the handsome-coloured leaves during the winter and spring months. Instead of choosing a shady spot I selected a fully exposed one, and here two plants have been for over a year, one in peat and the other in sandy loam. Both are vigorous." It succeeds well in various soils as described, and is hardy. N. America.

SIBTHORPIA (*Cornish Moneywort*).—*S. europæa* is a little native creeping plant, with slender stems and small round leaves. In summer it forms a dense carpet on moist soil, and should always be grown in the bog-garden. The variegated form is more delicate than the wild plant, and rarely succeeds in the open air.

SILENE (*Catchfly*).—Tufted alpine herbs, or herbaceous plants, of the Pink order, often of much beauty, and not difficult to grow.

Silene acaulis (*Cushion Pink*).—Tufted into dwarf light-green masses like a widespreading moss, but quite firm, this plant defies the storms, snows, and Arctic cold of numerous mountain climes in northern regions of the globe, from the White Mountains of New Hampshire to the Pyrenees, covering the most dreary positions with glistening verdure. In summer it becomes a mass of pink-rose flowers barely peeping above the leaves, and making lovely carpets where all else is branded with desolation. Many places on the mountains of Scotland, Northern Ireland, North Wales, and the mountains in the Lake District of England, are sheeted with its firm flat tufts, often several feet in diameter. This plant is indispensable for our purpose, and those who can, would do well to transfer patches from the mountains to humid but sunny slopes on the rock-garden, in peaty or sandy soil. It is, however, not a slow grower, and is easily increased by division. There are several varieties : *alba*, the white one ; *exscapa*, with the flower-stems even less developed than in the usual form, and *muscoides*, dwarfer still ; but none of them are far removed from the wild plant.

S. alpestris (*Alpine Catchfly*).—This has beauty of bloom, perfect hardiness, dwarf and compact habit, growing only from 4 to 6 inches high, and a constitution that enables it to flourish in any soil. It flowers in May, the flowers being of a polished whiteness, with the petals notched, and abundantly produced over the shining green masses of leaves. Like most high-mountain plants, it should have perfect exposure to the full sun; it should never be elevated amongst burrs or stones in such a position that a dry wind may parch the life out of the tiny roots, so unwisely cut off from the moist earth. I once regretted to see a colony of ants take up their abode under a tuft of this plant, and begin to raise the soil amongst its tiny leaves; but as the ants built their hill, the plant expanded its leaves, and finally grew to be a little mound of starry snow. Alps of Europe; readily increased by seed or by division.

Silene Elisabethæ (*Elizabeth's Catchfly*). —A remarkably distinct and rare alpine plant, the flowers looking more like those of some handsome but diminutive *Clarkia* than those of a Catchfly. They are large, of a bright rose colour, and with the base of the petals white, from one to seven being borne on stems 3 or 4 inches high. It is rare in a wild state, occurring in the Tyrol and Italy, where I had the pleasure of gathering it on Monte Campione, growing amidst shattered fragments of rock, and in one case in a flaky rock without any soil. It grows freely enough in sandy soil in a warm nook, as I observed in M. Boissier's garden, in Switzerland. Flowers in summer, rather late, by seeds.

S. maritima (*Sea Catchfly*).—A British plant, not uncommon on sand, shingle, or rocks by the sea, or on wet rocks on mountains, forming carpets of smooth glaucous leaves, from which spring generally solitary flowers about an inch across, and white, with purple inflated calyces. The handsome double variety of this plant, *S. maritima fl. pl.*, is well worthy of culture, not only for its flowers but for the dense, sea-green spreading carpet of leaves which it forms, and which make it particularly suitable for the margins of raised borders, for hanging over the faces of stones. The flowers appear in June, and, in the case of the double variety, rarely rise more than a couple of inches above the leaves, which form a turf about 2 inches deep.

S. Pennsylvanica (*Wild Pink*).—The wild Pink of the Americans is a dwarf and handsome plant, with nearly smooth root-leaves, forming dense patches, and with clusters of six or eight purplish-rose flowers, about an inch across, notched, and borne on stems from 4 to 7 inches high, somewhat sticky, and hairy. A native of many parts of North America, in sandy, rocky, or gravelly places flowering from April to June, and very freely in deep sandy soil.

Silene pumilio (*Pigmy Catchfly*).—An interesting kind from the Tyrol, resembling the Cushion Pink of our own mountains in its dwarf firm tufts of shining green leaves, which are, however, a little more succulent and obtuse, and bearing much larger and handsomer rose-coloured flowers, rising taller than those of *Silene acaulis*, and yet scarcely more than an inch above the flat mass of leaves, so that the whole plant seldom attains a height of more than between 2 and 3 inches. It should be planted in deep sandy loam, on a well-drained and exposed spot, sufficiently moist in summer, facing the south, a few stones being placed round the neck of the young plant to keep it firm and prevent evaporation.

S. schafta (*Late Catchfly*).—A much branched plant, not compressed into hard cushions like the alpine, stemless, or dwarf *Silenes*, forming very neat tufts, from 4 to 6 inches high, and covered with large purplish-rose flowers from July to September, and even later. It comes from the Caucasus, is quite hardy, and a fine plant for almost any position. In planting it, it may be as well to bear in mind its late-flowering habit. Seed or division of established tufts.

S. virginica (*Fire Pink*).—A brilliant perennial, with flowers of the brightest scarlet, nearly 2 inches across, somewhat straggling in habit, hardy and perennial, and the colour as fine as that of the scarlet Lobelia. A native of America, increased by seeds and division, growing from 1 to 2 feet high, and therefore most suited for association with the *Aquilegias* and taller alpine plants.

Having in cultivation such brilliant and distinct plants as the preceding Catchflies, we must consider *Silene Zawadskii*, dwarf and with white flowers, the diminutive soft-tufted *S. quadridentata* (for which *S. alpestris* is often mistaken), the woody *S. arborescens*, a dwarf, shrubby, evergreen species, with rose-coloured flowers, and the dirty-white *S. Saxifraga* — only worthy of a place in very large collections or in Botanic Gardens. *S. rupestris*, a sparkling-looking, dwarf, white species, little more than 3 inches high when in bloom, and reminding one of a dwarf *S. alpestris*, is better worthy of a place.

SISYRINCHIUM (*Satin Flower*).— Iris-like plants, few species of which are worthy of culture on the rock-garden. *S. grandiflorum* is a beautiful perennial, flowering in early spring, with grass-like foliage and flowers borne on slender stems 6 to 12 inches high, bell-shaped and drooping, a rich purple and a transparent white in the variety album. Both are graceful, thriving in sandy peat. Division. North-West America.

SKIMMIA.—Handsome dwarf evergreen shrubs, and among the best for the rock-garden worth cultivating are *S. japonica*, and *S. Fortunei*.

The plant, known in gardens as *S. japonica*, is not Japanese at all, but a native of China. Mr Fortune met with it in 1848 in a garden at Shanghai, the Nurseryman from whom he obtained it informing him that the plant was brought from a high mountain in the interior, called Wang Shang. Of all the plants Fortune sent home only one reached England alive. The proper name of this species is *Skimmia Fortunei*. The true *S. japonica* is a Japanese plant, and did not find its way into British gardens for some years after *S. Fortunei*.

The *Skimmias* thrive under very varied conditions as regards soil, I have seen them thrive splendidly in strong clay, and also in poor sandy soil and peat.

SMILAX (*Green-Brier*). — These handsome, evergreen, and neglected trailing shrubs, should have a place in gardens. They are natives of South Europe, North Africa, and North America, some hardy enough for our country, but rarely planted, and yet, I think, very suitable for the more bushy parts of the rock-garden. For a description of the species see in the "English Flower Garden" an article by Mr Lynch, of the Cambridge Botanic Gardens, in the dry soil of which these plants are grown well.

SOLDANELLA.—Modest and refined true alpine plants that live near the snow-line on many of the great mountain-chains of Europe—not brilliant, but withal beautiful, in pale-bluish bell-shaped flowers, cut into narrow, linear strips, and springing from a dwarf carpet of leathery, shining, roundish leaves. If sound young plants are placed out of doors in a little bed of deep and very sandy loam, they will succeed, especially in moist districts, and in dry ones it will be easy to prevent evaporation by covering the ground near the plants with some cocoa-fibre mixed with sand to give it weight. I have seen a carpet, several feet square, of these plants growing on a bed of fine moist sandy earth on a flat spot in a rock-garden, in this country, and none I saw in the Alps equalled it in luxuriance. The best place for the plants is a level spot on the rock-garden near the eye.

They are readily increased by division, though, as they are starved too often from confinement in small worm-defiled pots, they are rarely strong enough to be pulled in pieces. The smaller kinds will thrive under the same conditions, but require more care in planting, and should be associated *with the most minute alpine plants*, in a mixture of peat and good loam, with plenty of sharp sand, and get abundance of water in summer, especially in dry districts.

According to Mr H. Correvon, who knows these plants well, writing in the *Garden*, there are five wild and two hybrid kinds, natives of the mountain chains of Middle and South Europe, Jura, Pyrenees, Apennines, Tyrol, Transylvania, Carpathians.

Soldanella Alpina known by its reniform, entire leaves, very sparsely toothed, with two ear-like drooping lobes at the base, and by its flower-stem of a height of 3 inches to 5½ inches; the pedicels are a little roughened by the presence of sessile glands; the scales of the corolla (abortive stamens alternating with the lobes of the corolla) are attached to the filaments. Alps and Pyrenees.

S. montana.—In this species the leaves are rounded instead of being kidney-shaped, more or less crenate, the underside often of a strong purple colour; the flower-stem has a height of 12 inches to 14 inches; the scales of the corolla are free; the leaves are indented, and with untoothed lobes; the pedicel, calyx, and petiole bear with glandular hairs.

S. pyrolæfolia.—Leaves orbicular, thick, and bright green; undersides strongly ribbed and regularly pitted above; flower-stem very long, glandular at the base. Easter Alps.

S. pusilla.—Plant very small, leaves minute, very slightly crenate, and a little pitted towards their base; flower-stalk 3 inches to 6 inches high, set with small glands; flower solitary, corolla narrow, long-shaped, reddish-violet, fringed for nearly one-third of the length. Alps and Carpathians on granite. Syn., *S. Clusii*.

S. minima.—The smallest kind, liliputian; leaves very small, quite round, and never indented at the base; flower-stems from 3½ inches to 4 inches high, slightly downy, one-flowered; lilac-white, with fringing barely a quarter of the length. Limestone Alps of Switzerland and Austria.

Soldanella Gauderi is intermediate between *S. alpina* and *S. minima*, but rather nearer the former; and *S. hybrida*. Syn., *Media*, is half-way between *S. alpina* and *S. pusilla*.

SPARTIUM JUNCEUM (*Rush, or Spanish Broom*).—A handsome flowering shrub, valuable on account of its blooming in July and August, when shrubs are usually flowerless. It is 8 or 10 feet high, and its Rush-like shoots have so few leaves as to appear leafless. It bears erect clusters of fragrant bright yellow flowers, shaped like Pea-blossoms. It is hardy, and is useful for dry, poor soils, railway banks, or dry rocky places. I have naturalised it abundantly on very rocky and shaly railway banks, by merely throwing the seed down the bank. South Europe.

SPIGELIA MARILANDICA (*Wormgrass*).—A distinct and beautiful plant; the flowers 1½ inch long, crimson outside and yellow within, from three to eight borne on a stem from 6 to 15 inches high, and as, when the plant is well grown, these stems come up very thickly and form close erect tufts, the effect, when in bloom, is brilliant. A native of rich woods in North America, from Pennsylvania to Florida and Mississippi, flowering in summer, and increased by careful division of the root. I have not seen it grown to perfection except in deep and moist sandy peat.

SPIRÆA (*Meadow Sweet*). — Some of the smaller of these handsome shrubs may well find a place in our bushy rock-garden, taking the dwarfest and neatest kinds, such as *bumalda*, *Thurnbergi*, *Bella japonica*, also *S.*

pectinata, which Mr A. K. Bulley describes as follows:

"At first sight this plant would be mistaken for a mossy *Saxifrage*. The tufts of bright green foliage are not more than 3 inches in height; the flowers, borne on numerous short spikes, are of a soft cream colour."

STATICE (*Sea Lavender*).—Plants of the Leadwort or Plumbago family, all dwarf perennials or annuals, chiefly natives of sea-shores and mountains. Most of them bear twiggy flower-stems, and bear myriads of small flowers, which are, for the most part, membraneous, and long retain their colour after being cut. The larger species require least care when in open places in sandy soil, while some of them are admirable for the rock-garden. The best of the larger kinds are *S. Limonium*, of which there are several varieties; *S. latifolia*, with wide-spreading flower-stems with many small purplish-blue flowers; and *S. tartarica*, a dwarfer species, with distinct red flowers. The smaller species, such as *S. minuta*, *S. minutiflora*, *S. caspia*, *S. eximia*, are good rock-plants.

STERNBERGIA (*Winter Daffodil*). —Bulbous plants of distinct beauty especially for the garden in autumn. The species, as described and arranged by Mr Baker, are as follows:—

Sternbergia colchiciflora, as possessing delicious fragrance, and perfuming the fields of the Crimea, and about the Bosphorus. The leaves are narrow, and appear with the fruit in spring. The flowers appear in autumn, and are nearly 1½ inch long, pale or sulphur-yellow. It is found on dry exposed positions on the Caucasian Mountains, Crimea, and is hardy in this country, treated in the same way as *S. lutea*. *S. dalmatica* and *S. pulchella* are varieties.

Sternbergia clusiana (Ker, not Boissier).—*Narcissus persicus* (Clusius), *Amaryllis citrina*, *A. colchiciflora*, *S. ætnensis* and *S. Schuberti* are synonyms.

S. Fischeriana is nearly allied, and has the habit of *S. lutea*, from which it differs chiefly in flowering in spring instead of autumn. It is a native of the Caucasus, hardy in this country.

S. lutea.—This is the autumn or winter Daffodil (*Narcissus autumnalis major*) of Parkinson. A plant that flowers freely in autumn; where not disturbed often effective in its sheets of yellow bloom. *S. lutea* has five or six leaves, each about ½ inch broad, about a foot long, and produced at the same time as the flowers in autumn and winter, and is supposed by some writers to be the Lily of Scripture, as it grows in Palestine. A colony of it on the warm side of a rock is worth having, and when the plant is at rest in the summer, the ground might be covered with stonecrops.

S. angustifolia.—Appears to be merely a narrow-leaved form of *S. lutea*. It is very free-flowering, and grows rather more freely than *S. lutea*.

S. græca. — From the mountains of Greece; has very narrow leaves and broad perianth segments.

S. sicula.—Is a form with narrower leaves and segments than the type, while the Cretan variety has considerably larger flowers.

S. macrantha. — This, introduced by Mr Whittall from the mountains of Smyrna, is a handsome species. The leaves are blunt, and slightly glaucous, about an inch broad when fully developed about midsummer, flowers bright yellow, in autumn. A native of Palestine and Asia Minor.

STYLOPHORUM (*Celandine Poppy*). —*S. diphyllum* is a handsome Poppywort, resembling Celandine, but is a finer plant. Its foliage is greyish, and it has large yellow flowers in early summer. A plant of easy culture, 1 to 2 feet high. N. America. Syn., *S. japonicum*.

SWERTIA PERENNIS (*Fellowort*).
—A curious perennial, with slender stems, 1 to 2 feet high, and erect spikes of flowers, greyish-purple spotted with black, in summer. It is interesting for the bog-garden, or for moist spots near the rock-garden. Seed or division.

SYMPHYANDRA. — *Campanula*-like plants, *S. pendula* being a showy perennial from the rocky parts of the Caucasus, with branched pendulous stems and large cream-coloured bell-like flowers, almost hidden in the leaves. It is hardy, and rarely more than 1 foot in height is best seen about the level of the eye in the rock-garden. The Austrian *S. Wanneri* rarely exceeds a foot in height, with deep mauve flowers borne freely on branching racemes, preferring a light, rich soil, and a half-shady place. Seed.

TCHIHATCHEWIA.—This beautiful alpine plant, *T. isatidea*, is a native of Asia Minor, hardy, and not particular as to soil, preferring to grow among rocks. From a tuft of oblong leaves, formed in the first year, appear the flowers in the second year; the leaves dark green, with shining silky hairs, from amongst which rises the thick flower-stalk of Syringa-like bright rosy lilac flowers, fragrant like vanilla. The bunch is over a foot across, and is in great beauty throughout the month of May.

TEUCRIUM MARUM (*Cat Thyme*).
—I should no more have included this in the present selection than the Oak, previous to one afternoon in July 1868. On a dry old wall in one of the islands on Lago Maggiore, I noticed a mass of lilac flowers, on a plant which, from the profusion of bloom, appeared to be a dwarf heath; but was only our old friend the Cat Thyme, that, flowerless and neglected, used occasionally to be seen in old greenhouses. Here it had become a mass of flowers. This suggested to me that its true home was not in the greenhouse, but on some dry old sunny wall, or in a chalk pit or very dry spot on the southern face of a rock-garden. And, indeed, the wall would seem to be the only way of preserving it from cats, for they are desperately fond of it. A native of Spain; readily increased by cuttings.

Teucrium polium (*Poly Germander*), with silvery foliage, is also worth growing, and perhaps others, but they are southern rather than northern plants.

THALICTRUM (*Meadow Rue*).—
Usually vigorous hardy perennials, a few of which are good in the rock-garden, not so much for their flowers as for the effect of their fern-like leaves.

Thalictrum anemonoides (*Rue Anemone*).—A delicate, diminutive species, with the habit and frondescence of *Isopyrum*, the inflorescence of Anemone, and the fruit of *Thalictrum*. These qualities, in addition to its dwarf stature, usually only a few inches high, make it a plant for the rock-garden. The flowers are white, nearly an inch in diameter, open in April and May, the flower-stem bearing a few leaves near the summit, in the form of a whorl round the flowers. A native of many parts of N. America, increased by seed or by the division of its tuberous roots. There is a pretty double variety, *T. anemonoides fl. pl.*, with smaller flowers than those of the single one. Being small and fragile in its parts, it requires a little care, a light, peaty, and moist soil, and to be associated with other delicate growers. Syn., *Anemone thalictroides*.

T. minus (*Maidenhair Meadow Rue*).—
A native of Britain, but also found on the Continent and in Russian Asia. By pinching off the inconspicuous blooms that appear in summer, the plant can be made to resemble, in outline, the Maidenhair

PART II.] ALPINE FLOWERS FOR GARDENS 323

Fern. And the finely-cut leaves are as good for mingling with cut flowers, and better in one respect, as they are of a pretty firm consistency, and do not fade quickly, like those of the Fern. It will thrive in any soil, and requires no trouble whatever after planting.

Thalictrum adiantifolium — Is probably a variety of this plant, and of like use.

T. alpinum (*Alpine Meadow Rue*).—A species with few flowers and four purplish sepals. The plant is rarely more than 8 inches or 10 inches high, and has the same use for the rock-garden. Native of Britain, and N. America.

T. tuberosum (*Tuberous Meadow Rue*). —This is about 9 inches high, and besides the usually graceful foliage which we find in all the dwarf forms of the genus, we have, in this instance, an additional beauty in the abundant mass of yellowish creamcoloured flowers which this plant produces. It is quite hardy, and thrives in deep peat soil. Spain.

THLASPI LATIFOLIUM (*Showy Bastard Cress*). — A dwarf, stronggrowing plant, with large indented root-leaves and corymbs of pretty white flowers, somewhat like those of *Arabis albida*, but a little larger, and of a paper-white; early in March. It is worth growing with the earlier and more vigorous spring flowers, comes from the Caucasian mountains, and is easily increased by division. A few other kinds are worth a place—*T. rotundifolium* and *T. violascens*, of easy culture in moist spots.

THYMUS (*Thyme*).—Dwarf, tufted perennials on mountains and open heaths, not showy in flower, but charming from their close, turfy growth and pleasant odour, often neglected, I think, for more showy things. Their easy culture, and the pretty little carpets they form, make them much valued in the rock-garden. Our native Wild Thyme and its varieties are as pretty as any other. Division in autumn or early spring.

Thymus lanuginosus (*Downy Thyme*).— This is usually considered a woolly variety of *T. Serpyllum*, our common British Thyme, but given the same conditions, it is a better plant, forming cushions of grey leaves in any soil exposed to the sun. Few plants are more suited for such places, in which many other plants will not thrive, though it spreads so quickly into wide dense cushions that it ought not to be near very minute alpine plants. Various other kinds of Thyme are worthy of a place on the dry arid slopes of the large rock-garden and on walls, but space forbids any more than the enumeration of them here. There is a variegated form of the common garden Thyme (*T. vulgaris*), which makes a pretty tufted bush, and many plants sold as alpine plants have not half the merits of the Lemon Thyme as rock-plants. Other species in cultivation are—*T. azoricus, azureus, bracteosus, Zygis, thuriferus, carmosus, micans, nummularius, rotundifolius chamædrys*, and *villosus*, most of which are of easy culture and increase in poor soil. The white and highly coloured forms of our common Thyme are good rock or wall-plants.

TIARELLA CORDIFOLIA (*FoamFlower*).—A dwarf perennial plant of some beauty, both of leaf and flower; the little starry flowers creamy white, the buds tinged with pink, a mass of the white flowers seen a few yards off resembling a wreath of foam. The young leaves are of a tender green, spotted and veined deep red, while the older ones at the base of the plant are of a rich red-bronze. Whether planted in rock-garden or border, it is beautiful, and needs only division every two years, the plants being at their best the second year.

TRIENTALIS EUROPŒUS (*Starflower*).—A graceful perennial, living in woody and mossy places, with erect slender stems, rarely more than

6 inches high, bearing a whorl of leaves, from the centre of which arise from one to four slender flower-stems, each supporting a star-shaped white or pink-tipped flower. A native of Northern and Arctic Asia, America and Europe, and found in the Scotch Highlands and North of England. With healthy well-rooted plants to begin with, it is not difficult to establish among bog shrubs in some half-shady part of the rock-garden, or in the shade of Rhododendrons, in peat soil. It is best for association with *Linnæa*, the *Pyrolas*, and *Pinguiculas*, among mossy rocks. Flowers in early summer, and is increased by division of the creeping root-stocks.

TRIFOLIUM (*Clover*). — Notwithstanding the immense number of kinds, there are but few, excepting the alpine *Trifolium*, that are of consequence for the rock-garden; and there are so many pretty plants from the same Pea-flower order that we are never short of a like kind of beauty. The alpine Clover is a rather showy plant of easy culture.

TRILLIUM (*Water Robin*).—Singularly formed North American perennial plants of value and interest for the moist parts of the rock-garden, and also for the marsh-garden, thriving best in rich and moist sandy soil or peat, or, if in loam, with added leaf soil. They are natives of moist woods and thickets, and, therefore, if we wish to see them at their best, partial shade is a help, but they should not be robbed by hungry shrubbery roots.

Trillium grandiflorum (*White Wood Lily*).—One of the most singular and beautiful of hardy plants, so named from the larger parts being usually arranged in threes. When in good health, each stem bears a lovely, white, three-petalled flower, fairer than the white Lily, and almost as large when the plant is strong. It thrives in a free deep soil, full of vegetable matter, and a shady position. If placed in a sunny or exposed position, the large soft green leaves will not develop. At Biddulph Grange I saw it forming bushes of the healthiest green, more than 2 feet high, and spreading out as freely as any border-plant. It was planted in a moist spot, shaded and sheltered by high banks and shrubs. In such positions it may be grown as well as in its native woods.

Trillium erectum is a curious species, with broad leaves 2 to 6 inches wide, and brown-purple or white flowers. It is also found in East Siberia, and is nearly allied to the plant found in Japan, if not identical with it. It is figured in Salisbury's "Paradisus," t. 35, as *T. fœtidum*. Flowers in May and June, and is found from Canada to North Carolina.

T. **erythrocarpum** is a shy flowerer, and not easy to keep in health. It is called the Painted Lady, and surpasses all the others in the beauty of its flowers, which are white, with bright purple streaks. The flowers are, however, small, appearing in May and June. Georgia, on high mountains, or in cold damp woods.

T. pusillum, recurvatum, stylosum, nivale, ovatum, petiolatum, and *undulatum* are rare in gardens, and more worth growing. *T. sessile*, with brown flowers and mottled leaves, is best known through the variety *Californicum*, which has large rose-coloured or white flowers, and is a useful, easily grown plant.

TROLLIUS (*Globe Flower*).—Stout and handsome perennials, inhabiting alpine and northern pastures. Although plants of the semi-marshy sub-alpine pastures and copses, they will thrive in exposure if kept moist at the roots,—that is to say, planted in a deep, rich soil, as then the roots are less affected by drought. The best time to propagate the Globe-flower is in September, when the roots may be lifted and divided to almost any extent. If

ALPINE FLOWERS FOR GARDENS

left, as is often the case, until the end of March, they are almost sure to suffer. They may also be propagated by seeds, which should be sown quickly, as if kept for any length of time the germination becomes uncertain. If liberally treated, the seedlings will flower the second year, attaining their full strength during the third and fourth years.

They are too vigorous in growth to go with the dwarfer rock-plants, but if we grow the mountain shrubs in association with the rock-plants, then such handsome plants may be grown between them with good effect.

Trollius acaulis.—A native of the higher Himalayas, and one of the most charming of dwarf bog-plants, rarely exceeding 4 to 6 inches in height, bearing in early April its bright golden-yellow flowers, suffused with purple-brown on the outside. It is hardy, and will be found useful for the moist spots of the rock-garden, in moist peat.

T. Asiaticus, which also includes *chinensis, Fortunei,* and other forms, has deep, orange-yellow flowers, and bright, orange-red anthers. It has a wide distribution both in China and Japan, and is hardy even in exposed positions. It differs from the European Globe-flower chiefly in the flowers being orange, and less globular, and in the small and finely-divided foliage, and taller growth. This, and its varieties, form a valuable group, and when grown in moist places bear brilliant orange flowers.

T. Europæus is an extremely variable plant, and so widely spread that almost every locality has its particular form. Raised from seed, it also gives much variety, particularly in habit, and often in flowers and foliage. Many of the names in Catalogues are for slight forms of this. Some few of these, of course, are distinct varieties, such as *T. e. aurantiacus.* It is, like its parent, of strong constitution, flowers freely, and bears its flower-stems well above the handsome foliage.

The known species of *Trollius,* according to the "Hortus Kewensis," are *T. altaicus, americanus, asiaticus, caucasicus, dschungaricus, emarginatus, europæus, Ledebouri* (this has pale yellow flowers, and is a strong grower), and *patulus,* but whatever differences these may show botanically, a few species give us the best effects of the plants.

TROPÆOLUM (*Indian Cress*).—A few of these tuberous and fragile climbers of great beauty may well take a place among the shrubs near the rock-garden; their fine colour and distinct form being most precious. Where any shelter or background of Holly or evergreen shrub is used, they are admirable, planted beneath the bushes in rather open leaf-soil, and let alone.

Tropæolum polyphyllum (*Indian Rock-Cress*).—A distinct plant, whether in or out of flower; the leaves glaucous, densely crowded on a stem a quarter of an inch thick, and when planted on a warm sunny part of the rock-garden, the stems creep about, snake-like, through the vegetation around, some to 3 or 4 feet in length bearing yellow flowers. It is tuberous-rooted, quite hardy in dry spots and on sunny banks, where it should not be often disturbed; springs up early, and dies down at the end of summer. Cordilleras of Chili.

T. speciosum (*Flame Nasturtium*).—A splendid creeping plant, with long annual shoots, gracefully clothed with six-lobed leaves, and such brilliant vermilion flowers that a long shoot of the plant is startlingly effective. It is impossible to find anything more worthy of a position in which its shoots may fall over or climb up the face of some high bank in the rock-garden or among Hollies or other shrubs near. It thrives in deep, rich, and rather moist soil, best in cool places, or in those near the sea, and not so well in a dry atmosphere. When a position is selected for it, the soil should be made light, and deep, and free, by the addition of leaf-mould, peat, fibry loam, and sand, as the nature of the ground may require, and the surface should be mulched in summer with an inch or two of leaf-mould. It will also enjoy a bed of manure beneath the roots, and put below the soil in which the young plants are first placed, and is best planted in spring, the roots inserted 6 or 8 inches in the soil, and the young plants well watered. It is best planted where the shoots may ramble among the spray of shrubs, or trailers; and it is much better to let them have their own way, than to resort to any kind of staking or support, except that afforded by shrubs or low trees near. It ripens its pretty blue seed in early autumn, and the seeds come up the next spring, if sown in light sandy mould in pots, and placed in a greenhouse or pit.

Tropæolum tuberosum.—A handsome trailing plant, but tender on cold soils, and a shy bloomer in many places where it has been tried. It is a tall climber with succulent stems, leaves about 2 inches or 3 inches across, and rather small red and orange flowers. The colour of the flowers is beautiful, the calyx, with the exception of the green tip of the spur, being a deep red; and the entire petals, which scarcely exceed in length the lobes of the calyx, are of a rich golden-yellow, veined with black. Plant in warm loam on the sunny side of a rock.

TULIPA (*Tulip*).—Much attention is now being called to these splendid plants; not merely old garden kinds, but wild kinds from many countries, including countries not far away, as Savoy. Though they do not require rock-garden cultivation as a rule, still, so long as kinds are new and rare, the variety of surface and aspect of the rock-garden will often give us a home for them until they become plentiful.

Tulipa celsiana (*Dwarf Yellow Tulip*).— A species having slightly concave glaucous leaves, the largest nearly an inch across, and yellow flowers, smaller than those of the common Tulips, and, when in clumps and fully open, sometimes reminding one of a yellow Crocus; the outside of the petals is tinted with reddish-brown and green. It begins to flower about the first of May, and usually attains a height of 6 to 8 and sometimes 12 inches. The bulbs emit stolons after flowering. Southern Europe.

Tulipa Clusiana (*Clusius's Tulip*).— Usually our Tulips are great, bold, showy flowers, but in this species we have one, humble in stature, and modestly pretty. The bulbs are small, the stem reaching from 6 to 9 inches high, seldom more, and sometimes flowering when little more than 3 inches high. The flower is small, with a purplish spot at the base of each petal; the three outer divisions of the petals stained with rose, the three inner ones of a pure transparent white. A native of the South of Europe, a little more delicate than most of its family, and requiring to be planted in good, light, vegetable earth, in a warm, sheltered, and well-drained position, to succeed to perfection. Although so small, it will be the better of being planted rather deeply, say at from 6 to 9 inches, and of being placed in some snug spot, where it need not be disturbed too often.

TUNICA SAXIFRAGA (*Rock Catchfly*).—A small plant of the Pink order, with narrow leaves and wiry stems, bearing elegant rosy flowers, small, but numerous, thriving without particular care on most soils, and forming tufts a few inches high. A native of stony places on the Pyrenees and Alps, often descending into the low country, where I have found it on the tops of walls. It will grow in like positions in this country, and is a neat, free-growing plant for the rock-garden. It is easily raised from seed, and thrives in poor soil.

UVULARIA. — Slender perennials allied to the Solomon's Seal, bearing yellow blossoms. There are four cultivated species, *U. chinensis, grandiflora, puberula,* and *sessilifolia.* Of these, *U. grandiflora* is the finest plant; it attains a height of from 1 foot to 2 feet, and the numerous slender stems form a dense tuft, the flowers long, yellow, gracefully drooping. It is a good peat border plant, and thrives best in a moist peaty soil, in a partially shaded place, and in the bog-garden. It is a native of N. America, as are all the others except *U. chinensis.*

VACCINIUM VITIS-IDÆA (*Red Whortleberry*) is a dwarf British evergreen, with box-like foliage, but of a paler green, and with clusters of pale rose flowers, which appear in summer, followed by berries about the size of Red Currants, like those of the Cranberry, on wiry stems from 3 to 9 inches high. It forms a neat little bush in peat soil. The Marsh Cranberry (*V. Oxycoccos*), a native of wet bogs in Britain, with very slender creeping shoots and drooping darkrose flowers, requiring wetter soil than the preceding, is also worthy of a place where bog-plants are grown. The American Cranberry (*V. macrocarpum*), a much larger plant, distinguished from the preceding by its much larger fruit, is also worthy of a place in moist sandy peat, associated with bog shrubs. Some of the American kinds are too large for the rock-garden proper, though a few may come in well among the shrubs, among them *V. pennsyllvanicum,* if only for its fine colour in autumn.

VERONICA (*Speedwell*).—Herbaceous perennials, evergreen, alpine, rock and half-shrubby plants. An enormous genus of plants, many of the herbaceous kinds of which are too large for the rock-garden, and among the northern kinds this leaves a limited choice. The more beautiful of the half-shrubby kinds come from the southern hemisphere, and, unfortunately, are not hardy everywhere, so that these are less precious for our rock-gardens than the northern kinds.

Veronica chamædrys (*Germander Speedwell*).—A well-known and much-admired little native plant, with ovate, or heartshaped, hairy leaves, and with hairs curiously arranged in two opposite lines down the stem, while the other portions are bare. The flowers are bright blue, produced in great numbers. It is abundant in nearly all parts of Britain, and may be allowed to crawl about here and there in the less important parts of the rock-garden. Easily increased by seed or division.

V. prostrata. (*Prostrate Speedwell*).—A dwarf spreading plant, forming darkgreen tufts, under 6 inches high, the leaves lance-shaped or linear; the stems covered with a short down, forming circular tufts, and nearly woody at the base; flowers of a deep blue, but varying a good deal, there being several varieties with rosecoloured and white blooms, appearing in early summer, somewhat earlier than *V. Teucrium.* A hardy and pretty plant, flowering so freely that, when in full perfection, the leaves are often quite obscured by the flowers. A native of France, Central and Southern Europe, occurring on stony hills and in dry grassy places, and, in cultivation, succeeding in dry sandy soil, though by no means fastidious, and easily increased by seeds or division.

V. repens.—Clothes the soil with a soft carpet of bright green foliage, covered, in spring, with pale bluish flowers. It thrives well on moderately dry soil, but delights in moist corners of the rock-garden, and is an admirable little rock-plant.

V. saxatilis (*Rock Speedwell*).—A brilliant, dwarf, bush-like plant, a native of alpine rocks in various parts of Europe, and also in a few places in the Highlands of Scotland, forming close tufts, 6 or 8 inches high. The flowers are a little more than ½ inch across, and of a blue, striped with violet, with a narrow but decided ring of crimson near the bottom of the cup, its base being pure white; appearing in May and June, is increased by seed or cuttings, grows in ordinary soil, and should be in every rock-garden.

Veronica Taurica (*Taurian Speedwell*). — A dwarf, wiry, and almost woody species, forming neat dark-green tufts, under 3 inches high; the flowers a fine gentian-blue. Perhaps the neatest of all rock *Veronicas* for forming spreading tufts in level spots, or tufts drooping from chinks, hardy, growing in ordinary well-drained garden soil; flowering in early summer, and suitable for association with the dwarfer alpine shrubs. Tauria; increased by division or by cuttings.

V. teucrium (*Teucrium Speedwell*). — A continental plant, the stems forming spreading masses from 8 inches to a foot high, and covered with flowers of an intense blue in early summer. The flowers are at first in dense racemes, which afterwards become much longer, lower ones pointed. It is an excellent plant for the rock-garden, easily increased by seeds or division, and thriving in ordinary garden soil.

V. Bidwillii, Guthriana Telephifolia, V. Nummularia, of the Pyrenees, *V. aphylla*, the neat little bushy *V. fruticulosa, V. satureifolia*, and *V. candida*, with silvery-white leaves, are also worthy of a place; though, generally, the bloom of the rock Speedwells is not prolonged enough to make them of the first importance in the rock-garden.

NEW ZEALAND VERONICAS. —

The dwarfer kinds of these are scarcely so precious as the taller kinds. In our country away from the sea-shore, even in southern mild districts, they are not hardy, and although they give pretty evergreen effects in the winter, and are distinct and often good in habit, the flower rarely seems worthy of the plant. In fact, in our country they seem to be, with few exceptions, not nearly as well fitted for our rock-gardens as the plants of the Alps and the Rocky Mountains of America.

Undoubtedly, around the coasts, a good many of the bushy New Zealand kinds can be grown, as this coast climate suits them well. But our rock-gardens should be made for plants that will stand any weather; and in this case we should only try the hardier kinds, and those not much until we have proved them. From experiments made at the Royal Botanic Garden in Edinburgh, in 1892, the following appeared to be hardy species; but it should be noted that Edinburgh is under the sea influence, and that its soil is perhaps the most excellent in Britain for outdoor plants.

V. Hectori	V. Godefroyana
loganioides	glaucocœrulea
lycopodioides	Colensoi
cupressoides	Traversi
Armstrongi	rakaiensis
carnosula	monticola
pinguifolia	pimeleoides
amplexicaulis	linifolia
buxifolia	anomala

VESICARIA UTRICULATA. — A half-bushy perennial, with large yellow flowers, not unlike the alpine Wallflower, but with bladder-like pods. It usually grows from 10 inches to a foot high, a vigorous plant, though it perishes in winter on cold soils. A native of mountains in France, Italy, and Southern Europe generally, usually on calcareous rocks, and most likely to flourish and endure on dry sunny spots or on walls. It is very easily increased from seed.

V. græca is a handsome plant, the flowers opening in succession. It is a hardy evergreen perennial, a native of Dalmatia and other places in South Europe. Increased by cuttings placed in soil under a hand-glass and also by seeds.

VICIA (*Vetch*). — Perennial and annual plants, several of which are natives, and, as I think, worthy of more care than they often get. *V.*

Cracca, V. Orobus, V. sylvatica, V. Sepium, and *V. argentea* are among the best. *Vicias* grow freely in almost any soil, and are raised from seeds, and increased by careful division.

Vicia argentea (*Silvery Vetch*) has silvery leaves, and of prostrate habit, but without tendrils, and rarely more than 8 inches high, spreading about freely in light and well-drained soil; the rather large whitish flowers are veined with violet in the upper, and spotted with purple in the lower, part. It is not a brilliant plant in flower, but the elegant foliage makes it worthy of a place in the rock-garden. Pyrenees, rare in gardens; easily increased by division or seed.

V. onobrychus is a lovely Vetch, bearing long and handsome racemes of flowers in summer on the Alps of France and Italy, and giving an effect like that of some of the purple Australian Pea-flowers. It is best grouped or scattered in a colony or grassy bank in the rock-garden.

VINCA (*Periwinkle*).—Hardy, wiry, trailing perennials, easily grown, free —almost too free—but nevertheless useful for bare banks, and welcome for their bloom in spring.

Vinca major is useful on masses of rootwork, near cascades, etc., and also in rocky places or banks. There is a variety called *elegantissima*, finely blotched and variegated with creamy white, and several other variegated varieties. The lesser Periwinkle (*V. minor*), a much smaller plant, is also useful for like positions; there are several varieties of it well worthy of cultivation, a white-flowered one (*V. minor alba*), one with reddish flowers, one or two double varieties, and also, as of the larger, several variegated forms.

V. herbacea is a plant much less frequent than the common Periwinkles, and more worthy of culture on rocks, as it is not rampant in habit. A native of Hungary, flowering in spring and early summer, the stems dying down every year, it thrives best in an open position.

VIOLA (*Violet*).—Dwarf, growing perennials of the mountain, woodland, and pasture, many kinds of which are alpine flowers.

Some *Violas* are among the most beautiful which bedeck the alpine turf; and even the common Violet may almost be claimed as an alpine plant, for it wanders along hedgerow and hillside, copses and thin woods, all the way to Sweden. From all kinds of *Violas* the world of wild flowers derives a precious treasure of beauty and delicate fragrance; and no family has given to our gardens anything more precious than the numerous races of Pansies, and the various large, sweet-scented Violets. Far above the faint blue carpets of the scentless wild Violets in our woods and heaths, thickets and bogs, and the miniature Pansies that find their home among our lowland field-weeds; far above the larger Pansy-like *Violas* (varieties of *V. lutea*) which flower so richly in the mountain pastures of northern England, and even on the tops of stone walls; and above the large free-growing Violets of the American heaths and thickets, we have true alpine Violets, such as the yellow two-flowered Violet (*V. biflora*), and the large blue Violets, such as the *V. calcarata* and *V. cornuta*. It would be difficult to exaggerate the beauty of these alpine *Violas*. They grow in a turf of high alpine plants not more than an inch or so in height. The leaves do not show above this densely-matted turf, but the flowers start up, waving everywhere thousands of little banners. *Violas* are of the easiest culture; even the highest alpine kinds thrive with little care, *V. cornuta* of the Pyrenees thriving even more freely than in its native uplands. Slow-growing compact kinds, like the American Bird's-Foot Violet, from their stature and their comparative slowness of growth, are entitled to a place in the rock-garden.

Tufted Pansy.

Viola biflora (*Two-Flowered Yellow Violet*).—This is a bright little Violet, widely distributed. From its delicate condition in gardens, few would suspect what a lovely little ornament it is on the Alps, in many parts of which every chink between the moist rocks is clothed with it. It even crawls far under the great boulders and rocks, and lines shallow caves with its fresh verdure and little yellow stars. In our gardens its home will be on the rock-garden, running about among such plants as the yellow annual Saxifrage, and Sandworts, in moist spots. If obtained in a weakly condition, it may seem difficult to establish, but this is not by any means the case; and once fairly started in a moist and half-shady spot, it soon begins to creep about, and may then be readily increased by division.

V. calcarata (*Spurred Violet*).—This is a pretty plant on the Alps, usually in high situations, amidst dwarf flowers, sometimes so plentiful that its large purple flowers form sheets of colour, the leaves being scarcely seen amidst the other dwarf plants that form the turf. There is a yellow variety, *flava* (*V. Zoysii*). In some high pastures the flowers vary in colour every step one takes, and yet every variety in colour is delicate and lovely. Try it among a short turf of Sandworts or any dwarf plants. Alps.

V. cornuta (*Horned Pansy*). — A fine Pyrenean Violet, with pale-blue or mauve-coloured and sweet-scented flowers. Generally speaking, it does poorly on dry soils and in warm districts, and exceedingly well in wet places. I have rarely seen anything to equal its appearance in the cold wet climate of East Lancashire, while it looks poor indeed in many gardens in the South. It is easily propagated by division, cuttings, or seeds.

Viola cucullata (*Large American Violet*), bears some resemblance to the common Violet, though without its scent. It flowers more freely, and its foliage is bold and sometimes variegated. It belongs to a section which contains some good varieties, such as *V. primulæfolia*, *semperflorens*, *blanda*, *obliqua*, *sagittata*, *delphinifolia*, *canadensis*, *pubescens*, *striata*, and others. All these varieties are worthy of culture in a botanical collection. N. America.

V. gracilis is a remarkably pretty dwarf species, never failing to produce in spring an abundance of deep purple blossoms in dense tufts. It is hardy in light soil. Mount Olympus.

V. lutea (*Mountain Violet*).—This is one of our native Violets classed by Bentham as a variety of *V. tricolor*, but considered distinct by other botanists, and is distinct for garden purposes. Being called *lutea*, one is surprised to find the flowers of nearly every wild plant of it a fine purple, with a yellow spot at the base of the lower petal. In cultivation the yellow form is a neat plant, rising from 2 to 6 inches high, and flowering from April onwards, the flowers of a rich yellow, the three lower petals striped with thin lines of rich black.

V. munbyana.—One of the prettiest of Violets, abundant in flower, free and robust in growth, and quite hardy. Generally it begins to bloom about the end of February, but it attains its greatest beauty in May. The deep purple-blue flowers resemble those of *V. cornuta*; and there is also a yellow variety. Algeria.

V. odorata (*Sweet Violet*).—This well-known plant is, in a wild state, widely spread over Europe and Russian Asia, and common in various parts of Britain. Its odour distinguishes it immediately from the numerous other *Violas*. The Sweet Violet and most of its varieties may be used in many places where few things but weeds succeed; it will form carpets for open groves or the fringes of woods, or in open parts of copses, or on hedgebanks, demanding in such positions no care, and rewarding the planter by filling the cold March air with sweetness; and in the garden, instead of confining it to a solitary bed for cutting from, as is often the case, it should be used on the rock-garden, and it grows well on dry walls.

The newer seedling forms, like La France, are so good that if used more as carpets in the rock-garden and near, all the better. It will grow in almost any soil, but succeeds best in free sandy loams, and should be put in such when there is any choice.

The varieties of the Violet are numerous. We have the Single White and the Single Rose, the Double White, the Czar, the Queen of Violets, Admiral Avellan, La Grosse Bleue, La France, California, Princess of Wales, Luxonne, Belle de Chatenay, White Czar, Marie Victoria Regina, *Wellsiana*, and the perpetual blooming Violet — well known in France as La Violette des Quatre Saisons. It differs slightly from the Sweet Violet, but is valuable for flowering long and continuously in autumn, winter, and spring. It is the variety used by the cultivators round Paris. The Neapolitan Violet comes from a different and more delicate species, and its varieties are not fitted for open-air culture, save in very favoured districts.

Viola pedata (*Bird-Foot Violet*).—The most beautiful of the American Violets, with handsome flowers, an inch across, pale or deep lilac, purple or blue, the two upper petals sometimes deep violet, and velvety like a Pansy; the leaves deeply divided, like the foot of a bird, and the plant dwarf. In a wild state it inhabits sandy or gravelly soil in the Northern States of America, flowering in summer, and increased by seeds or division. It is best adapted for the rock-garden, where the soil is sandy and moist.

V. rothomagensis (*Rouen Violet*).— A handsome plant belonging to the tricolor group, dwarf, and with low creeping stems which bear in spring numerous purple and white blossoms. It is a free grower, but, being a native of Sicily, is not so hardy as some Violets, and should be grown in a light soil and a warm spot.

Viola tricolor (*Heartsease*).—The common Pansy is usually included under the head of *V. tricolor*, though it is more likely to have descended from *V. altaica*; in any case, from some kinds nearly allied to that species. But the kinds are so numerous, so varied, and, withal, so distinct from any really wild species of Violet in cultivation, that little can be traced of their origin. Of one thing we may be certain: the parents of this precious race were true mountaineers. Only alpines could give birth to such rich and brilliant colour and noble amplitude of bloom, considering the size of the plant. Its season never ends, it blooms often cheerfully enough at Christmas, and is sheeted with delightful gold and purple when the Hawthorn is whitened with blossoms. Such a flower must not be forgotten on our rock-gardens, even though it thrive in almost any soil and position. It may be treated as an annual, biennial, or perennial, according to climate, position, and soil. Good varieties are quickly and easily raised from seed, while the plant may be raised freely from cuttings or by division. Only the most delicate colours are worthy of the rock-garden.

In addition to the Violets here described, other species are worthy of cultivation in large collections, for example: *V. striata, V. canadensis, V. obliqua, V. palmata, V. blanda, V. pennata, V. palmaensis*; but most of these are all exceeded in size and beauty of flower by those described, and surpassed in odour by the Sweet Violet.

Hybrids of Viola.—The common Pansy of our gardens is a hybrid *Viola*. Of late years a beautiful race of plants has been raised by crossing this with other *Violas*, giving us the plants I call Tufted Pansies, which are of the highest value for the rock-garden or any other flower-garden use. The delicate colours, facility of increase, and almost perennial character make them more precious than the older race of Pansies, which are rather of a biennial character, and not easy to perpetuate. For a full account of these plants, see the "English Flower-Garden."

VITTADENIA TRILOBA (*New Holland Daisy*).—A pretty Australian plant, bearing an abundance of flowers with yellowish disks and rosy-white rays, somewhat like those of a Daisy; the plant has a spreading diffuse habit, and forms neat little bushes about a foot high. The plant may be raised as freely as any annual, sown in frames or on a gentle hot-bed, in March or early in April; when put out in April in free sandy soil in a sunny spot, it flowers abundantly from early summer to late autumn. I probably should not have mentioned it in this book, had I not met with it in North Italy adorning some rocks on which it had become naturalised. Although often treated as an annual, it is a perennial on soils and in positions where not destroyed by wet and frost.

WAHLENBERGIA. — Dwarf and pretty alpine plants of the harebell family, but a little more alpine in nature, and perhaps a little more difficult of cultivation, as, to succeed well, they require some of the choicest spots on the rock-garden. Mr F. W. Meyer, of Exeter, who has been very successful with this family, writes of them in the *Garden*:—

"According to my experience, none of them succeed if planted on flat ground, but if planted into an upright or sloping fissure, with the roots in a horizontal, instead of a vertical position, success is certain, if the plants receive an abundance of sunshine. There are fast-growing and slow-growing varieties, but, with the exception of planting the dwarfest kinds closer together, I make no difference in the treatment.

"The rock on which I grew them best, which is facing south-east, was composed of pieces of limestone so arranged as to leave between them long, almost perpendicular, crevices 2 inches or 3 inches wide, and from 2 feet to 2½ feet in depth. These crevices were filled with plenty of broken stones for drainage, and before filling in the soil the lowest visible or outward part of a crevice was closed up by a small wedge-shaped stone, held in place by a kind of mortar made of clay and Sphagnum Moss, mixed with a very small quantity of soil. The small stones, acting as drainage, would be on a lower level and in the inside part of the crevice. By means of more 'mortar' and more small stones, the outside part of the fissure is now built up to the height where it is desired the first plant should be, and simultaneously the inside part of the crevice is filled to the same height with a mixture of loam, leaf-mould, small broken stones (limestone), and stony grit. The plant is then inserted with its roots in a horizontal position, and more of the stony soil is filled in and rammed around and between the roots with a small stick. On each side of the neck of the plant a small stone is next driven into the crevice in such a manner as not in any way to injure the roots, but to take the pressure of other small stones used for building up the front of the crevice above the first plant, say to the height of 10 inches or a foot in precisely the same way as was done below the first plant; the second plant is then introduced, and in the same way a third or fourth plant may be added, according to the height of the fissure or the size of the plants, but care must be taken not to use the clay mixture as mortar above the last plant, as the more or less impervious clay would check the free access of water to the roots. I use soil and Moss only as a 'mortar' for small stones above the last plant. If the tiny crevices between the small stones are not filled up, they become a harbour for slugs and other pests."

Wahlenbergia, or any other plants requiring to be grown sideways (*i.e.* with their roots in a horizontal position), succeed remarkably well if planted in the manner just described, as water can never rest on the foliage of the plants to any dangerous extent, while free access of water and perfect drainage are assured to the roots. The native home of most *Wahlenbergia* is in South-Eastern Europe and Asia Minor. Syn., *Edraianthus*.

Wahlenbergia dalmaticus.—One of the best, robust in growth, and the easiest to cultivate. In planting, the plants should be kept at least a foot apart. The large flowers form clusters or heads, each consisting of from eight to twelve flowers, of a violet-blue, and white at the base in May and June. The height of the plant is seldom more than 4 inches or 5 inches, as the stout flower-stems do not stand up erect, but lie on the ground or stones. Dalmatia.

Wahlenbergia graminifolius and W. dalmaticus in the rock-garden at Abbotsbury, Newton Abbot. (Engraved from a photograph sent by Mr F. W. Meyer.

Wahlenbergia dinaricus.—It is one of the smallest, and more compact than the robust *W. dalmaticus*. The flowers are nearly as large, of a more purplish shade of colour, more bell-shaped in form, singly, or two or three on a stem. The leaves are very small and narrow, covered with very minute hairs on the upper surface. May and June.

W. pumilio.—A very small kind, the flowers solitary and 1 inch in length, and about ¾ of an inch in diameter, of a bright purplish blue. The upper surface of the leaves is covered with minute hairs to such an extent as to have quite a silvery appearance, which in all plants, as a general rule, is a sure indication of the requirement of a sunny position. But though the plant itself grows best when its foliage is moderately dry, its roots, though well drained, should never want for moisture.

W. Kitaibeli is a robust kind. It is a native of Bosnia, and growing about 6 inches high, the flowers large, purplish blue.

W. serpyllifolius.—A gem for the rock-garden, and, planted sideways into an upright fissure, does remarkably well. The flowers are very much like those of *W. pumilio*, but of a deeper bluish shade.

W. tenuifolius.—A native of the mountainous districts bordering on the Adriatic from Trieste to Montenegro.

WALDSTEINIA FRAGARIOIDES
(*Strawberry Waldsteinia*).—A showy plant from North America, with creeping bright-red, hairy stems, growing about 6 inches high, bearing in summer bright-yellow blooms about ½ inch across, and thriving in ordinary soils.

Waldsteinia trifolia (*Three-Leaved W.*). —A dwarf vigorous plant, spreading about with stout stubby strawberry-like runners. The trifoliate and rich yellow flowers in April, on dwarf stems, with a dense brush of golden stamens in the centre. A hardy plant, good for any kind of rock or wall gardening. Division.

WULFENIA.—*W. carinthiaca* is a dwarf, almost stemless, evergreen herb, 12 to 18 inches high, bearing in summer spikes of drooping purplish-blue flowers, and found only on one or two mountains in Carinthia. It is a plant for rock-gardens or borders, thriving in a light moist sandy loam. *W. Amherstiana* from the Himalayas is similar to the Carinthian species, but more showy and rare, and we have seen it only in Kew Gardens. It is hardy, grows freely in any position in the rock-garden, but prefers a shady spot and light rich soil.

XEROPHYLLUM ASPHODELOIDES (*Turkey's-Beard*). — A tuberous-rooted plant with the aspect of an *Asphodel*, beautiful, forming a spreading tuft of grassy leaves, and bearing on a flower-stem, from 2 to 4 feet high, a raceme of numerous white blossoms. It grows well in a moist sandy peaty border, or in the drier parts of the bog-garden. A common plant in the Pine barrens in North America.

YUCCA (*Adam's Needle*). — Evergreen plants of good and distinct form, which, although used much as lawn-plants, are best for the rock-garden or dry banks, coming as they do from arid and sandy regions in North America.

Their varieties really hardy in our climate are *Y. gloriosa, recurva, filamentosa, flaccida*. In damp localities Yuccas are apt to form soft growths, easily pulped by severe frosts. Planted on dry mounds, or in sand and stones, and lime rubble, or among sheltered rocks by the sea, they are quite at home, and flower well. Starvation is the best treatment for them, especially in cold inland places.

In the rock-garden the best way is to keep to the dwarfer free-flowering kinds, which have the merit also of flowering annually. Their effect, even in winter, on a knoll is good, and there is nothing one could plant on a dry poor bank that would be likely to do or look as well. A little fringe of some small-leaved Ivy surrounding them looks well.

ZAUSCHNERIA CALIFORNICA (*Californian Fuchsia*).—A distinct and bright perennial, hardy in warm soils,

ZENOBIA.—*Z. speciosa* is one of the most beautiful of rock shrubs of the Heath family, about a yard high, with small pale green leaves. In the variety *pulverulenta*, the leaves are covered with a mealy glaucescence. The flowers are white and wax-like, resembling those of Lily of the Valley, in summer, in loose drooping clusters. A well-flowered plant is most charming, and lasts for some weeks in beauty,

A group of Yuccas. (Engraved from a photograph by Mrs Henderson, Sedgwick Park, Horsham.)

12 to 18 inches high, with an abundance of bright vermilion flowers during summer and autumn. It thrives in sandy loam in the rock-garden, and grows well on an old wall, but on heavy and moist soils does not thrive. Where any difficulty is found in cultivating it, it will certainly succeed in a "dry" wall.

doing best in a peaty soil or a sandy loam. It comes from the Southern United States, but is hardy in the southern countries. In Nurseries it is known as *Andromeda speciosa*, and *A. pulverulenta*.

ZEPHYRANTHES (*Zephyr Flower*). —Pretty bulbous plants requiring a warmer climate than Britain for their fullest beauty, and in our land requiring the warmest positions and light well-drained soils. The grassy leaves appear in spring with or before the Crocus-like flowers, which are white or rose-pink, and, for the most part, handsome. *Zephyranthes* require rest during winter, and at that season are best kept dry. In spring they should be planted out in the full sun in very sandy soil.

Zephyranthes atamasco (*Atamasco Lily*).—A beautiful, lily-like plant, bearing handsome white flowers tinged with purple, 3½ inches across, on stems from 6 to 12 inches high. Although growing abundantly in North America, this fine plant is too rare in our gardens, where it is well worthy of culture, thriving in light, rich, sandy soil, and flowering in early summer. Dotted over a turf formed of some carpet-plant like the Lawn-Pearlwort, it is seen to great advantage when its great bell-like flower opens. Division of established tufts.

Zephyranthes carinata.—This lovely plant has narrow leaves, and its flower-stem, which is about 6 inches high, bears a rosy flower, 2 or 3 inches long. It thrives in the open border if kept dry in winter in light sandy loam.

Zenobia speciosa pulverulenta.

ZIETENIA.—*Z. lavandulæfolia* is a dwarf, creeping, half-shrubby perennial of a grayish hue, 6 to 12 inches high, with purple flowers in summer, borne in whorls, forming a spike about 6 inches long, with a slender downy stalk. Suitable for the rougher parts of the rock-garden. Division. Caucasus.

INDEX

[ILLUSTRATIONS IN ITALICS]

A

Abbotsbury, Newton Abbot, Wahlenbergia on the rock-garden at, 333
Acæna, 147
Acantholimon, 147
—— *venustum*, 148
Achillea, 149
Acis, 150
—— *autumnalis*, 150
Aconite, Winter, 218
Adonis, 150
Æthionema, 150
Ajuga, 153
Allium, 153
Allosorus, 153
Alpine and Rock plants, watering, 92
—— *Flowers at home, facing title-page.*
—— —— *a ledge of*, 24
—— —— for gardens, Part II. 147
—— —— in borders and beds, rock and, 34; in pans or baskets, 79; in pots, 82; the rocky mountains, 141
—— —— *small bed of*, 32
—— —— wall-gardens of rock and, 38
—— gardening in adverse conditions, 68; planting, 73; soil, 72
—— gardens, trees and, 50
—— *Larchwood*, 124
—— Marsh garden, the, 51
—— *plant growing between stones in a pot*, 83
—— *plant on border surrounded by half-buried stones*, 36
—— plants from seed in the open ground, 87; raising, 77, 87
—— —— *frontispiece of a book on*, 95
—— —— *growing in a level border*, 83; *on the level ground*, 25
—— —— raised from seed in pots, 90
—— *View, an*, 112
—— *Village, an*, 117
—— *Stream, an*, 20
—— *Waterfall*, 118
Alpines, frames for, 78
Alps of Europe and the Rocky Mountains of N.W. America, some notes of a journey in the, 111

Alsine, 153, 172
—— *laricifolia*, 172
Alyssum, 154
—— *montanum*, 154
Anagallis, 155
Andromeda, 155
Androsace, 155
—— *lanuginosa in the Rock-Garden, The Friars, Henley-on-Thames*, 157
—— *villosa*, 158
Androsaces, pot for, 83
Anemone, 159, 239
—— *blanda*, 161
—— *Greek, the*, 161
—— *vernalis*, 163
Annuals for the Rock-Garden, 85
—— some dwarf and more refined, 86
Antennaria, 162
Anthemis, 163
—— *macedonica*, 164
Anthericum, 164
Anthyllis, 164
Antirrhinum, 165
Aquilegia, 165
—— *cærulea*, 168
Arabis, 169
Arch, Rustic, 95
Arctostaphylos, 170
Arenaria, 153, 171
—— *laricifolia*, 172
Arethusa, 172
Armeria, 173
Arnebia, 173
Artemisia, 174
Arum, Bog, 182
Asarum, 174
Asperula, 174
Asphodel, Bog, 265
Aster, 174
—— *Stracheyi*, 175
Astragalus, 175
Atragene, 176
Aubrietia, 176
Auricula, 284
Avens, 234
—— Mountain, 217
Azalea, 177

Y
337

338   INDEX

[ILLUSTRATIONS IN ITALICS]

B

Barbary Ragwort, 270
Barberry, 179
Barrenwort, 217
Baskets, Alpine flowers in pans or, 79
Bearberry, 170
Board Tongue, 273
Beauty of the Rocks, 274
Bed kept saturated by perforated pipes, 84
—— *of Alpine flowers, small, 32*
Bellium, 179
Berberis 179
Bergenia, 180
Betula, 180
Bindweed, 193
Birch, 180
Bitter Root, 251
Bleeding Heart, 213
Bletia, 180
Bloodroot, 300
Bluets, 240
Bog bed, the cemented, 59; beds without cement, 59
—— —— the partly cemented, 60
Bogs, artificial, 55
Border, Alpine plant growing in a level, 33
—— —— *on, surrounded by half-buried stones,* 36
—— *rough stone-edging to,* 35
Borders and beds, rock and Alpine flowers in, 34
Boretta, 180
Box, 181
Brachycome, 180
Bramble, 299
Bridge, stepping-stone, with water-lilies and water-plants, 27; *plan of,* 27
Bridges and Cascades, 27
Brookfield, Hathersage, Sheffield, part of Rock-Garden at, 5
Broom, Rock, 228
—— Spanish, 320
Bruckenthalia, 180
—— *spiculifolia,* 180
Bryanthus, 181
Buckbeam, 260
Bugle, 153
Bulbocodium, 181
Bulrush, 312
Butcher's Broom, 300
Buttercup, 295
—— *Lady,* 295
Butterwort, 278
Buxus, 181

C

Calamintha, 181
Calandrinia 181
Calla, 182
Calluna, 182
Calophaca 182

Caltha, 182
Camomile, 163
Campanula, 182
—— *garganica,* 184
—— *turbinata,* 186
Campion, 255
Candytuft, 242
Cardamine, 186
Cascade in a high wood, 126
Cascades, Bridges and, 27
Cassandra, 155
Cassiope, 155, 186
Catchfly, 255, 317
—— Rock, 326
Cat's-Ear, 162
Cave for Killarney Fern, entrance to, 30
Centaury, 223
Cerastium, 187
Chickweed, Mouse-Ear, 187
Chieranthus, 188
Chimaphila, 188
Chiogenes, 188
Christmas Rose, 237
Cinquefoil, 281
Cistus, 189
—— *formosus,* 190
Clark, Mr Latimer, on forming the Rock-Garden, 55
Claytonia, 191
Clematis, 176, 191
Clover, 324
Colchicum, 191
Columbine, 165
—— *flower of blue,* 168
Conandron, 193
Concrete, Rocks formed of, 30
Convallaria, 193
Convolvulus, 193
Coptis, 194
Coris, 194
Cornus, 194
Coronilla, 195
Cortusa, 195
Corydalis, 195
Cotoneaster, 196
Cotyledon, 197
Cowslip, 288
—— American, 213
Cranesbill, 238
Creeping Jenny, 257
Cress, Indian, 325
—— rock, 169; purple, 176
—— showy bastard, 322
—— silvery, 150
—— violet, 243
Crocus, 197
Crowberry, 217
Cyananthus, 199
Cyclamen, 199
Cypripedium, 203
Cystopteris, 205
Cytisus, 205

INDEX

[ILLUSTRATIONS IN ITALICS]

D

Daffodil, 265
—— Winter, 321
Daisy, New Holland, 332
—— Rock, 179
Dalibarda, 205
Daphne, 206
—— *Blagayána*, 206
Darlingtonia, 208
Dentaria, 208
Desfontainea, 209
Dianthus, 209
—— *alpinus*, 210
—— *cæsius*, 211
—— *neglectus*, 212
Diapensia, 212
Dicentra, 213
Diphylleia, 213
Dodecatheon, 213
Dogwood, 194
Dondia, 214
Draba, 214
Dracocephalum, 216
Dragon's Head, 216
Drosera, 216
Dryas, 217
Dutchman's Breeches, 213
Dyer's Greenweed, 229

E

Echinocactus, 217
Edelweiss, 249
Edelweiss, 250
Edging to border, rough stone, 35
Edraianthus, 333
Elmet Hall, Leeds, part of Rock-Garden at, 19
Emmotts, Ide Hill, Sevenoaks, Kent, part of Rock-Garden at, 37
Empetrum, 217
Epigæa, 217
Epilobium, 217
Epimedium, 217
Epipactis, 218
Eranthis, 218
—— *hyemalis*, 218
Erica, 180, 182, 219
Erigeron, 220
Erinus, 220
Eriogonum, 221
Eritrichium, 221
Erodium, 221
Erpetion, 222
Eryngium, 222
Erysimum, 222
Erythræa, 223
Erythronium, 223
Evening Primrose, 265
Everlasting, Yellow, 237

F

Fellwort, 322
Fern, Bladder, 205
—— *Killarney, entrance to Cave for*, 30
—— Parsley, 153
Fernery, Rock-Garden, 29
Ferns on an old wall, 58
Feverfew, Alpine, 249
Fissure, horizontal, 21
Fissures, right and wrong, 21, 22, 23
Flag, 244
—— *Algerian*, 245
Flax, 253
—— Toad, 252
Fleabane, 220
Foam-flower, 323
Fog-fruit, 254
Forget-me-not, 263
—— antarctic, 263
—— creeping, 267
—— Fairy, 221
Fota, Co. Cork, Water-Garden at, 68
Fountain and Rockwork, what to avoid, 97
Fragaria, 226
Frankenia, 226
Fritillaria, 226
Frog-bit, 241
Fruiting Duckweed, 265
Fuchsia, Californian, 335
Fumitory, 195

G

Galanthus, 227
Galax, 227
Garland Flower, 206
Gaultheria, 227
Genista, 228
Gentian, 229
Gentiana, 229
—— *decumbens alba*, 231
—— *macrophylla*, 232
Gentianella, 229
Geographical arrangements of rock-plants, 26
Geological aspects of Rockwork, on the, 99
Geranium 233
Geum, 234
Ginger, Wild, 174
Glacier, a, 123
Globe Flower, 324
Globularia, 235
Gold Thread, 194
Golden Club, 270
—— Drop, 267
Goodyera, 235
Granite tor, 100
Grape Hyacinth, 262
Grass of Parnassus, 272
—— Whitlow, 214

INDEX

[ILLUSTRATIONS IN ITALICS]

Green Brier, 319
Grit, 17
Gromwell, 254
Ground Pine, 257
Gypsophila, 235

H

Habenaria, 235
Haberlea, 235
Habranthus, 236
Hairbell, 182
Hawkweed, 239
Hawthorn, Japanese, 297
Heartsease, 331
Heath, 219
—— Arctic, 186
—— Rocky Mountain, 181
—— Sea, 226
Heather, 220
—— Whin, 228
Hedysarum, 236
Helianthemum, 236
Helichrysum, 237
Helleborus, 237
Helonias, 239
Hemiphragma, 239
Hepatica, 160, 161, 239
Herniaria, 239
Heronsbill, 221
Hesperochiron, 239
Hieracium, 239
Hippocrepis, 239
Honeysuckle, 255
—— Swamp, 177
Horminum, 239
Hottonia, 239
Houseleek, 315
Houstonia, 240
Howth, Co. Dublin, Rhododendrons among natural rocks at, 47
Hutchinsia, 151, 240
Hyacinth, 240
Hyacinthus, 240
Hydrocharis, 241
Hydrocotyle, 241
Hypericum, 241
—— *polophyllum*, 241

I

Iberidella, 151, 241
Iberis, 242
Incarvillea, 243
Ionopsidium, 243
—— *acaule*, 243
Iris, 244
—— *stylosa*, 245
Isopyrum, 245

J

Jankæa, 245
Jasione, 245
Jasmine, 246

Jasminum, 246
Jeffersonia, 246
Juniperus, 246

K

Kalmia, 246
Kernera, 247
Knotweed, 280

L

Lady's Slipper, 203
—— Smock, 186
Lake Maggiore, margin of, Island in, 28
Larch-wood, Alpine, 124
Lathyrus, 247
Laurel, Mountain, 247
Lavender, Sea, 321
Ledge, rocky, 110
Ledges, well-formed, sloping, 20
Ledum, 249
Leiophyllum, 249
—— *buxifolium*, 248
Leontopodium, 249
—— *alpinum*, 250
Leucanthemum, 249
Leucojum, 150, 249
Leucothoe, 155, 251
Lewisia, 251
Libertia, 251
Lilies, Lenten, 238
—— *Water and water plants, stepping-stone bridge with,* 27 *plan of,* 27
Lilium, 251
Lily, 251
—— Atamasco, 336
—— of the valley, 193 ; twin-leaved, 257
—— St Bruno's, 164, 271
—— water, 265 ; *the white,* 64
—— White Wood, 324
Limestone, 103
Limestones, 108
Linaria, 252
Ling, 220
Linnæa, 252
Linum, 253
Lippia, 254
Lithospermum, 254
Lloydia, 254
Loiseleuria, 254
Lonicera, 255
Lungwort, 293
—— Smooth, 260
Luzuriaga, 255
Lychnis, 255
Lychnis 256
Lycopodium, 257
Lyndhurst, Sussex, rocky path at, 11
Lyonia, 155
Lysimachia, 257

INDEX 341

[ILLUSTRATIONS IN ITALICS]

M

Madwort, 154
Maianthemum, 257
Mallow Rock, 257
Malvastrum, 257
Marjoram 269
Marsh garden, the Alpine, 51 ; the, 52
—— Marigold, 182
—— plants, a selection of, 55
Masterwort, dwarf, 214
Mayflower, 217
Mazus, 257
Meadow Beauty, 298
—— Saffron, 191 ; Spring, 181
—— Rue, 322
—— Sweet, 320
Meconopsis, 258
—— *aculeata*, 259
Megasea, 259
Melittis, 259
Menyanthes, 260
Menziesia, 260
Merendera, 260
Mertensia, 260
Mozereon, 207
Milkwort, 279
Mimulus, 261
Mitchella, 261
Mocassin flower, 204
Modiola, 261
Mœhringia, 261
Moneywort, Cornish, 317
Monkey-flower, 261
Monte Campione, 128
Morisia, 261
Mound of earth with exposed points of rock, 8
Mountain, flank in process of degradation, 2
—— vegetation in America, 137
—— *Woods of California*, 139
Mountains, miniature 96
Muhlenbeckia, 262
Mullien, Rosette, 293
Muscari, 262
Mutisia, 262
Myosotidium, 263
Myosotis, 263
Myrica, 264

N

Narcissus, 264
Narthecium, 265
Nasturtium, flame, 325
Nertera, 265
Nierembergia, 265
Nymphæa, 265

O

Oak Lodge, Rocky bank at, 31 ; *water margin at*, 28
Œnothera, 265
—— *cæspitosa*, 266

Omphalodes, 267
Ononis, 267
Onosma, 267
Ophrys, 267
Opuntia, 268
Orchid, 268
Orchis, 268
—— Bee, 267
—— *maculata superba*, 269
—— Rein, 235
Origanum, 269
Orontium, 270
Othonna, 270
Ox-eye, 150
Oxalis, 270
Oxlip, 288
Oxytropis, 271
Ozothamnus, 271

P

Pans or baskets, Alpine plants in, 79
Pansy, horned, 330
—— tufted, 330
Papaver, 271
Paradisia, 271
Parnassia, 272
—— *palustris*, 272
Parochetus, 272
Paronychia, 272
Parsley Fern, 153
Partridge Berry, 261
Pasque flower, 162
Passerina, 272
Path, rocky, at Lydhurst, Sussex, 11
Pathway, ascending, in Rock-Garden, Warley Place, 10
—— *stone, in Rock-Garden, Warley Place*, 9
Pathways, 10
Pea, Everlasting, 247
—— Shamrock, 272
Pear, prickly, 268
Pearlwort, lawn, 300
Pelargonium, 272
Pennywort, 241
Pentstemon, 273
Pepino, 274
Periwinkle, 329
Pernettya, 274
Petrocallis, 274
Philesia, 274
Phlox, 274
—— *divaricata*, 276
—— *stellaria*, 277
Phyteuma, 277
Pieris, 155, 277
Pimpernel, 155
Pincushion flower, 310
Pines, limit of the, 132
Pinguicula, 278
Pink, 209
—— *Alpine*, 210

[ILLUSTRATIONS IN ITALICS]

Pink *Cheddar*, 211
—— Cushion, 317
—— Sea, 173
Pinxter-flower, 178
Pitcher plant, 301
—— —— Californian, 208
Plantain Rattlesnake, 235
Planting, (1) *Right*; (2) *Wrong*, 94
Plants for dry walls, 42
Platycodon, 278
Plumbago, 278
Polemonium, 279
Polygala, 279
Polygonatum, 279
Polygonum, 280
Pontederia, 281
Poppy, 271
—— Bush, 299
—— Celandine, 321
—— Satin, 258
Pot, Alpine plant growing between stones in a, 83
—— *for Androsaces, etc.*, 83
Potentilla, 281
—— *nitida*, 282
Pots, Alpine flowers in, 82
—— *plunged in sand, bed of small Alpine plants in*, 84
—— raising Alpine plants from seed in, 90
Pratia, 282
Prickly Pear, 268
—— Thrift, 147
Primrose, 282
—— Evening, 265
Primula, 282
Prophet-flower, 178
Prunella, 292
Pulmonaria, 293
Puschkinia, 293
Pyrola, 293
Pyxidanthera, 293

R

Ragwort, 316
Ramondia, 293
—— *pyrenaica*, 294
Rampion, 277
Ranunculus, 295
—— *amplexicaulis*, 295
Raphiolepis, 297
Rest Harrow, 267
Rhamnus, 298
Rhexia, 298
Rhodiola, 298
Rhododendron, 298
Rhododendrons among natural rocks at Howth, Co. Dublin, 47
Rhodora, 298
Rhodothamnus, 299
Rock and Alpine flowers in borders and beds, 34; planting, 93; wall-gardens of, 38

Rock cress, purple, 176
Rockfoil, 302
—— *broad-leaved the*, 306
Rock-Garden, annuals for the, 85; some dwarf and more refined, 86
—— *at Elmet Hall, Leeds, part of*, 19
—— construction, 12
—— cultural, 1
—— Fernery, 29
—— *half-buried stone in*, 18
—— Japanese dwarfed trees for the, 50
—— materials, 6
—— Mountain Shrubs for the, 47
—— Mr Latimer Clark on forming the, 55
—— *on level ground at Emmotts, Ide Hill, Sevenoaks, Kent*, 37
—— *part of, at Brookfield, Hathersage, Sheffield*, 5
—— position for the, 6
—— small, the, 31
—— soil, 13
—— *Warley Place, Essex, ascending pathway in*, 10; *stone pathway in*, 9
—— water plants in the, 61
—— Gardens on level ground, 36; various, 18
—— *hidden natural*, 8
—— *mound of earth, with exposed points of*, 8
—— *near water, suitable for bold vegetation*, 28
—— *on which plants do not thrive, artificial*, 20
—— plants, a wall made for, 46
—— —— dry stone walls for, 46
—— —— *established on an old wall*, 4
—— —— geographical arrangements of, 26
—— —— *hollow wall for. Plan and Section of*, 44
—— —— *on sloping wall of local sandstone*, 39
—— Rose, 189
—— *with base buried, showing ascending*, 23
Rockeries, stone for, 75
Rockery, barrow-shaped, the, 70
—— Facing, 71
—— Sunk, the, 71
Rocks formed of concrete, 30
—— *in a Sussex garden, unearthed*, 8
—— *trees on*, 50
Rook-spray, 196
Rockwork in Villa in Hammersmith, 9
—— What to avoid, 97, 98, 99
Rocky Mountains, Alpine flowers in the, 141
—— —— of N.W. America, some notes of a journey in the Alps of Europe and the, 111
—— —— *Isolated rocks in the*, 188
—— —— *Scene in the*, 137

INDEX 343

[ILLUSTRATIONS IN ITALICS]

Rodgersia, 299
Romanzoffia, 299
Romneya, 299
Rosa, 299
Rose, 299
Rosemary, 299
Rosmarinus, 299
Rubus, 299
Rue, 300
Ruscus, 300
Rush, 320
Ruta, 300

S

Saas Valley, the, 116
Sagina, 300
Salix, 300
Sandstone, old red, 105
Sandwort, 171
Sanguinaria, 300
Sanicle, Alpine, 195
Santolina, 301
Saponaria, 301
Sarracenia, 301
Satin-flower, 319
Savin, 246
Saxifraga, 259, 302
—— *cordifolia,* 303
—— *Juniperina,* 307
Saxifrage, home of the purple, 135
Scabiosa, 310
Scabious, Sheep's, 245
Schists and Shales, 109
Schivereckia, 310
Schizocodon, 311
Scilla, 311
Scirpus, 312
Scutellaria, 312
—— *indica,* 312
Sea Holly, 222
Sedum, 312
Selaginella, 315
Self-heal, 292
Sempervivum, 315
Senecio, 316
Shales, Schists and, 109
Sheffieldia, 317
Shortia, 317
Shrubs, Mountain, for the rock-garden, 47
Sibthorpia, 317
Silene, 317
Sisyrinchium, 319
Skimmia, 319
Slugs, 25
Smilax, 319
Snakeshead, 226
Snapdragon, 165
Snowberry, creeping, 188
Snowdrop, 227
Snowflake, 249
Soapwort, 301
Soldanella, 319

Soil for certain plants, need of poor, 17
Solomon's Seal, 279
Sowbread, 199
Sparrow-wort, 272
Spartium, 320
Speedwell, 327
Spigelia, 320
Spiræa, 320
Spring Beauty, 191
Squill, striped, 293
St John's Wort, 241
Starflower, 323
Starwort, 175
Statice, 321
Steps, rocky, 11
Sternbergia, 321
Stone for Rockeries, 75
Stonecrop, 312
Strawberry, 226
Stream, an Alpine, 20
Stubwort, 270
Studflower, 239
Stybarrow Crag, Ullswater, 107
Stylophorum, 321
Sundew, 216
Sunrose, 237
Sweet Gale, 264
Swertia, 322
Symphyandra, 322

T

Tchihatchewia, 322
Tea, Labrador, 249
Teucrium, 322
Thalictrum, 322
Things to avoid, 94
Thlaspi, 323
Thrift, 173
—— *on the hills at Anglesey,* 17
Thyme, 323
—— Cat, 322
Thymus, 323
Tiarella, 323
Toothwort, 208
Trees and Alpine gardens, 50
—— Japanese dwarfed, for the Rock-Garden, 50
—— *on Rocks,* 50
Trientalis, 323
Trifolium, 324
Trillium, 324
Tropæolum, 325
Tufted Bur, 147
Tulip, 326
Tulipa, 326
Tunica, 326
Turkey's-beard, 334

U

Uvularia, 326

[ILLUSTRATIONS IN ITALICS]

V

Vaccinium, 327
Valerian, Greek, 279
Veronica, 327
Veronicas, New Zealand, 328
Vesicaria, 328
Vetch, 328
—— Bitter, 248
—— Crown, 195
—— Kidney, 164
—— Milk, 175
Vicia, 328
Village, an Alpine, 117
Vinca, 329
Viola, 329
—— hybrids of, 331
Violet, 329
—— Dog's tooth, 223
—— New Holland, 222
—— Sweet, 330
—— Water, 232
Vittadenia, 332

W

Wahlenbergia, 332
—— *graminifolius and W. dalmaticus in the Rock-Garden at Abbotsbury, Newton Abbot*, 333
Waldstenia, 334
Wall for rock plants, hollow. Plan and Section of, 44
—— gardens of rock and Alpine flowers, 38
—— made for rock plants, a, 216
—— of local sandstone, sloping, 39; *sandstone blocks supporting earth banks*, 42
—— *old, rock plants established on*, 43
—— plants from seed, 40
Wallflower, 188
Walls for rock plants, dry stone, 41
—— plants for "dry," 42
Wand plant, white, 227

Warley Place, Essex, ascending pathway in Rock-Garden at, 10; *stone pathway at*, 9
Water Dock, the great, 65
—— *garden at Fota, Co. Cork*, 63
—— Leaf, Sitcha, 299
—— Lily, 265
—— plants, hardy, 64; in the Rock-Garden, 61
—— Robin, 324
Waterfall, an Alpine, 118
Watering Alpine and Rock plants, 92
Weed, Pickerel, 281
What to avoid, 95
Willow, 300
Willow-herb, 217
Windflower, 159
—— *Alpine*, 160
Winter green, 293 ; creeping, 227
—— —— spotted, 188
Wood plants, 125
—— Sorrell, 270
Woodruff, 174
Woods of California, Mountain, 139
Wormgrass, 320
Woundwort, 164
Wulfenia, 334

X

Xerophyllum, 334

Y

Yarrow, 149
Yucca, 334
—— *group of, a*, 335

Z

Zauschneria, 335
Zenobia, 155, 835
—— *speciosa pulverulenta*, 336
Zephyr-flower, 335
Zephyranthes, 335
Zietenia, 336

www.ingramcontent.com/pod-product-compliance
Lightning Source LLC
Chambersburg PA
CBHW031249230426
43670CB00005B/106